Nanotechnology for Mechanical Engineers

This book exclusively aims to deliver a basic understanding of nanotechnology from a mechanical engineering perspective. It begins with the history and fundamentals of nanotechnology and comprehension of the relationship between the properties and the structure. A brief overview of the several techniques available for the synthesis of various nanostructures and the techniques for size control is provided in the subsequent section. Further, it demonstrates applications of nanostructured materials in the field that are closely related to mechanical engineering.

- Presents exclusive discussion and elaboration on the nanomaterials in varied aspects of mechanical engineering.
- Covers machining techniques for nanostructure manufacturing, such as chemical grinding and additive manufacturing.
- Discusses advanced synthesis techniques of nanostructures and nanomaterials.
- Illustrates computational techniques relevant to mechanical properties of nanomaterials.
- Includes smart materials in the military, automobile, and aerospace applications.

This book is aimed at researchers and graduate students in mechanical engineering and nanotechnology.

Nanotechnology for Mechanical Engineers

Gibin George, Raghav G. R. and Jeetu S. Babu

CRC Press
Taylor & Francis Group
Boca Raton London New York

CRC Press is an imprint of the
Taylor & Francis Group, an **informa** business

First edition published 2025
by CRC Press
2385 NW Executive Center Drive, Suite 320, Boca Raton FL 33431

and by CRC Press
4 Park Square, Milton Park, Abingdon, Oxon, OX14 4RN

CRC Press is an imprint of Taylor & Francis Group, LLC

ISBN: 9781032329048 (hbk)
ISBN: 9781032329055 (pbk)
ISBN: 9781003317272 (ebk)

DOI: 10.1201/9781003317272

Typeset in Times
by Newgen Publishing UK

Contents

Preface

Nanotechnology has become essential to any engineering and technological development due to the possibility of manipulating materials near the atomic and molecular scales. This book aims to provide basic knowledge on various realms of nanotechnology within the robust field of mechanical engineering. Irrespective of whether you are a student, researcher, or engineer in practice, this book can help you with a strong understanding of how nanotechnology is revolutionizing science and engineering applications close to mechanical engineering. Through this book, one can understand how nanotechnology helps to develop new products for various applications.

Nanotechnology deals with materials and their applications at the nanoscale regime. It is not a new technology; it was developed during ancient civilizations, and its applications are unearthed from the remains of those civilizations. Nanotechnology can be a new way of thinking in which we can manipulate materials with predictable behaviors, enhance performances, and create previously unimaginable products.

Mechanical engineering is multifaceted, and a mechanical engineer is believed to be easily trained to perform well in any other science and engineering discipline. Nanotechnology is the most promising field in mechanical engineering. Mechanical engineers are accustomed to macroscopic elements of engineering. However, if they know how to manipulate the building blocks of materials, new solutions can be developed to revolutionize industries from manufacturing to medicine. A mechanical engineer with good knowledge of nanotechnology can have many opportunities to create more efficient, durable, and innovative materials and machines.

This book is an introduction to nanotechnology and a call to innovation. It encourages you to think beyond traditional engineering solutions, innovate, and integrate nanoscale technologies into your projects and research. This book begins with an introduction to an overview of nanotechnology and various nanomaterials, including their unique properties and the physical principles governing their behavior. The nanomaterials' size, shape, and morphology are important for the intended applications. The second chapter discusses the synthesize techniques for nanomaterials with different sizes, morphology, and composition. The various synthesis techniques and apparatus used for the synthesis are also covered in the chapter. The nanoscale fabrication techniques are important for creating various nanostructured architectures, especially for the fabrication of MEMS devices, sensors, electronic components, etc. These techniques are discussed in the succeeding chapter. The instrumentation for nanomaterial characterization and the various applications of nanotechnology in the sectors related to mechanical engineering, such as energy, transportation, and military, are also discussed in detail in this book.

In creating this book, we have collaborated with experts in both mechanical engineering and nanotechnology to provide comprehensive and authoritative content. We hope this book will serve as a reliable resource for understanding nanotechnology and inspire you to participate in shaping the future of engineering.

Gibin George
Raghav G. R.
Jeetu S. Babu

Acknowledgments

First and foremost, the authors thank all those who have supported us throughout this journey.

Our deepest thanks go to our families, friends, and colleagues for their unwavering encouragement and love. Your belief in us has been our greatest motivation and advice.

We are immensely grateful to Dr. Gagandeep Singh, Senior Publisher at CRC Press, for constantly pushing us beyond our boundaries and inspiring us to strive for excellence. Your guidance and constructive feedback were invaluable.

We especially thank Prof. Pramod P. Thevannoor, Vice Chairman of SCMS Group, for inspiring us to pursue our passions.

Lastly, we thank the readers for engaging with our work and supporting us.

Author Biographies

Gibin George is an Associate Professor at the Department of Mechanical Engineering, SCMS School of Engineering and Technology, Karkutty, Ernakulam, Kerala, India. He was a Postdoctoral Fellow in the Department of Chemistry and Physics at Fayetteville State University, North Carolina, before joining the SCMS School of Engineering and Technology. Gibin received his Ph.D. in 2015 from the National Institute of Technology Karnataka, India. His research interest is in electrospinning, oxide nanomaterials for energy storage and production, photoluminescence, sensors, and scintillators.

Raghav G. R. is currently serving as an Associate Professor in the Department of Mechanical Engineering at SCMS School of Engineering and Technology, Karukutty, Ernakulam, Kerala, India. He completed his Doctoral Degree from Anna University, Chennai, India with a specialization in Tribology and Corrosion resistant materials for dental applications. He has more than 12 years of teaching experience. He has contributed significantly to academia, with several of his research papers being published in prestigious SCIE and Scopus-indexed journals. In addition to his research, he is actively engaged as a reviewer for international journals and is recognized as a research supervisor by APJ Abdul Kalam Technological University. Dr. Raghav G. R.'s extensive academic and professional background underscores his expertise in the field of Mechanical, Metallurgy and particularly in Nanotechnology.

Jeetu S. Babu currently serves as an Assistant Professor at the Department of Material Science and Engineering, National Institute of Technology Calicut, Kerala, India. He received his Ph.D. in Computational Nanotechnology from NIT Calicut. Later, he worked as a Postdoctoral Fellow at the TIFR Centre for Interdisciplinary Sciences, Hyderabad, and IIT Kharagpur.

Author Biographies

Introduction

This book is an introduction to nanotechnology and a call to innovation. It encourages you to think beyond traditional engineering solutions, innovate, and integrate nanoscale technologies into your projects and research. This book begins with an introduction to an overview of nanotechnology and various nanomaterials, including their unique properties and the physical principles governing their behavior. The nanomaterials' size, shape, and morphology are essential for the intended applications. The second chapter discusses the synthesis techniques for nanomaterials with different sizes, morphology, and composition. The various synthesis techniques and apparatus used for the synthesis are also covered in the chapter. The properties and characteristics of the nanostructured materials and architectures are important factors in the respective applications; many advanced techniques are available to evaluate the nanostructured material and the architectures. The techniques used for evaluating the chemical, physical, and spectroscopic properties of nanostructured materials and architectures are discussed in detail in the third chapter. The manufacturing/fabrication techniques for nanoscale structures are essential for creating various nanostructured architectures, especially for fabricating MEMS devices, sensors, electronic components, etc. These techniques are discussed in the succeeding chapter. The various applications of nanotechnology in the sectors related to mechanical engineering, such as tribology, energy, transportation, and military, are discussed in detail in the following chapters of this book. Overall, this book will be a valuable resource for anyone interested in learning about nanotechnology and its potential applications in mechanical engineering. It offers a solid foundation in nanotechnology fundamentals while providing practical insights into its real-world uses.

1 Nanotechnology
An Introduction

Nanotechnology or Nanoscience can be defined as the technology or science of small things. The term nano has a Greek origin and represents the physical quantities in the order 10^{-9} or one billionth of the reference units such as meters, seconds, farads, etc., as nanometers, nanoseconds, and nanofarads. Nanotechnology involves studying the properties of nanostructured materials, devices, and applications. Generally, nanomaterials are those structures or elements with at least one dimension less than 100 nm and contain approximately 10 to 10,000 atoms. However, from a commercial perspective, the dimensions up to 500 nm are considered nanomaterials. Therefore, one cannot firmly define the nanoscale range, and the right to define the nanoscale is assigned to the end users based on the quality of the product.

The nanoscale dimension refers to the dimension along any axes within the nanoscale regime. Based on these dimensions, nanomaterials can be classified as zero-dimensional (0D), one-dimensional (ID), and two-dimensional (2D) materials. A nanosheet is a 2D nanomaterial because its dimensions along length and breadth are in the range of bulk (in the order of microns and above), while the thickness is in nanometers; the length and breadth may be in larger units. A nanowire is a 1D nanomaterial with a nanometer diameter and a micrometer length. A nanoparticle has a radius or dimension along all the directions within the nanoscale regime. For a nanostructured material, the dimension at the nanoscale determines its characteristics, which is called the characteristic dimension. For example, nanosheets have thickness as their characteristic dimension. This means any change in the breadth or length of the nanosheet will not make any change in the properties of the nanosheets.

Several nanostructures exist in nature, and they exhibit unique characteristics. For instance, the self-cleaning lotus leaves are one of the common examples of superhydrophobic surfaces with a natural origin. The tiny nanostructures on the lotus leaves lead to the exceptionally superhydrophobic nature of the lotus leaves, also called the lotus effect. Another example is the gecko's feet. (Gecko is a wall-climbing lizard) Geckos' feet are covered with microscopic hairs called setae, setae split into smaller nanostructures known as spatulas. This hierarchical arrangement allows the gecko foot to adhere to surfaces without using any sticky liquids. Such tiny structures can be observed in the case of small insects such as spiders, flies, etc., as

DOI: 10.1201/9781003317272-1

FIGURE 1.1 Nanostructures behind nature's wonders (a) lotus leaf, (b) spider's legs, and (c) gecko's feet.

well. A few examples of nanostructures of natural origin with unique properties are shown in Figure 1.1. The natural nanostructures can be mimicked to reproduce their properties for real-time applications. Mimicking the nanostructures of biological origin for practical applications is called biomimicry. Like a lotus leaf surface, the nanotextured coatings will exhibit a superhydrophobic nature, and these coatings can find applications in textile, automotive, and biomedical fields. Nanostructures comparable to the gecko's feet can provide bonding without any adhesives for biomedical applications.

1.1 WHY NANOTECHNOLOGY

Nanotechnology applications are widespread, and they cover all the fields of science and engineering applications. The multidisciplinary nature of nanotechnology is due to numerous reasons; the most unique one is its size regime. The nano-regime is 10,000 times smaller than the width of a human hair. The size chart in Figure 1.2 compares the artificial nanomaterials with naturally existing nano/microstructures such as building blocks of cells, cells, microorganisms, human hair, pollen, etc.

The properties of nanomaterials differ from bulk materials with the same chemical composition; the unique properties of the nanomaterials originate from their high surface area to volume ratio. The nanostructured materials have more atoms on their surface than the structure's interior. As the materials become at the nanoscale, the surface area tremendously increases. One can easily estimate the increase in surface area with the size reduction. For instance (Figure 1.3), let us consider a cube of side 1 cm; its surface area is 6 cm^2. If the cube is cut into eight cubes of sides 0.5 cm, the total surface area of the cube becomes 12 cm^2. If the cube is cut into 64 cubes of side 0.25 cm, the total surface area becomes 24 cm^2. Imagine if the cube is cut into cubes of 1 nm side; the total surface area of the cubes becomes 60,000,000 cm^2. Such a dramatic increase in the surface area is proportionate to the number of atoms available on the surface of the nanostructured materials.

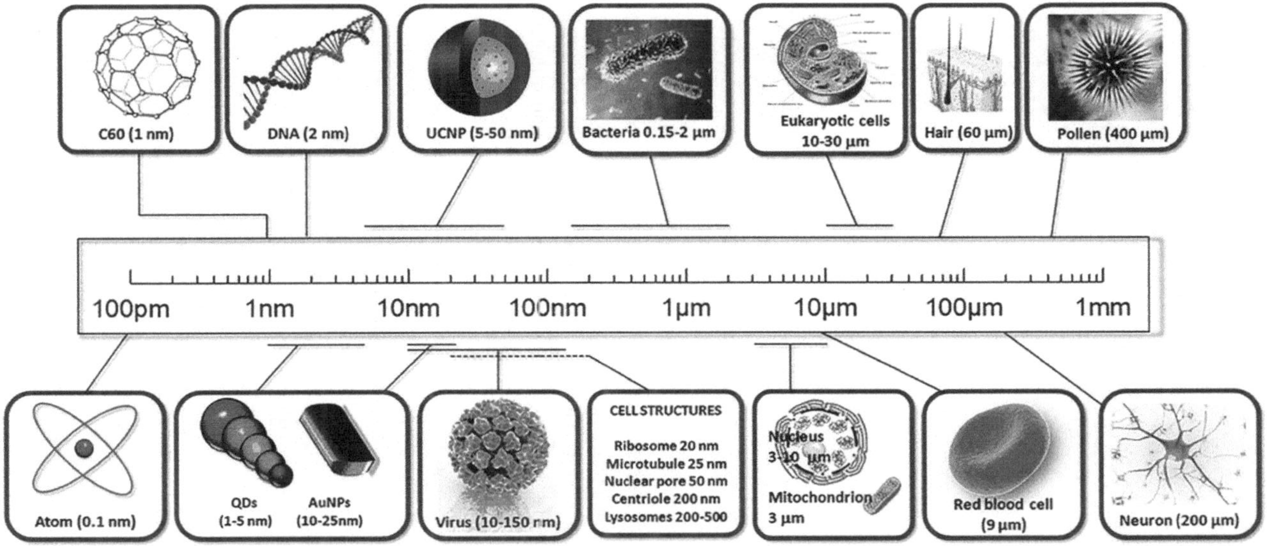

FIGURE 1.2 Compares natural micro/nanostructures with man-made nanostructures with length scale. (Reproduced with permission of the Royal Society of Chemistry, Gnach et al., 2015)

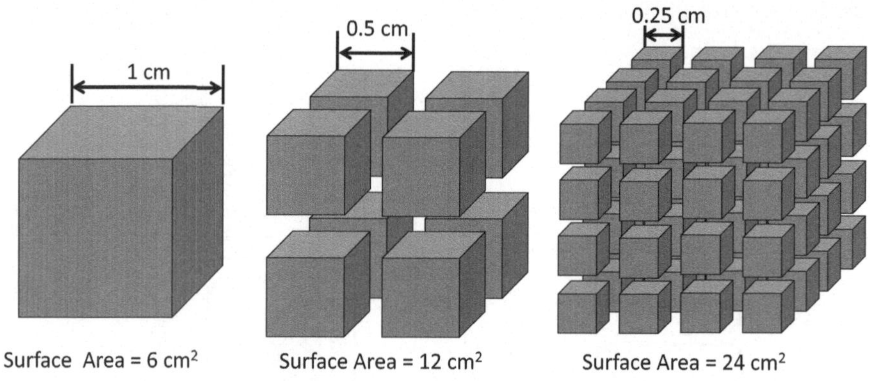

FIGURE 1.3 Increase in surface area with size reduction.

Due to the large number of atoms on the surface of the nanomaterials, the structure and properties of nanomaterials fall between that of atoms and materials in bulk. Tuning the properties at an atomic scale is impossible, as the atoms are the fundamental units. However, one can tune the properties of nanomaterials by combining atoms of different elements. As a result, we can achieve unique formulations with special properties, and often, such nanomaterials exhibit better properties than the bulk counterparts when employed for the same applications. In the case of nanomaterials, most of the atoms are at the surface; these surface atoms have dangling (unsaturated) bonds. The surface atoms have coordination numbers/nearest neighbors less than that of atoms at the center of the particles. As a result, they carry a partial charge and constitute the high surface energy of nanostructures. The nanomaterials reduce the free energy at the surface by active reaction, neutralizing the dangling bonds on the surface and thus eliminating the surface energy. The surface energy is also reduced by surface restructuring. Additionally, agglomeration, functionalization, adsorption, etc., reduce the surface energy. Therefore, the reactivity of the nanomaterials will be high as the particle size decreases. The essence of nanoscience and nanotechnology is that the nanoscale's chemical, physical, and biological properties significantly differ from the bulk material, as shown in Figure 1.4.

A mechanical engineering student needs to learn about nanotechnology, as nanotechnology is one of the new and rapidly growing frontiers in engineering. Any knowledge of nanotechnology may help students obtain new skills and develop new ideas that will expand the boundaries of the mechanical engineering discipline. Moreover, nanotechnology is an interdisciplinary approach, and understanding nanotechnology may provide basic knowledge in many disciplines so that one can explore the new possibilities in materials science, manufacturing, heat and mass transfer, fluid mechanics, robotics, and many others.

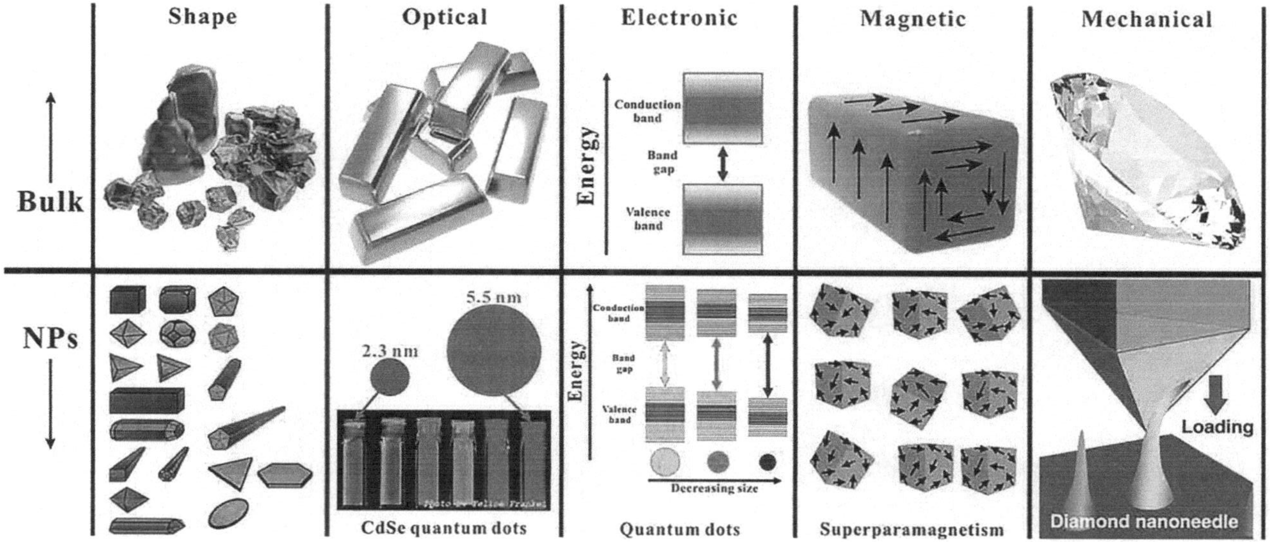

FIGURE 1.4 Comparison of different properties of bulk materials with NPs, optical, electronic, magnetic, and mechanical properties (left to right). (Huang et al., 2020)

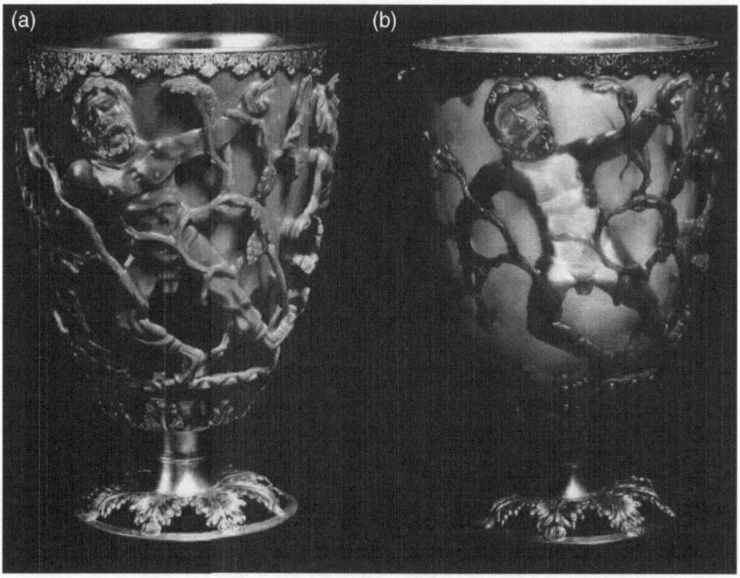

FIGURE 1.5 The Lycurgus cup, (a) green under reflected light and (b) red-purple under transmitted light. (Reproduced with permission of Springer Nature, Leonhardt, 2007)

1.2 HISTORY OF NANOTECHNOLOGY

The superior qualities of nanostructured materials were exploited during medieval times and by ancient Romans and Greeks, and they understood the size-property relationship remarkably well. One example is the Lycurgus cup, which was displayed in the British Museum. Romans made the Lycurgus cup in the 4th century, and it is the earliest example of dichroic glass; a glass displays two different colors when lit from the front, and the light passes through it. The Lycurgus cup is red when the light is transmitted through it and green when a direct light lights it. The dichroism of the Lycurgus cup is due to the presence of gold-silver (3:7) alloy nanoparticles with 50–100 nm diameter in glass. The absorption of red light by the gold nanoparticles results in the red color of the cup, and the scattering of light by the silver particles results in the green color (Freestone et al., 2007). The color change in the Lycurgus cup is shown in Figure 1.5. The nanoparticles in the Lycurgus cup are the oldest synthetic nanomaterial known till today.

A similar approach can be observed in late medieval church windows, where gold and silver nanoparticles of different sizes are infused into glass to achieve various shades as the light passes through the window. The color change in gold nanoparticles with the size is shown in Figure 1.6. Gold nanoparticle suspension<40 nm exhibits red color, and 40–100 nm transforms from red to blue. The silver nanoparticles also exhibit a size-dependent absorption and color change. Later, between the 9th and 17th centuries, Ag and Cu nanoparticles were commonly used to give a glittering "luster" to ceramic pottery artworks. During the same period, the Ottoman produced

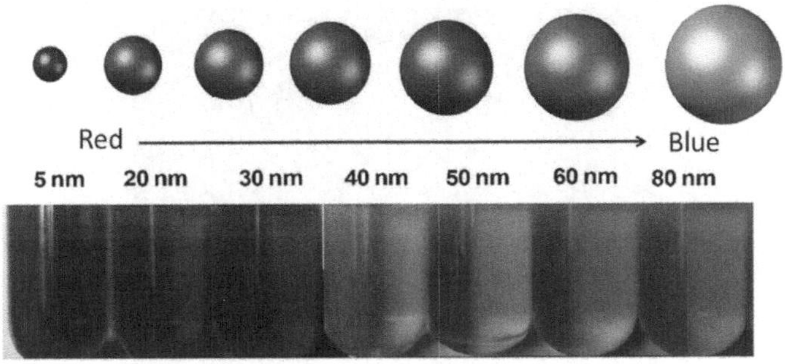

FIGURE 1.6 Color change with particle size of gold nanoparticles as colloidal suspension (size of the gold colloids: 5 nm, 20 nm, 30 nm, 40 nm, 50 nm, 60 nm, and 80 nm). (Reproduced with permission of Springer Nature, Abrica-González et al., 2018)

"Damascus" blades with keen edges. The cementite nanowires and carbon nanotubes were reinforced in steel to provide strength, resilience, and the ability to hold a keen edge. Interestingly, the artists or smiths in the respective era never knew the causes of these astonishing properties.

The colloidal form of gold attracted many enthusiasts during the 1600s and 1700s. The colloid form of gold was then used for dyeing and proposed as a medicine with some curative properties. In India's history, colloidal gold was a component of many Ayurvedic medicines. Faraday's first report on the systematic preparation of gold colloid was done in 1856, and he called it a "divided metal."

The potential of nanostructured objects in the modern era is envisaged by the Nobel Prize-winning American physicist Richard Feynman in his famous talk at the APS annual meeting in 1959, "There's plenty of room at the bottom." Richard Feynman is considered as the father of nanotechnology. Feynman proposed assembling atoms to form nanostructured materials with different properties. The manipulation or assembly of atoms was made possible by IBM's introduction of a scanning tunneling microscope (STM) in 1981. The first manipulation of atoms was carried out using STM in 1990 at IBM, where 35 xenon atoms were placed on a nickel surface to form the letters of the IBM logo (Figure 1.7).

Yet another remarkable discovery in nanotechnology occurred in 1985: carbon in the form of stable spheres called fullerenes or buckyballs. These carbon balls have a chemical formula, C60 or C70, derived from graphite when evaporated in an inert atmosphere. C60 has the exact physical appearance of a football or a soccer ball, as shown in Figure 1.8. The structure of the C70 molecule slightly deviates from the shape of a perfect sphere.

In 1991, Iijima et al. discovered the existence of carbon in a tubular form called carbon nanotubes (CNTs), another allotrope of carbon. The existence of carbon dots was identified in 2004. In the same year, the discovery of graphene sheets took place, and it was a breakthrough in the history of nanostructured materials. The defect-free structure of the nanostructured carbon materials improves their strength and flexibility, making them suitable for a wide range of applications. Carbon-based nanomaterials

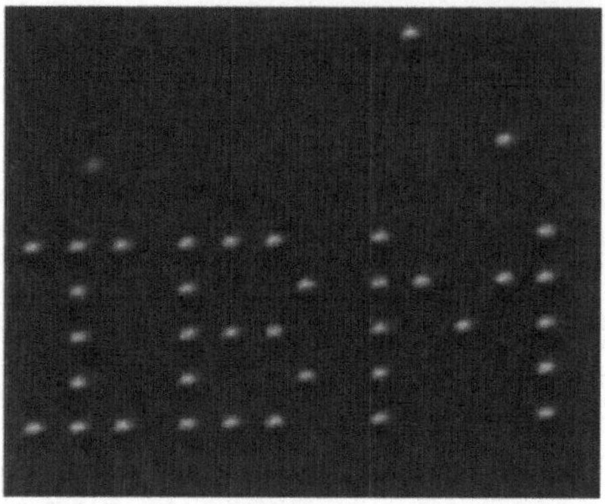

FIGURE 1.7 The IBM logo was created by positioning 35 Xenon atoms on a nickel (110) substrate using an STM. (Reproduced with permission of Springer Nature, Eigler and Schweizer, 1990)

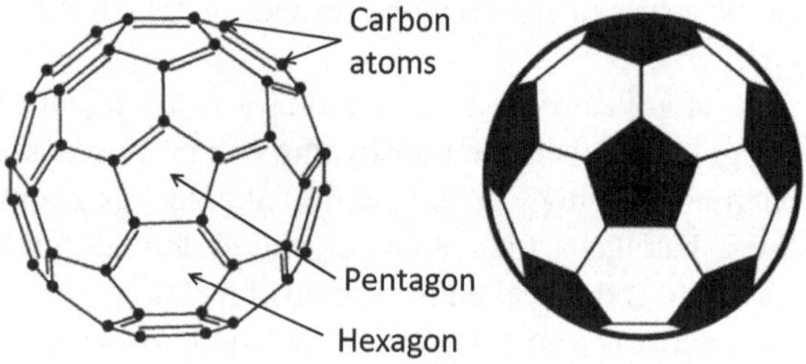

FIGURE 1.8 Comparison of C60 molecule with a football/soccer ball.

are used extensively as field-emitting transistors, energy storage materials, catalysis, electronic components, sensors, etc. Carbon atoms are successfully replaced by metal atoms in allotropic structures to create new formulations with superior properties.

With the introduction of nanotechnology, an immediate effect is observed in reducing the size of electronic components. The growth in the technology drastically reduced the size of the devices. The dimension reduction is expected to follow Moore's law, proposed by Gordon Moore, co-founder of Intel, in 1965 (Figure 1.9). According to Moore's law, the number of transistors in an integrated circuit (IC) doubles every two years (Schaller 1997). The extrapolated version of Moore's law is still followed on the number of transistors in a chip. Still, it is likely to end very soon as the semiconductor process technology reaches the molecular limits.

The advancement in nanotechnology in the different fields of science and technology is remarkable. Figure 1.10 shows the significant milestones in nanotechnological developments in various fields.

The never-ending search for better properties and performance replaced many conventional materials with nanostructured materials, since the latter surpasses the former. Nanomaterials with different compositions and structures are explored

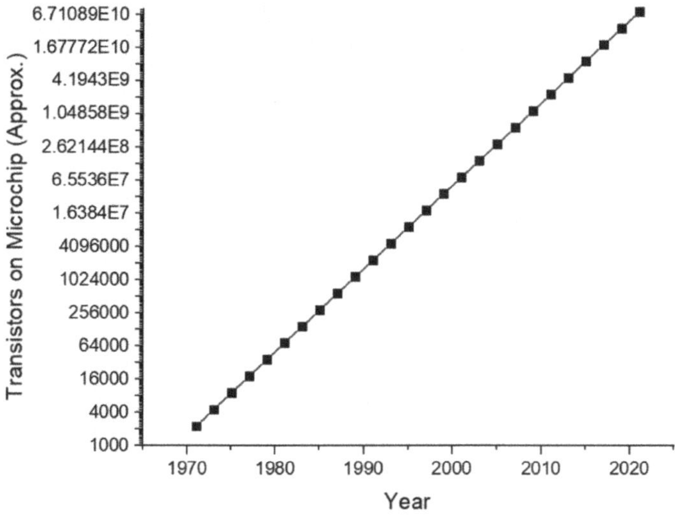

FIGURE 1.9 Moore accurately predicted in 1965 that the number of transistors on an IC would double yearly for the next ten years, and the law still holds true.

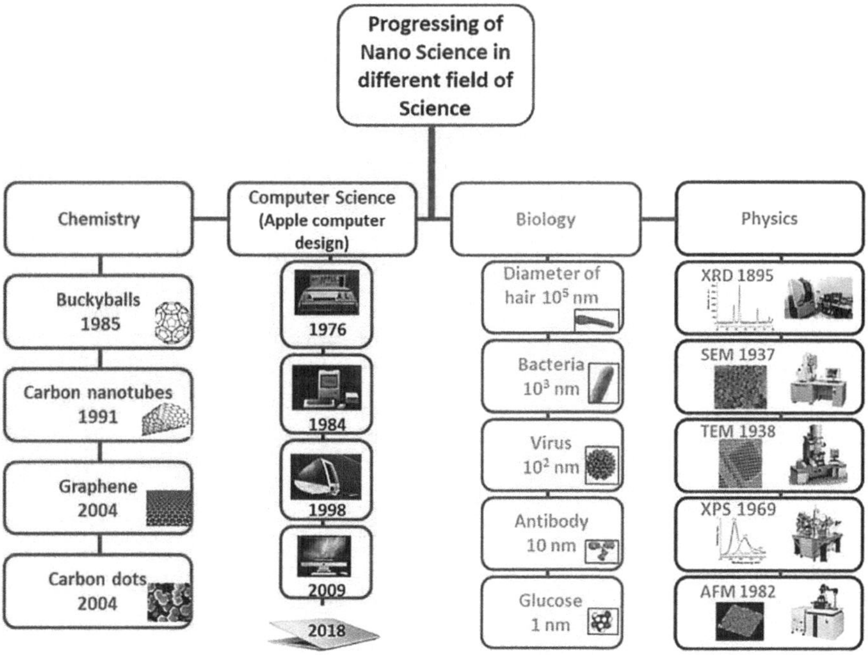

FIGURE 1.10 The advancement in nanotechnology from its origin. (Bayda et al., 2019)

tremendously. The current scenario is such that at least one new nanostructured material is being introduced day by day worldwide. Besides the health and environmental aspects, nanomaterials are doing wonders in all developments and technologies, from a single nanofiber for a field emission transistor to paints, tires, coolants, etc., filled with nanomaterials.

1.3 TYPES OF NANOMATERIALS

The nanostructured materials have different dimensionalities and compositions. The dimensionality and composition significantly affect the properties of the nanomaterials. As the morphology changes, so do the surface area and its properties. The composition determines the functionality of the nanomaterials. Often, nanomaterials are multifunctional. We can classify the nanomaterials based on their composition and dimensionality.

1.3.1 Classification by Composition

First, let us classify the nanomaterials based on their composition. The nanomaterials can be classified as organic, inorganic, carbon, and organic-inorganic hybrid materials, as listed in Figure 1.11.

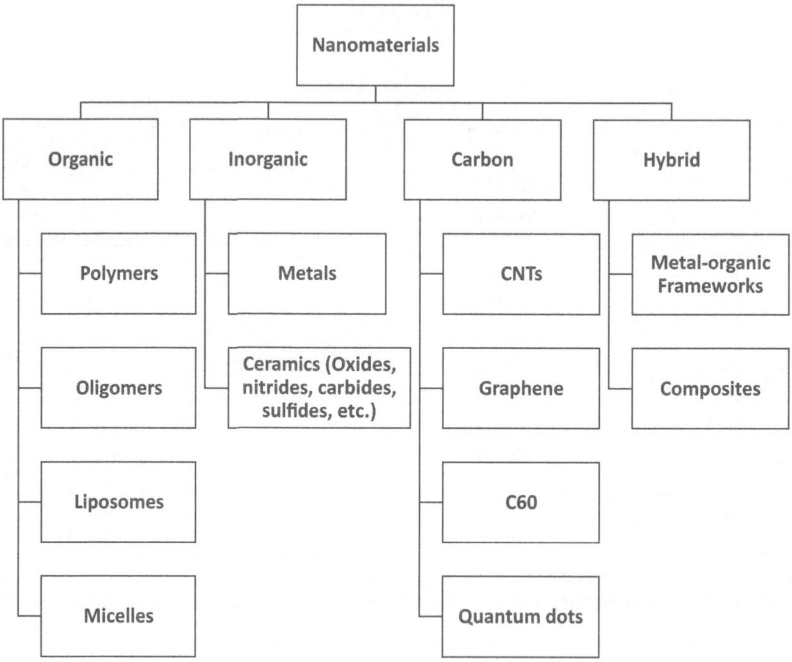

FIGURE 1.11 Classification of nanomaterials based on their composition.

1.3.1.1 Organic Nanomaterials

Organic materials represent carbon compounds with non-metallic elements formed by covalent bonds. Non-covalent interactions are essential in forming nanostructures of these compounds. Many organic nanostructures are fabricated by a bottom-up approach where atoms or molecules are assembled to form the nanostructure. In the case of organic materials, the binding force is Van der Waal's force, also known as hydrogen bonding. Van der Waal's forces lead to the self-organization of a wide range of molecules, forming micelles, vesicles, liposomes, and polymersomes. Controlled synthesis of dendrimers, polymer nanoparticles, and polymeric nanofibers is also possible. Additionally, high-aspect-ratio organic nanostructures can be fabricated using template-based synthesis, electrospinning, etc. These organic molecules are used for biomedical applications such as tissue engineering, targeted drug delivery, etc.

Micelles, vesicles, liposomes, and polymersomes (Figure 1.12): These organic nanoparticles are similar in construction with a symmetric hollow structure. Usually, they adopt a spherical shape in the respective solvent. The micelles and liposomes are formed by surfactants and lipids, respectively, whereas the polymersomes constitute the block polymer chains as the separating layer. Micelles can adapt to any shape and are formed by the orientation of surfactants' hydrophilic or hydrophobic heads towards or away from the solvent. The nanoparticles above are generally non-toxic. The micelles and liposomes can change size and shape with time or pH of the medium. These nanoparticles can be loaded with drug molecules by conjugation on the surface or in the core, enabling them for drug delivery and biomedical applications.

Dendrimers: Dendrimers are nanostructured organic materials with spherically symmetric structures with tree-like arms or branches, also known as "cascade" molecules. Dendrimers are considered hyperbranched macromolecules with tree-like branching grown around the core of a small molecule or a linear polymer chain. "Dendrimer" represents the structure, not a compound. Dendrimers can be functionalized to obtain the required physical, chemical, or biological properties. Dendrimers hold many functional groups in a confined structure, making them ideal candidates for drug delivery, biological imaging, etc.

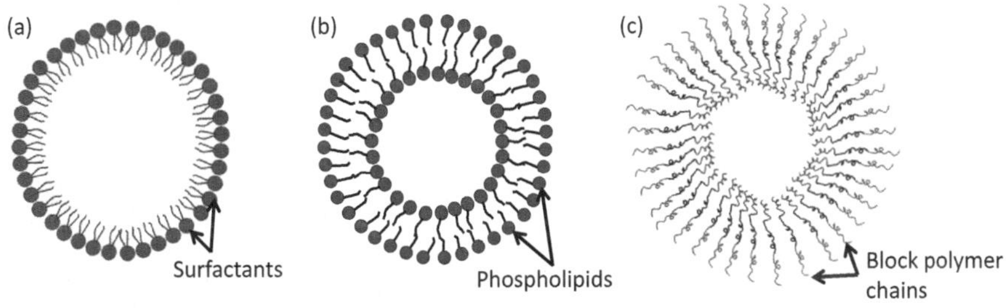

FIGURE 1.12 Micelles, liposomes, and polymersomes in water.

Polymer Nanoparticles: Polymeric nanoparticles are colloids of natural or synthetic polymers with a size ranging from 10 to 1,000 nm. These particles can be formed either from the preformed polymeric chains or by the controlled polymerization from the monomer. Polymeric nanoparticles are advantageous over liposomes or micelles for drug-controlled delivery because of their stability in various biological fluids.

Polymer nanofibers: Polymer nanofibers are high-aspect-ratio nanostructures with diameters in the nanometer regime and lengths up to meters. They can be fabricated using any of the following techniques: drawing, centrifugal spinning, electrospinning, self-assembly, template-assisted synthesis, and phase separation. Most polymers can be converted to nanofibrous morphology. These nanofibers are used in many applications, such as sensors, separators in supercapacitors and batteries, filters, tissue engineering, wound healing, drug delivery, etc.

1.3.1.2 Carbon-Based Nanomaterials

Carbon can form covalent bonds with other carbon atoms with different hybridization states, such as sp, sp^2, and sp^3, forming various nanostructures. These carbon nanostructures are allotropes of carbon and collectively represent this group. Figure 1.13 shows the arrangement of carbon atoms in various carbon allotropes, and Table 1.1 lists their properties and applications.

Based on the hybridization, the carbon-based nanomaterials exhibit different morphologies. The most prominent forms of carbon nanomaterials are carbon nanotubes (CNTs), graphene, carbon dots, fullerenes (C60, C70), carbon nanofibers, and carbon black. Among these allotropes, the most important ones are CNTs and graphene.

Carbon nanotubes (CNTs): CNTs are sp^2 hybridized carbon allotropes with a tubular structure (1D nanomaterial). It has the appearance of rolled graphene sheets

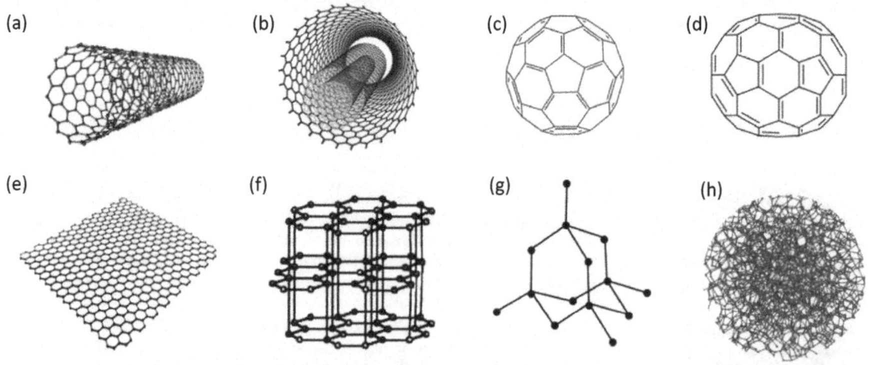

FIGURE 1.13 The structure of various allotropes of carbon: (a) Single-walled carbon nanotube, (b) Multi-walled carbon nanotube, (c) Fullerene (C60), (d) C70 molecule, (e) Graphene sheet, (f) Graphite, (g) Diamond, and (h) carbon dot.

TABLE 1.1
Nanostructured Allotropes of Carbon and Their Properties

Carbon Nano Allotrope	Features	Applications
Carbon nanotubes	• sp^2 Hybridized • 1 Dimensional structure • High mechanical strength • Good electrical conductivity	Transistors, electron-field emitters, reinforcement in composites, chemical/electrochemical sensors, biosensors, lithium-ion batteries, hydrogen storage cells, supercapacitors, and electrical shielding devices.
Graphene	• sp^2 Hybridized • 2 Dimensional structure • High mechanical strength • Good electrical and thermal conductivity	Solar cells, batteries, sensors, electronics, coatings, fuel cells, composites, drug delivery, luminescence, flexible screens, superconductors, paints, infrared vision, and radiation shielding.
Carbon dots	• Amorphous • Fluorescent • Size <10 nm in all the directions	Bioimaging, optical imaging, drug delivery, biosensing, disease detection.
Fullerenes	• Formed by 60 or 70 number of carbon atoms. • Cage-like hollow structure	Absorbent, lubricant, antimicrobial applications, solar cells, batteries, paints, antioxidants.
Carbon nanofibers	• High-aspect-ratio nanostructures	Li-ion batteries, drug delivery, catalyst support, reinforcement in composites, electrodes.

(Figure 1.13 a&b), is light in weight (density is 1/6th of steel), and has a high surface area (~1,000 cm^2/g. CNTs can be single-walled (SWCNT) with a diameter as small as 2 nm or multiwalled (MWCNT) with diameters ranging from 10–100 nm, and they are fabricated using the chemical vapor deposition (CVD) technique. CNTs are high-aspect-ratio nanomaterials with an aspect ratio (length-to-diameter ratio) of ~1,000. SWCNTs are CNTs with a single layer of carbon sheet (graphene) rolled in the form of a tube. The MWCNT has multiple rolls of graphene layers that are rolled coaxially. The CNTs exhibit high tensile strength, better than steel, and can be used as a reinforcement in composites without altering the properties of the matrix. SWCNTs exist in three forms: armchair, chiral, and zigzag nanotubes, based on how the graphene sheets are rolled (Dresselhaus et al., 2004). Zigzag and chiral CNTs are semiconducting, while armchair CNTs are metallic. Along the tube direction, the electric conductivity of armchair CNTs is 1,000 times that of copper. Due to their unique properties, CNTs and graphene are the most popular nanomaterials.

Graphene: Graphene is a single layer of carbon atoms with a hexagonal lattice structure. This 2D nanostructure is an allotrope of carbon with sp^2 hybridization. Graphene is the carbon layer separated form of graphite in which multiple layers of carbon atoms are stalked together by hydrostatic bonding. A single layer of carbon atoms forms the graphene layer with a thickness equal to the diameter of the carbon atom. Graphene is the lightest carbon allotrope, weighing 0.77 mg for 1 m^2. Graphene exhibits a remarkable tensile strength of 130 GPa and Young's modulus of 1 TPa (300 times than steel), high thermal conductivity ~5.0×10^3 $W \cdot m^{-1} \cdot K^{-1}$, and good electrical conductance. A single layer of carbon atoms is transparent to visible light.

Carbon dots (C-dots):. C-dots are amorphous carbon allotropes with a size of less than 10 nm. They can be produced at a low cost as compared to CNTs and graphene. C-dots are luminescent materials, meaning they produce visible light under UV excitation.

Fullerenes (C60, C70) are another nanostructured allotrope of carbon in which carbon atoms are covalently bonded to their neighbors by single and double bonds. Fullerenes have a hollow cage-like structure with 60 or 70 carbon atoms. The number of carbon atoms can be higher than 70. However, a perfect spherical shape is observed in the case of C60 molecules. The presence of the pentagonal and hexagonal rings in fullerenes makes the ideal spheres with a structure similar to a soccer ball. These structures are also known as buckyballs and are closely connected to CNTs.

Carbon nanofibers: Carbon nanofibers (CNFs) comprise layers of graphene stalked in different orientations. Various synthesis techniques are used to fabricate carbon nanofibers, such as electrospinning, templated-assisted synthesis, chemical vapor deposition, etc.

1.3.1.3 Inorganic Nanomaterials

Inorganic nanomaterials include metals and ceramic nanostructures. Ceramic means a compound of a metal and a non-metallic element. Most ceramics are inert; however, their nanostructured counterparts exhibit multifunctionality, meaning properties suitable for different applications. The nanostructures of metal oxides/hydroxides, carbides, nitrides, fluorides, selenides, tellurides, etc., are ceramic materials. Ag, Au, Fe, and Cu nanoparticles are the most common and stable metallic nanomaterials. TiO_2, ZnO, Co_3O_4, NiO, CaF_2, BaF_2, TiC, etc., are a few examples of ceramic nanomaterials. Table 1.2 lists the various nanostructured ceramics and their applications. These nanomaterials are synthesized as particles, wires, sheets, and other shapes. Figure 1.14 shows the different morphologies of Co_3O_4 nanomaterials prepared under various conditions. Controlled synthesis procedures are essential to tailoring the morphology and size of these materials.

1.3.2 Classification by Dimensionality

Nanomaterials can be classified based on their dimensionality as zero-dimensional (0D), one-dimensional (1D), two-dimensional (2D), and three-dimensional (3D) nanomaterials, as shown in Figure 1.15. The classification of nanomaterials based

TABLE 1.2
Some Ceramic Nanostructures and Their Applications

Sl. No	Composition	Application
1	TiO_2	Photocatalysis, wastewater treatment, drug delivery, implants
2	MgO	Photocatalysis, water filtration, sensor, absorbent
3	ZnO	Sensor, optoelectronics, antibacterial, photocatalysis
4	CuO	Sensing, infrared filters, catalysis, superconductors, and thermoelectric materials
5	Al_2O_3	Adsorbent, antimicrobial applications
6	Fe_3O_4	Diagnostic and drug delivery, sensing, MRI imaging, cancer treatment
7	Co_3O_4	Sensor, electrocatalysis, drug delivery, solar selective absorbers, anode materials in lithium-ion batteries, energy storage, pigments, and dyes
8	NiO	Sensor, capacitors, thermistors and varistors, batteries, electrochromic devices, chemical or temperature sensing devices
9	SnO_2	Sensor
10	WO_3	Sensor
11	ZrO_2	Antioxidants, absorbents, antibacterial applications
12	MoO_3	Catalysis, OLED, electronics applications
13	MnO_2	Electrodes for batteries and supercapacitors, catalysis, adsorbent, sensors
14	CeO_2	Catalysis, luminescence
15	IrO_2	Electrocatalysis
16	V_2O_5	Sensor
17	In_2O_3	Sensor
18	Nb_2O_5	Corrosion resistance coatings, battery, supercapacitors
19	Ta_2O_5	Antireflection coating, an insulating layer, gate oxide, corrosion-resistant material, and sensitive layer
20	La_2O_3	Optical glasses, phosphors for fluorescent lamps, capacitors, X-ray intensifying screens
21	$BaWO_4$	X-ray scintillation, photoluminescence, photocatalysis, sensors
22	$CaWO_4$	Phosphors, temperature sensing, catalysis
23	$SrWO_4$	LEDs, phosphors, photocatalysis
24	YVO_4	Polarizers, fluorescent emitters, lasers
25	$ZnWO_4$	Photocatalysis, scintillators
26	$MgWO_4$	Photocatalysis, scintillators, fluorescent probe
27	$GdVO_4$	Sensors, Photocatalysis
28	$LaMnO_3$	Photocatalysis, electrodes for batteries and supercapacitors, water splitting catalysis
29	$LaCoO_3$	Catalysis, sensing, fuel cells
30	$LaFeO_3$	Adsorbent, Photocatalysis
31	$BaTiO_3$	Optical imaging, piezoelectric devices, pyroelectric sensors, semiconductive ceramics, varistors, and electro-optic devices

(continued)

TABLE 1.2 (Continued)
Some Ceramic Nanostructures and Their Applications

Sl. No	Composition	Application
32	$SrTiO_3$	Gas sensors, solar cells, UV detectors, capacitors, and photocatalysis
33	$YAlO_3$	Sensors, Photocatalysis
34	$CuAlO_2$	Solar cells, sensors, electrocatalysis
35	$ZnFe_2O_4$	Photocatalysis, electrodes for batteries, and supercapacitors

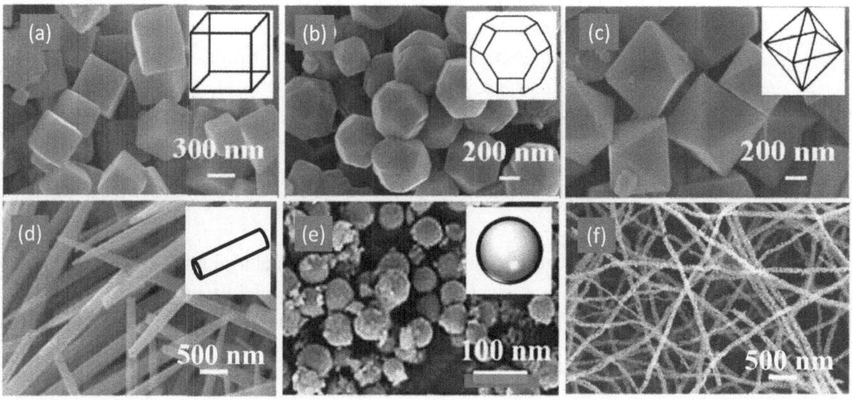

FIGURE 1.14 Co_3O_4 nanostructures with different morphologies. (Reproduced with permission of John Wiley & Sons and Elsevier, Xiao et al., 2012; Han et al., 2018)

on dimensionality is discussed at the beginning of this chapter. Zero-dimensional nanomaterials are nanostructures with all dimensions (x, y, z) at the nanoscale, meaning, along all directions, the dimensions should be less than 100 nm. Quantum dots are zero-dimensional nanomaterials with all dimensions (x, y, and z) <10 nm.

One-dimensional nanomaterials have two dimensions (x and y directions) in the nanoscale regime and one dimension outside the nanoscale. As shown in Figure 1.15, nanofibers, nanotubes, nanorods, and nanowires are a few examples of 1D nanomaterials. Two-dimensional (2D) nanomaterials have one of their dimensions (x-direction) in the nanoscale, and the other two dimensions are outside the nanoscale–the 2D nanomaterials' sheet-like structures with nanometer thickness. Graphene, layered double hydroxides, MXene, etc., are a few examples of 2D nanomaterials. Three-dimensional nanomaterials have dimensions above the nanoscale; however, the pores in the material or the 3D structure are composed of nanostructured units. Clay is a 3D nanomaterial consisting of two-dimensional sheets of tetrahedral (SiO_4) and octahedral (M_2O_3, M = Mg, Al), as shown in Figure 1.16. Each set of these layers is stacked together by interlayer metals such as Na^+ and Ca^{2+} by Van der Waal's forces. The layers can be separated by exfoliation or intercalation; the processes are discussed

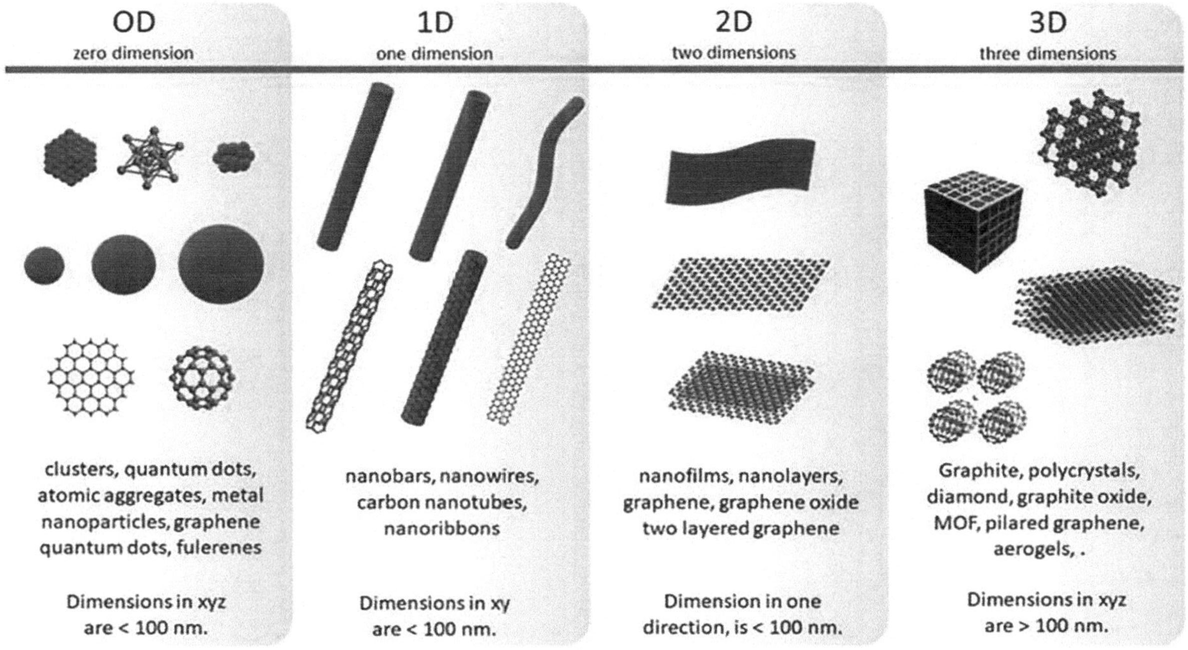

FIGURE 1.15 Classification of nanomaterials based on the dimensionality. (García-Betancourt et al., 2020)

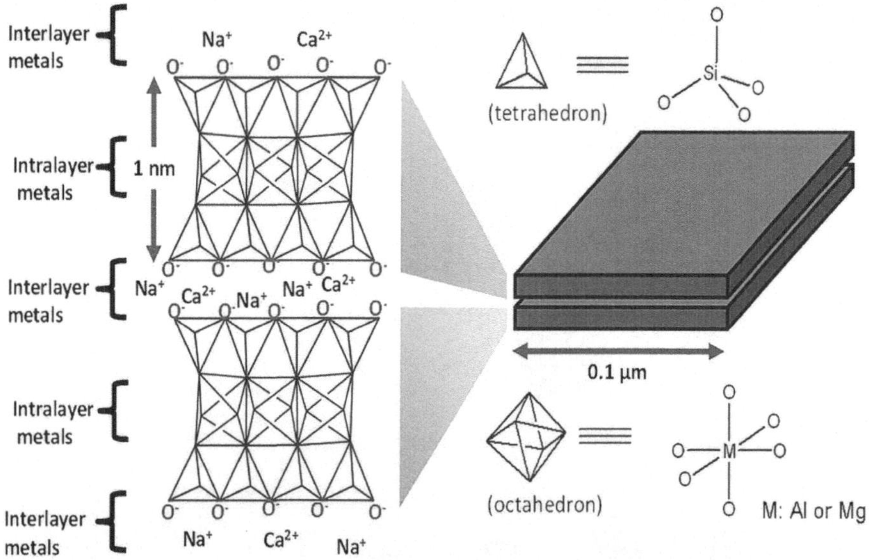

FIGURE 1.16 The nanolayered structure of clay. (Reproduced with permission of the Royal Society of Chemistry, Zhu et al., 2012)

in Chapter 2. Layered double hydroxides also adopt a similar structure. Metal-organic frameworks are 3D structured nanomaterial with nanosized pores in them.

REFERENCES

Abrica-González, P., Zamora-Justo, J.A., Chavez-Sandoval, B.E., Vázquez-Martínez, G.R., Balderas-López, J.A., 2018. Measurement of the Optical Properties of Gold Colloids by Photoacoustic Spectroscopy. Int. J. Thermophys. 39, 93. https://doi.org/10.1007/s10765-018-2412-1

Bayda, S., Adeel, M., Tuccinardi, T., Cordani, M., Rizzolio, F., 2019. The History of Nanoscience and Nanotechnology: From Chemical–Physical Applications to Nanomedicine. Molecules 25, 112. https://doi.org/10.3390/molecules25010112

Dresselhaus, M.S., Dresselhaus, G., Jorio, A., 2004. Unusual Properties and Structure of Carbon Nanotubes. Annu. Rev. Mater. Res. 34, 247–278. https://doi.org/10.1146/annurev.matsci.34.040203.114607

Eigler, D.M., Schweizer, E.K., 1990. Positioning single atoms with a scanning tunneling microscope. Nature 344, 524–526. https://doi.org/10.1038/344524a0

Freestone, I., Meeks, N., Sax, M., Higgitt, C., 2007. The Lycurgus Cup — A Roman nanotechnology. Gold Bull. 40, 270–277. https://doi.org/10.1007/BF03215599

García-Betancourt, M.L., Jiménez, S.I.R., González-Hodges, A., Salazar, Z.E.N., Escalante-García, I.L., Aparicio, J.R., García-Betancourt, M.L., Jiménez, S.I.R., González-Hodges, A., Salazar, Z.E.N., Escalante-García, I.L., Aparicio, J.R., 2020. Low Dimensional Nanostructures: Measurement and Remediation Technologies Applied to Trace Heavy Metals in Water, in: Trace Metals in the Environment – New Approaches and Recent Advances. IntechOpen. https://doi.org/10.5772/intechopen.93263

Gnach, A., Lipinski, T., Bednarkiewicz, A., Rybka, J., Capobianco, J.A., 2015. Upconverting nanoparticles: assessing the toxicity. Chem. Soc. Rev. 44, 1561–1584. https://doi.org/10.1039/C4CS00177J

Han, D., Ji, Y., Gu, F., Wang, Z., 2018. Cobalt Oxide Nanorods with Special Pore Structure for Enhanced Ethanol Sensing Performance. J. Colloid Interface Sci. 531, 320–330. https://doi.org/10.1016/j.jcis.2018.07.064

Huang, C., Chen, X., Xue, Z., Wang, T., 2020. Effect of Structure: A New Insight into Nanoparticle Assemblies from Inanimate to Animate. Sci. Adv. 6, 1321. https://doi.org/10.1126/sciadv.aba1321

Leonhardt, U., 2007. Invisibility Cup. Nat. Photonics 1, 207–208. https://doi.org/10.1038/nphoton.2007.38

Schaller, R.R., 1997. Moore's Law: Past, Present, and Future. IEEE Spectrum. 34, 52–59. https://doi.org/10.1109/6.591665

Xiao, X., Liu, X., Zhao, H., Chen, D., Liu, F., Xiang, J., Hu, Z., Li, Y., 2012. Facile Shape Control of Co_3O_4 and the Effect of the Crystal Plane on Electrochemical Performance. Adv. Mater. 24, 5762–5766. https://doi.org/10.1002/adma.201202271

Zhu, W., Lu, C.-H., Chang, F.-C., Kuo, S.-W., 2012. Supramolecular Ionic Strength-Modulating Microstructures and Properties of Nacre-Like Biomimetic Nanocomposites Containing High-loading Clay. RSC Adv. 2, 6295–6305. https://doi.org/10.1039/C2RA20523H

2 Synthesis of Nanostructured Materials

Various techniques are available to synthesize nanostructured materials with different sizes, shapes, dimensionality, aspect ratios, etc. The selection of the proper synthesis technique is based on the purity, quality, ease of production, availability of the precursors, and their reactivity in addition to size, shape, etc. Any synthesis technique used for nanomaterials comes under the broad category "top-down" and "bottom-up" approaches. The meaning of the top-down approach is from the top (large) to the bottom (small). In the top-down approach, the bulk material is broken down into nanostructured materials. This approach is similar to making gravel from a large piece of rock or machining tiny mechanical components from a large metal block. Ball milling, electron beam, laser beam, chemical etching, etc., can fabricate nanostructures from bulk materials. The bottom-up approach means bottom to top, meaning combining atoms and molecules to achieve the nanostructure. In a bottom-up approach to nanomaterial fabrication, the building blocks are molecules, molecular complexes, atoms, or aggregates that vary in size from tenths of nanometers to tens of thousands of nanometers, and the mortar is their assembly technique. Most of the chemical synthesis methods are examples for the bottom-up approach. Figure 2.1 shows the schematic drawings of top-down and bottom-up approaches.

The "top-down" and "bottom-up" synthesis methods can be solid-state, liquid-phase, and gas-phase. The important characteristics of top-down and bottom-up approaches are compared in Table 2.1. The selection of the synthesis technique is vital for tailoring the functionality of the nanomaterials, as the morphology and surface characteristics vary with the synthesis technique, resulting in different properties. Therefore, the appropriate selection of synthesis technique is crucial for the high performance of the nanomaterials in the intended applications. For instance, the catalytic properties of nanomaterials depend on the particle size and shape. To prepare highly active nanocatalysts, one has to select a synthesis technique that may provide nanoparticles with smaller sizes and a large surface area. Likewise, applications such as photovoltaics and microelectronics require thin films on different substrates, so we have to choose a fabrication technique suitable for producing thin films. This chapter delivers the important "top-down" and "bottom-up" approaches for fabricating nanostructured materials for different applications.

DOI: 10.1201/9781003317272-2

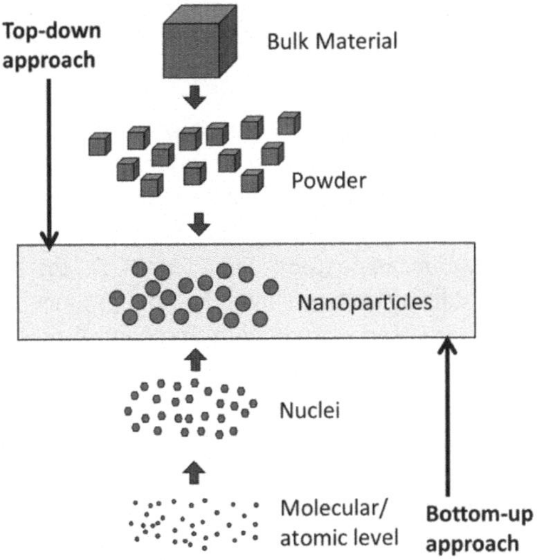

FIGURE 2.1 Schematic representation of top-down and bottom-up approach.

TABLE 2.1
Comparison of Top-Down and Bottom-Up Approach

Top-Down Approach	Bottom-Up Approach
Fabrication starts with bulk materials.	Fabrication starts with molecules and atoms.
The starting material is always a solid.	The starting material can be a solid, liquid, or gas.
The size reduction is limited.	Ultrafine nanoparticles, nanosheets, nanorods, nanotubes, etc. can be fabricated.
The size and morphology control is difficult; therefore, the size distribution is non-uniform.	Controlling size and morphology is easy.
No purification is required if the final product is the same composition as the starting materials.	Purification is needed as the starting materials and the end product are different.
Due to enormous forces, defects and imperfections are introduced to the nanomaterials.	Fewer defects and imperfections.
Suitable for large-scale production.	Large-scale production is challenging.
Expensive equipment is needed.	Cheaper than a top-down approach.
E.g., lithography, ball milling, etching, grinding, and exfoliation.	Any solution-based synthesis.

2.1 TOP-DOWN APPROACHES

2.1.1 Solid-State Reaction

This is one of the most straightforward techniques to produce nanostructured materials by heating the salts of the final compound at a high temperature; the process is called calcination. This technique can prepare nanostructured ceramics such as oxides, nitrides, and carbides of transition metals. The starting materials for solid-state synthesis can be nitrates, carbonates, oxides, and acetates. As the name implies, in the solid-state reaction, the starting materials and the final product are in the solid state. The reaction allows for large-scale production, but control over the size and purity of the final product is difficult. However, this method is helpful for complex ceramic nanomaterials with multiple transition elements. During the process, the solid starting materials are finely ground in a stoichiometric ratio and mixed well to increase the surface area of the starting materials. If the composition of the nanomaterial is complex, pelletizing is preferred to ensure contact between precursor particles. The powdered/pelletized precursors are then heated to a high temperature to remove the volatile part or the precursor to get the respective oxide nanomaterials.

For a complex ceramic composition, the interdiffusion of cations is necessary to obtain the final product. The interdiffusion of cations takes place at very high temperatures (above 1,000 °C) for 12–24 hours). In addition, the precursors such as carbonates and nitrates can accelerate the process as they form structural defects. The calcination temperature can be determined by the "Tamman's Rule." The rule suggests that a solid-state reaction occurs in a reasonable amount of time only when the temperature of the reaction is about two-thirds of the melting point of the lower melting reactant. At these temperatures, the nucleation of the final product phase is intended to occur, and the crystal continues to convert the reagents to the final product. The reaction rate can also be accelerated by choosing the reactants with similar crystal structures. If the final product has a structure identical to reagents, it is called topotactic transition. In the epitaxy transition, the product crystal growth occurs on a particular reagent crystal plane, so the reagents and products have the same crystal structure. This processing technique is successfully employed in fabricating simple, layered, complex, and nanostructured ceramics (oxide, fluoride, nitride, carbide, boride, etc.) by the appropriate selection of calcination/annealing temperature.

2.1.2 Ball Milling

Ball milling is a valuable top-down approach for large-scale production of nanomaterials. In ball milling, the bulk materials undergo impact and attrition between the balls used in the process, leading to the mechanical breakdown into fine powder under a centrifugal force. One can achieve a uniform dispersion of particles by using a ball milling operation. Ball milling is suitable for nanostructured metallic and ceramic materials. A typical ball mill consists of a powerful motor, a concealable rotating drum/container lined with abrasion-resistant lining, and abrasion-resistant balls. The materials for balls are stainless steel, agate, hardened steel, tungsten

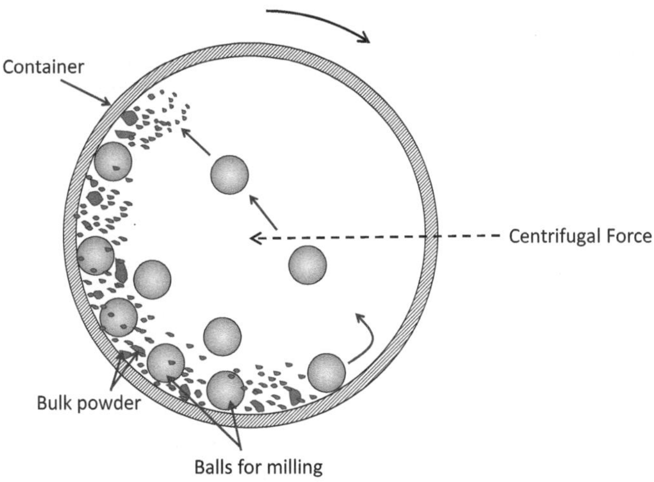

FIGURE 2.2 Milling operation inside a planetary ball mill.

carbide, zirconia, Al_2O_3, etc. The drum is usually lined with the same material as the balls. The bulk material will be powdered, and the balls will be placed in a concealed container rotated about an axis by the motor. The number of balls and amount of material are selected based on the parameters listed below. The axis of rotation can be horizontal, vertical, or at a certain angle with respect to the vertical or horizontal axis, depending on the type of ball milling used.

The bulk powder must be subjected to high-energy collision from the balls to prepare nanomaterials from bulk. The type of ball milling used for such applications is high-energy planetary ball mills. In planetary ball mills, cylindrical grinding drums/jars are placed as a planet with respect to an axially rotating wheel called a sun wheel. The bulk material and grinding balls are placed in the concealed planetary jars, which rotate in orbit around the sun wheel so that, as the sun wheel spins, the planetary jars are self-rotating in the opposite direction. The resulting centrifugal and coriolis forces produce strong grinding effects. The grinding action in a ball milling operation is schematically shown in Figure 2.2. The size of the nanoparticles obtained using ball milling operation depends on several parameters.

Parameters of Ball Milling Operations are

Material and Size of Milling Balls: The material for milling balls should be denser than the bulk material to be milled. The size of the balls also significantly impacts the grinding operation. Milling balls of 10 to 30 mm are typically used to prepare nanomaterials.

Ball-to-Powder Ratio: The ball-to-powder ratio is the weight of balls to that of the bulk material to be milled. One must optimize the ball-to-powder ratio to achieve the required nanoparticle size.

Occupied Milling Drum Volume: The free space in the drum after filling the material and the balls must be optimal for the free movement of the balls. Overfilling may lead to inefficiency in milling and, at the same time, increase the risk of wear on the milling balls and the drum's inner lining.

Milling Time: Bulk powder materials are gradually broken down into nanopowder form during ball milling. The process requires a specific amount of time to attain the anticipated particle morphology and size. Long milling time results in the deterioration of milling balls and the drum liner, while short milling time will not produce the required result. Milling time for nanomaterials ranges from 4 hours to 24 hours.

Milling Speed: The milling speed significantly affects the operation of the ball mills. The ball mill exerts a centrifugal force on the milling balls, colliding with the material and resulting in the desired breakdown. At the critical speed of a ball mill, the balls defy the force of gravity. As a result, the balls rotating along the liner of the jars at the starting will detach from the rotating jar liner and start impacting the materials under milling, as shown in Figure 2.2. The optimal milling speed is approximately 70% of the critical speed of the ball mill. A high speed can deteriorate the balls and jar material, and a lower speed is inefficient.

2.1.3 Laser Ablation

Laser ablation removes material from a solid substrate using a high-powered laser beam. It is also known as photoablation. The laser ablation technique synthesizes nanoparticles by ablating or vaporizing a target material using a high-energy pulsed laser. The target material is immersed in a solvent; usually, water is used as a solvent. The schematic of the laser ablation technique is presented in Figure 2.3. The target material is a solid, such as a metal plate, foil, or a mixture of materials. When the high-energy laser pulse hits the target, it rapidly heats a small volume of the material, causing it to vaporize or form a plasma. This process ejects atoms, ions, and clusters of atoms from the target surface into the surrounding liquid. The ejected material rapidly cools and condenses in the solvent, forming nanoparticles. The size and

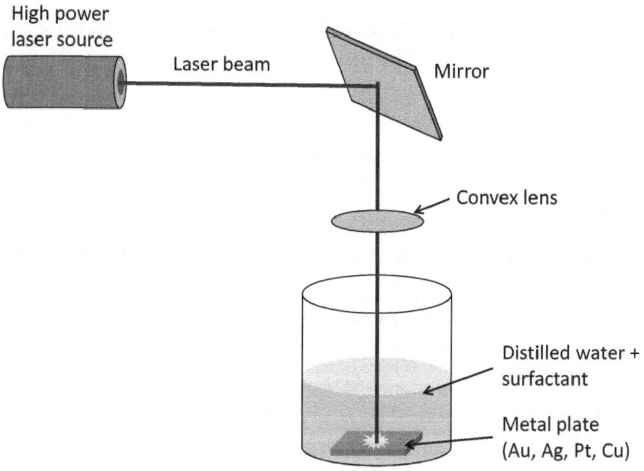

FIGURE 2.3 Laser ablation synthesis of nanomaterials.

composition of the nanoparticles can be controlled by adjusting parameters such as laser energy, pulse duration, repetition rate, and the properties of the surrounding medium (solvent). A suitable surfactant is added to the solvent to prevent aggregation of the nanoclusters. The nanomaterials synthesized using laser ablation have high purity and uniform size distribution, shape, and composition. Nanoparticles of stable metals, such as Au, Pt, Si, Cu, etc., can easily be fabricated using laser ablation techniques. Ceramic nanoparticles such as TiO_2, ZnO, and Fe_3O_4, semiconductor quantum dots such as cadmium selenide (CdSe), lead sulfide (PbS), and indium phosphide (InP) are successfully fabricated using laser ablation technique.

2.1.4 Exfoliation and Intercalation

Exfoliation is commonly used to produce two-dimensional (2D) nanomaterials from bulk materials formed by Van der Waals and ionic forces. This process is a top-down approach in which the bulk material is disintegrated into 2D nanomaterials. During exfoliation, the bulk material is expanded along the axis perpendicular to the layers, forming the bulk structure. The exfoliation technique can produce high-purity nanomaterials, and it is used for high-quality nanomaterials such as graphene, layered double hydroxide, MXene, nano clay, transition metal dichalcogenides (TMDs), etc.

Generally, the Van der Waals or ionic forces holding the layer together are overcome through a series of operations. Chemical exfoliation is the most effective technique for synthesizing 2D layered structures from bulk materials. The first step in chemical exfoliation is to identify a suitable solvent that can reduce the effect of binding forces, such as Van der Waals and ionic forces. The next step is to intercalate foreign ions/molecules between the layers. Metallic ions or surfactant molecules are used for this purpose, as shown in Figure 2.4. The separation of graphene from

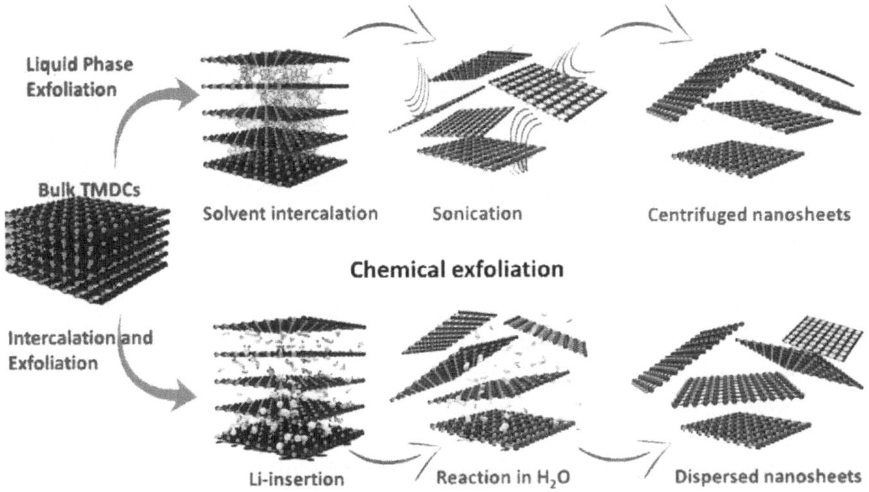

FIGURE 2.4 A typical chemical exfoliation of 2D nanolayers from bulk materials. (Reproduced with permission of the John Wiley & Sons, Raza et al., 2021)

bulk graphite mostly follows intercalation-assisted exfoliation. However, some bulk materials have covalent, ionic, and metallic bonding between the layers. Such materials can be exfoliated only after etching the bulk material using a strong acid. For instance, the Ti_3C_2 MXene is initiated by etching the Ti_3AlC_2 MAX phase using hydrofluoric acid (HF) along the edges.

The exfoliation process can also be achieved using mechanical and electrochemical exfoliations. Mechanical exfoliation is suitable for large-scale production of 2D nanolayers. Sonication (treating under ultrasonic waves), ball milling, and high shear mixing are commonly used to separate the 2D layers mechanically. The product quality may not be as good as chemically exfoliated 2D layers. Electrochemical exfoliation is an intercalation-assisted process during which a voltage is applied to intercalate ions from the electrolyte into the bulk material, for instance, graphite. The ions then expand and exfoliate individual graphene sheets. The type of the applied potential and magnitude determines the exfoliation reaction rate.

2.1.5 Physical Vapor Deposition

Physical vapor deposition (PVD) deposits a thin film onto substrates in a high vacuum. PVD is used in electronics, optics, and coatings to produce thin films with precise and uniform thickness, specific composition, and properties. In PVD, the physical transport of material from a source (usually a solid target) to a substrate takes place, where it condenses to form a thin film.

2.1.5.1 Pulsed Laser Deposition

Pulsed laser deposition (PLD) is a PVD process in which a high-power pulsed laser beam vaporizes the target material to be deposited. Pulsed laser beams, typically 10–30 ns pulses, are used during this process. The entire deposition is performed under a high vacuum for metals; however, the chamber can be filled with reactive gases depending on the thin-film composition. For instance, oxygen is used as the background gas for oxide films, nitrogen for nitrides, and acetylene for carbides. As the laser beam strikes the target material, the target is evaporated, forms a plume, and ultimately deposits on a heated substrate facing the target. A schematic of the PLD process is shown in Figure 2.5.

The film growth mechanism in PLD is complex, as several forms of energy conversion occur during the process. As the target absorbs the laser light, a thin layer on the target material is superheated to form a luminous plume resulting from evaporation, ablation, and plasma. The plume is at a high temperature and contains ionic species, atoms, molecules, electrons, particulates, etc. The deposition thickness varies from ~0.25–1.0 nm/sec under typical deposition conditions. The stoichiometric simultaneous ablation of multiple targets or the target exchange systems allows the achievement of the desired chemical composition of the film. The precise control of the target exchange with individual growth rate regulation can also prepare heterostructures and superlattices. The quality of the films depends on the quality of the laser source and the frequency of pulses. Even though the PLD is simple and versatile, the area of the depositions is limited up to 2 cm^2. Moreover, the high kinetic energies of the target particles impacting the substrate can result in a sputtering effect.

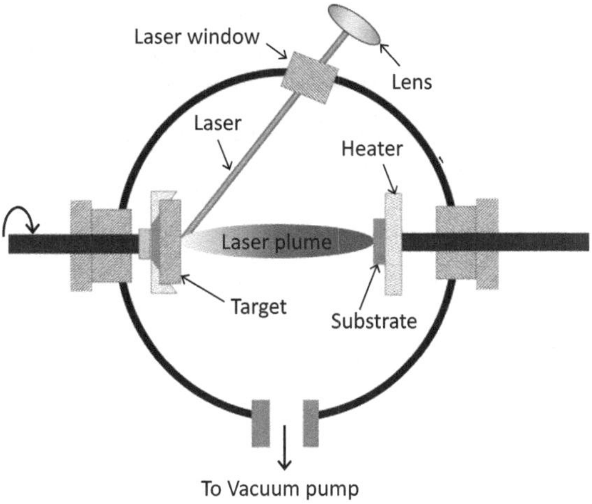

FIGURE 2.5 Schematic diagram of a PLD process.

The laser ablation in PLD can be compared to laser machining, where the material from the target surface is removed by converting energy from photons to electronic excitations and then into thermal, chemical, and mechanical energy. (Miotello and Kelly 1999). In the PLD process, the laser's wavelength/energy should be absorbed in a very shallow layer near the surface of the target to avoid subsurface boiling, which otherwise can lead to a large number of particulates at the film surface. (Christen and Eres 2008). For synthesizing materials with complex composition, retaining the stoichiometry is challenging as the stoichiometry depends on the material removal rate from the target. Suppose the vaporization temperatures of the component elements are dissimilar; the elements with high volatility must be evaporated in excess, or the substrate must be placed at an angle from the axial direction of the plume.

2.1.5.2 Sputtering

Sputtering is a thin-film deposition process where the films are stoichiometric to the specific target. During the process, the atoms are ejected from the target with the help of energized ions to form a plasma directed to the substrate under a high vacuum. Argon is the commonly used sputtering gas, carried out with the help of a DC power source, radio frequency alternating current, or ion-assisted deposition. In reactive sputtering, a reactive gas, such as oxygen for oxides and nitrogen for nitrides, is also passed to the reaction chamber along with argon. These reactive gases react with the target atoms in the plasma to form the desired composition. The schematic of the sputtering system is shown in Figure 2.6.

While forming multi-compositional films, multiple targets with different elements are simultaneously sputtered and deposited as the desired film. Suppose oxygen is used as a reactive gas due to the difference in electronegativity of the target elements and oxygen; the ions formed can be negatively charged. Thus, they are accelerated toward the substrate due to the difference in the potential of the negatively charged target and the grounded substrate. These ionic fluxes can resputter the growing thin

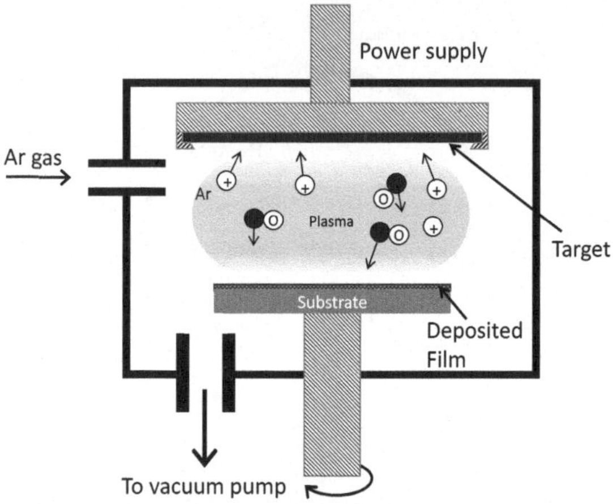

FIGURE 2.6 Sputtering of thin films.

film on the substrate to modify the composition of the films or etch the substrate. To omit such problems, avoid placing the substrate directly facing the target, minimize the discharge voltage, and design the target composition to compensate for the sputtering effects. (Habermeier 2007).

2.1.5.3 Molecular-Beam Epitaxy

In molecular-beam epitaxy (MBE), the respective elements to be deposited are sublimed by heating the target in individual cells called effusion cells or furnaces. The heating can also be carried out using a laser source or an electron-beam evaporator. The reaction chamber is always maintained under a very high vacuum. The elemental vapors are then condensed on the substrate, which is continuously rotated and heated. A precise control of the target and substrate temperature is necessary to control the rate of material deposition on the substrate. A molecular beam represents the evaporated atoms, which may or may not interact with each other depending on the composition of the thin film. A typical MBE process setup is shown in Figure 2.7. A typical MBE system is composed of four main parts: (1) an ultra-high vacuum epitaxial chamber with a background pressure of 1.3×10^{-8} Pa containing a four-target holder and a substrate holder, (2) a means of heating to vaporize the target, (3) a scanning device for the composition analyzing, and (4) a high-energy electron diffraction (RHEED) system and the charge-coupled device (CCD) camera.

A high-energy electron diffraction (RHEED) accompanies the MBE setup to monitor film growth continuously. The shutters control the release of vaporized elements from the effusion cell. This method is used for metal film deposition. However, injecting the respective reactive gases into the chamber achieves the oxide or nitride films. In reactive MBE, the pressure is maintained sufficiently low to guarantee the reaction between the vaporized atoms and the reactive gases. A hybrid MBE technique is used to deposition complex oxide films. In a hybrid MBE, one element is evaporated in the effusion cells, and the chemical beam produced by the thermal

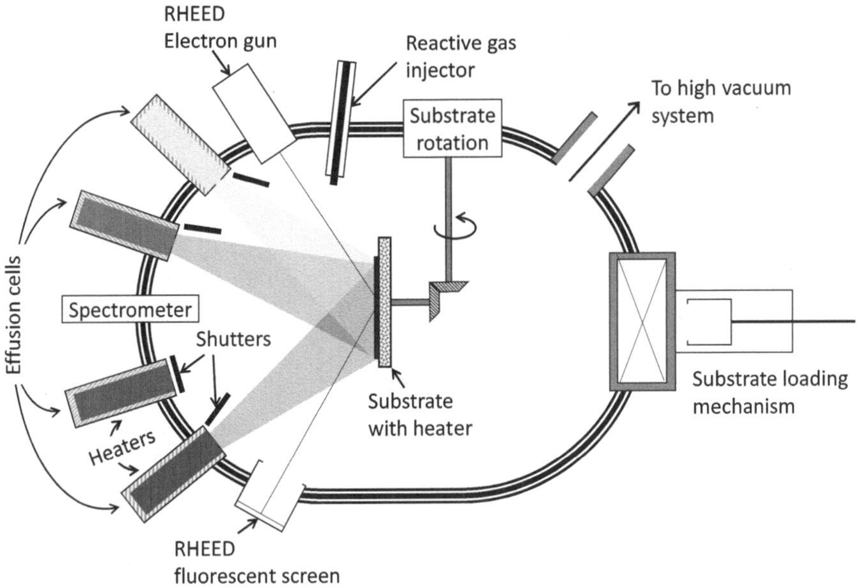

FIGURE 2.7 MBE reaction chamber.

evaporation of the respective precursor is used as the source for the second element. In reactive MBE, the pressure is reduced to 10^{-3} Pa; thus, the mean free path of the atoms evaporated from the cells is much larger than the source–substrate distance. For the effective deposition of complex oxides, retaining the stoichiometric ratios of the molecular beams of the different constituent elements and the corresponding fluxes is essential. Spectroscopic techniques such as atomic absorption spectroscopy, mass spectroscopy, quartz-crystal monitors, and electron microscopes are often employed to do the same.

The significant advantage of the molecular-beam epitaxy is the fabrication of thin films with few unit cell thickness. As with other thin-film deposition techniques, MBE requires a substrate; thus, the obtained films have a heterostructure. However, MBE deposits on any substrate that can withstand a high temperature. MBE deposits either a functional electrode on a conductive substrate, or the conductive layer on a functional electrode. A laser source can replace the effusion cell to vaporize the targets; such a technique is called laser MBE.

2.1.5.4 Thermal Evaporation

In thermal evaporation thin-film coating, the target material is physically evaporated, employing heat under vacuum. The evaporated target material will condense directly to the substrate in a solid state, similar to the condensation of water vapor on a lid. The thermal evaporation system consists of a sealed chamber connected to a vacuum pump with provisions to heat the target material resistively and holders for the substrate. A schematic of the thermal evaporation system is shown in Figure 2.8. The metallic target materials are fed into the "boat," the name of the evaporators used in

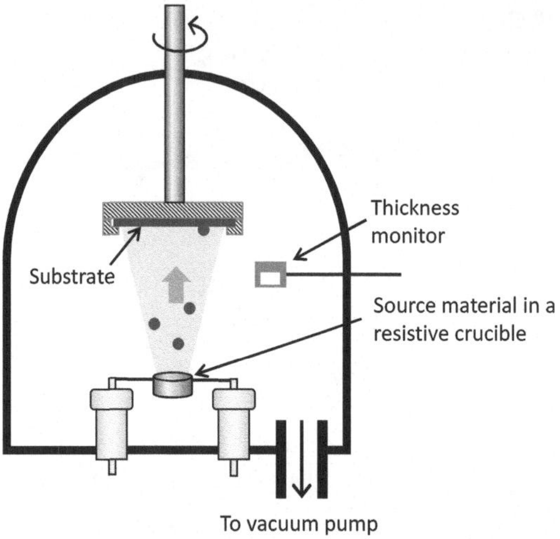

FIGURE 2.8 Thermal evaporation technique for thin films.

typical thermal evaporation due to their unique shape. The target materials are heated above the melting temperature and evaporated to deposit on the substrate. Often, the boats themselves are highly resistive, and the adequate power supply allows them to be heated above the melting temperature of the targets. Alternatively, the target material in a crucible is heated radiatively by an electric filament, or it is fed continuously onto a heated element, which allows evaporation.

In a typical thermal evaporation process, a high vacuum is mandatory because the presence of gas molecules in the chamber can either redirect the travel of vaporized molecules toward the substrate or interact with the vapors to affect the purity of the films. However, in reactive thermal deposition, the evaporated metal atoms can interact with reactive gases to form the respective compound film.

2.2 BOTTOM-UP APPROACHES

2.2.1 Solution-Assisted Reactions

In a simple solution-assisted technique, the molecular level missing is achieved with the help of solvent, which is otherwise impossible to accomplish in a solid-state process. Once thoroughly mixed metal salt solution is prepared, the solvent is evaporated, and the steps followed in the solid-state reaction are followed. This method is commonly used for the synthesis of oxide nanomaterials. The calcination temperature can be reduced significantly using the solution technique compared to the solid-state reaction. This method can be modified to synthesize nanoparticles with tailored shapes and sizes. There are two steps in the process; in the first step, the precursor nanomaterials are prepared, and in the latter step, these precursor nanomaterials are heated at a high temperature to yield the respective ceramic nanomaterials. The surfactants and catalysts control the morphology and the size of the nanoparticles.

2.2.2 Coprecipitation

In a coprecipitation process, the nanomaterials are formed by nucleation and growth in a solvent containing the precursor. The solvents are selected based on the solubility of the precursors of the nanomaterials. After the precipitation reaction, the formed nanostructures must be insoluble in the chosen solvent. One can use a surfactant as a capping agent to control the nanomaterials' growth after nucleation. The resulting nanomaterials will have a uniform shape and size. In the absence of a capping agent, the growth of the nanoparticles continues through the Ostwald ripening processes. Additionally, one can control the growth rate of the nanoparticle using mixed solvents. Water/ethanol mixture at different ratios is commonly used to fabricate nanomaterials of various sizes and shapes. The shape and size of the nanoparticles formed by precipitation also depend on the precursors used for the precipitation reaction. Interestingly, the shape of the nanostructures formed during coprecipitation depends on the crystal structure of the nanomaterial composition. If the intended composition has a cubic crystal structure, the nanomaterial adopts a cubic shape, hexagonal or trigonal structure form rods, and orthorhombic structure form spheres.

A super-saturated mixture of precursors dissolved in a solvent is prepared first for a coprecipitation reaction. The precursor solution is then mixed with a precipitating agent, such as hydroxides, oxalate, citrate, carbonate, etc., to precipitate the intended nanostructured materials. Sometimes, the precipitates need further heat treatment after washing to form the final nanomaterials. During the precipitation process, one can easily control the shape and size of the obtained nanomaterials by adjusting the temperature and pH of the precursor solution before mixing with the precipitating agents such as oxalic acid, ammonium oxalate, ammonia, NaOH, ammonium bicarbonate, sodium carbonate, tetramethylammonium hydroxide, etc. A representative coprecipitation reaction is shown in Figure 2.9.

2.2.3 Hydrothermal/Solvothermal Synthesis

Hydrothermal/solvothermal synthesis is any heterogeneous chemical reaction occurring in the presence of a solvent media above atmospheric pressure and temperature. "Hydrothermal reactions" generally represent the reactions if the solvent is water, and a "solvothermal" reaction is carried out using any solvent. Frequently, such reactions are performed in air-tight volume vessels called hydrothermal/solvothermal autoclave reactors. During hydrothermal reactions, the reactors are heated to a fixed temperature; above the solvent's boiling temperature, the pressure builds up autogenously in the reactor by the vapor pressure of the solvent. Under such conditions, solid phases can crystallize from the reactants at the specific temperature and composition of the hydrothermal solution. Therefore, any hydrothermal process can be distinguished as a combination of two steps; the former is preparing the precursor solution in a liquid phase with the accurate stoichiometric composition of elements. Later, temperature treatment determines the morphology.

The hydrothermal synthesis of nanomaterials can be classified into two methods. In the first method, the reactants and a mineralization agent are dissolved in water,

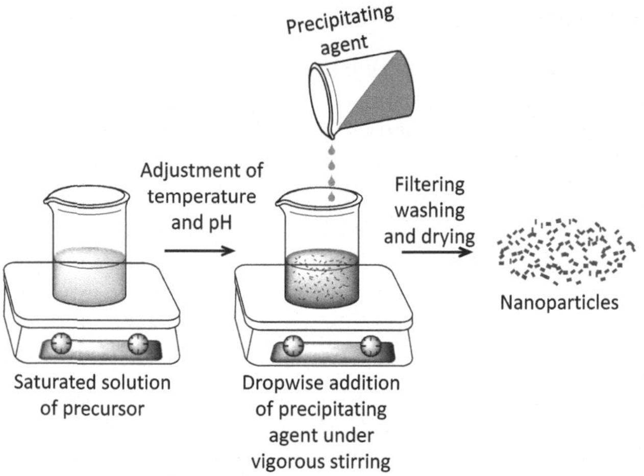

FIGURE 2.9 Coprecipitation reaction.

and the above solution is heated above the supercritical temperature of the water, > 350 °C. In the second method, autoclave vessels lined with Teflon are used; therefore, the temperature conditions for using such reactors are limited to 200 °C and pressures less than 10 MPa. Under severe reaction conditions that involve highly concentrated acidic or basic solutions, metallic reactor materials are not recommended due to possible rapid corrosion. The hydrothermal method involves the dynamic interaction of processing parameters such as solvent type, temperature, and pressure, which govern the chemical reactions by controlling ionic mobility. During the reaction, the particles are grown from liquids, and the size and shape of the particles depend on the number of nuclei formations and the subsequent particle growth in the controlled atmosphere. The supersaturation of the reactant solutions can significantly accelerate the nucleation rate and growth. The presence of surfactants, adjusting pH, and growth inhibitors can control the morphology of the particles formed during hydrothermal synthesis. The simultaneous precipitation, reduction, oxidation, hydrolysis, etc., also determines the morphology and composition of the final products.

Hydrothermal methods are the most straightforward processing does not require any specialized instruments or complicated processes; nonetheless, the method allows reasonable control over particle size, chemical composition, phase, and morphology of the resultant products. Hydrothermal processes are often combined with microwave, reverse micelle, microemulsion, colloids, or sol-gel processes to obtain characteristic nanostructures. In a typical hydrothermal synthesis, the precursor salts are dissolved in suitable solvents and heated to a temperature above the boiling temperature of the solvents and under pressure in sealed reactors. The surrounding medium controls the nucleation and growth of the nanostructures; often, the formed crystalline nanostructures are in a single phase.

The generalized synthesis of oxide nanomaterials using hydrothermal processes can be described as follows. The stoichiometric quantities of the precursor salts are dissolved in a suitable solvent, preferably at room temperature. Capping agents or surfactants are added to the above solution to control the growth or morphology. Capping agents also stabilize the nanostructured materials from agglomeration. Finally, an oxidizing or reducing agent is added to the precursor solutions to accelerate precipitation. Then, the reaction is further carried out in hydrothermal reaction vessels to obtain the precursor for the oxides. The product is then calcined at an elevated temperature at a controlled heating rate for several hours to obtain the final oxide nanomaterial.

2.2.4 Sol-Gel Processing

The sol-gel technique is used to synthesize ceramic nanomaterials, such as oxides, carbides, nitrides, etc. By definition, "sol" is a suspension of colloidal particles or molecules in a liquid, and "gel" is a three-dimensional network formed when the sol is mixed with a liquid, which can help to create the network. Advanced ceramic nanomaterials, such as spinel and perovskite oxides with high purity, homogeneity, and accurate composition for many sophisticated applications, are obtained using sol-gel processing. The sol-gel process needs the inclusion of polymers or in-situ polymerizations of organic or inorganic entities, which essentially contribute to the final composition of the materials.

In general, sol-gel processing involves four significant steps: (1) molecule mixing, which involves mixing the precursor chemicals for the desired ceramic material, a gel-forming medium (or polymer), a solvent, and a catalyst; (2) chemical treatment, which will accelerate the polymerization of the mixture to form a gel, (3) shaping of the polymeric gel from the previous step to the final morphology and (4) high-temperature treatment to remove the volatile and inorganic phases from the molded gel or xerogel from the amorphous to crystalline ceramic material of the desired shape. Electrospinning and freeze drying are examples of the shaping processes that transform the polymeric sol into a gel before it is calcined at high temperatures. If a dense material is desired, drying is followed by hot or cold isostatic pressing, and then sintering is required at a higher temperature. The starting material for a sol-gel process is a metal salt or ligand. Metal salts such as nitrates, acetates, formates, citrates, and tartrates, which are soluble in various solvents, are commonly used, and the more thermally stable salts such as chlorides and sulfates are avoided. Metal salts provide a viable source of oxides in sol-gel processing because they are readily converted to oxide by thermal or oxidative decomposition. The sol(gel processing requires a temperature lower than that is necessary for solid-state processes since reagent mixing occurs at an atomic/molecular level, and the interdiffusion of ions from one phase into the other is much easier.

The metal salts and a complexing medium are subsequently polymerized in a typical sol-gel process to form a hybrid material. Different morphologies can be obtained based on the polymeric gels' processing after hydrolysis and condensation, as shown in Figure 2.10.

The homogeneity of the solutions is essential in a sol–gel process. The metal salts should be uniformly dispersed, and subsequent gelation should freeze all elements in a gel network (Livage et al. 1988). Pechini's process is widely used in the

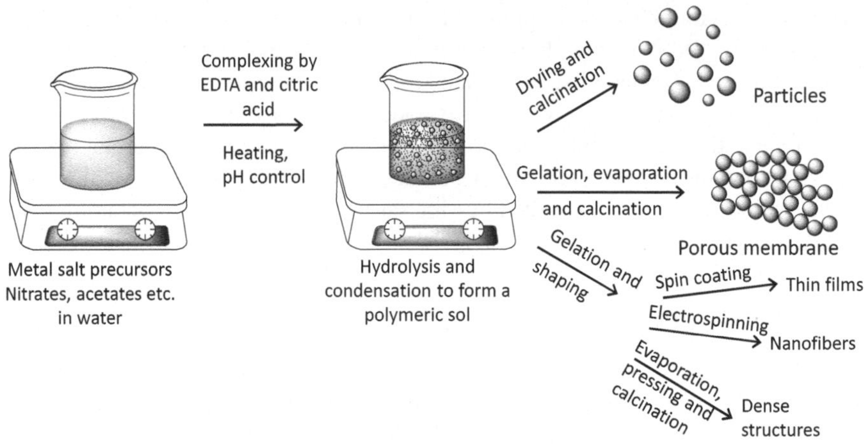

FIGURE 2.10 Steps in sol-gel processing.

sol-gel synthesis of oxides. In the typical fabrication of nano oxides using Pechini's method, the metallic ions are complexed using ethylenediaminetetraacetic (EDTA) and citric acid, and ethylene glycol is added as a stabilizing agent. The pH of the solution is maintained between 8–11 using ammonium hydroxide. The electron-donating carboxylic and aliphatic amine groups of EDTA are linked to the metal cation by six bonds, and the citric acid links with the metal cation by three bonds. Chelating agents like EDTA, citric acid, glucose, etc., arrest the movement of the metallic ions in the sol and prevent the partial segregation of cations in the final compound, possibly by the interaction between the metallic ions. Adding ethylene glycol or mannitol, polyhydroxy alcohols, under heating, will lead to polyesterification of the chelates, forming a cross-linked network of metal atoms and organic radicals. All the above processes take place in a basic medium. Therefore, a $pH > 7$ must be maintained during the reaction.

Since the number of chelating agents exceeds the number of metal ions, all the metallic ions undergo complexation. As a result, a stable chelate complex system is formed, which is essential to avoid the precipitation of any new phases. The addition of ethylene glycol promotes polymerization, which facilitates homogeneity. Citric acid's primary role is forming a polymeric network during the condensation reactions. With EDTA, glycine can be used as the chelating agent since glycine contains both carboxylate and aliphatic amine groups to participate in the complexation of metal ions and has a stronger chelating ability than citric acid, which contains only carboxylate groups. Therefore, the amount of glycine required for the fabrication is less than that of citric acid. (Liu et al. 2002). Additionally, glycine and citric acid undergo exothermic pyrolysis, increasing the temperature in the reaction zone to the surrounding temperature during the calcination. The Pechini method is a popular method adopted by many researchers working on complex oxides because of its simplicity, excellent control of the composition of the complex structures, and lower processing temperature requirements.

2.2.5 Microemulsion

Microemulsions are thermodynamically stable isotropic dispersion of two immiscible liquids stabilized by an interfacial film of surfactant molecules. Unlike emulsions with macroscopic domains, microemulsion domains are microscopic, with the domains 100 times smaller than emulsions, typically in the range of 1 to 100 nm. Such nanosized dispersions are called micelle, and the monolayer of surfactant molecules provides a consistent shape and dimensions. The microemulsions comprise at least three components: a dispersed non-polar solvent in a continuous polar solvent (water) in the liquid state or *vice versa* in the reverse micelle and a surfactant. The dispersed phases always carry the soluble precursors of the end product. In reverse microemulsions, in addition to the surfactant with long-chain hydrocarbons, typically with 8–18 carbon atoms, co-surfactants (short-chain aliphatic alcohols) are also added.

Surfactants are molecules with amphiphilic (polar) functional groups on one end and a hydrophobic tail. The hydrophobic hydrocarbon chains weakly interact with the polar solvent, while the polar functional groups interact strongly with the polar solvent. The surfactant molecules are oriented with the hydrophobic tail towards the non-polar phase (oil) and the hydrophilic part oriented towards the polar phase, forming a distinguishable boundary between the above phases. As a result, microscopic microheterogeneous domains of oil-in-water (O/W) or water-in-oil droplets (W/O) form. The microscopic emulsion O/W or W/O acts as tiny reactors that allow the nucleation and growth of the nanostructured products in a confined space. The schematic representation of the micelles and reverse micelles is shown in Figure 2.11.

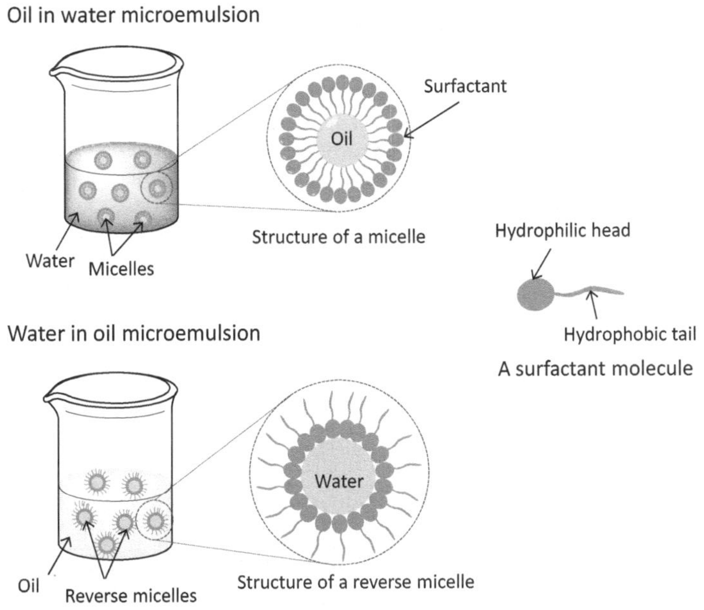

FIGURE 2.11 Structure of micelle.

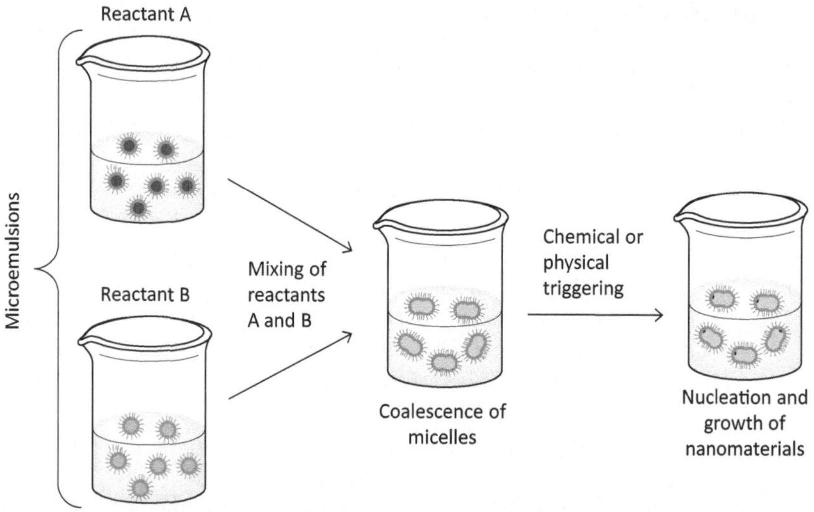

FIGURE 2.12 Microemulsion synthesis of nanomaterials.

Suppose we have two separate microemulsions of reactants that are mixed; the coalescence of the microemulsion takes space to form a worm-like structure, as shown in Figure 2.12, and the formation of resulting nanostructures. Sometimes, the reaction occurs under a high temperature and pressure in a hydrothermal autoclave or a microwave.

The conventional formation of micelles or reverse micelles microemulsions is associated with the hydrophobic and hydrophilic interaction between dispersed and continuous phases. The formation of micelles is a total function of the surfactant concentration determined by the critical micelle concentration. The microemulsion method is recognized for synthesizing inorganic nanoparticles, such as metals, metal oxides, and other inorganic materials. The significant advantage of synthesis through microemulsions is the homogeneity and monodispersion of the nanoparticles. The microemulsion technique helps to synthesize nanosized particles, spheres, cubes, wires, etc., with uniform size distribution.

2.2.6 Combustion Synthesis

In combustion synthesis, a fuel is added to the precursor material to increase the reaction rate by raising the local temperature in the reaction zone. Combustion synthesis combines the advantage of exothermic, rapid, and self-sustaining chemical reactions between metal salts and suitable organic fuels. As a result, the synergistic reaction supplies most of the heat required for oxide formation, and the mixture of reactants need only be heated to a temperature significantly lower than the actual phase formation. The highly exothermic and self-sustaining reaction in a combustion synthesis can be achieved using metal nitrates as precursors (since the nitrate groups act as oxidants) and an organic fuel that produces CO_2, H_2O, and N_2 during the combustion

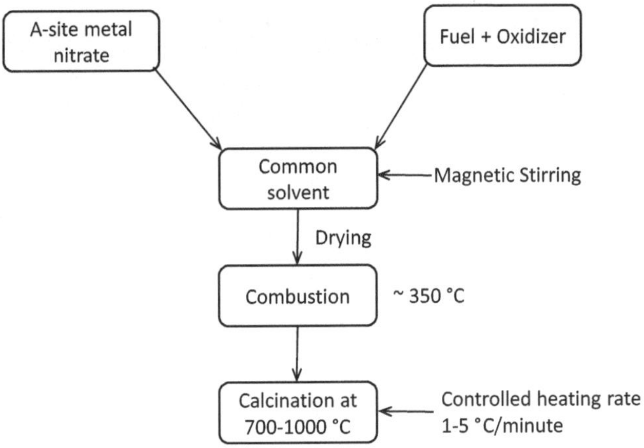

FIGURE 2.13 Steps in combustion synthesis.

reaction (Civera et al. 2003). The organic fuel molecules such as glycine, citric acid, urea, etc. play a double role: they react with the precursors, especially metal nitrates, and form complexes with metal cations to improve the homogeneity, at the same time, avoid the precipitation of new species (Specchia et al. 2004). The flow chart of combustion synthesis is shown in Figure 2.13.

The precursors susceptible to exothermic reactions can lead to a self-propagating combustion reaction, and the local temperature in the reaction zone can be considerably higher than the actual temperature used in the process. As a result, the reaction can occur rapidly at a lower operating temperature, and the obtained products will be highly pure and single phase. Ammonium dichromate is a precursor that undergoes exothermic decomposition to Cr_2O_3, N_2, and H_2O. Metallic materials such as gallium, magnesium, and aluminum can also undergo self-propagating combustion reactions due to their ability to undergo exothermic oxidation under high temperatures. An oxidizing agent, such as sodium perchlorate, is added to promote oxidation. Besides heating, the combustion reaction can be triggered using electrical means. Several organic molecules can be a potential fuel for the combustion synthesis in the presence of a metal nitrate; they are glycine, alanine, glycerol, tetraformal triazine, hexamethylenetetramine, urea, citric acid, etc.

2.2.7 Template Synthesis

A hard or soft template can fabricate metallic, polymer, and ceramic nanosized structures in template-assisted synthesis. A template is identical to a stencil, and the final product will have the size and morphology of the pores of the templates. Porous anodic alumina, porous silica, carbon, etc., are a few examples of hard templates and amphiphilic molecules, such as block copolymers and surfactants for soft templates. When a hard template is used, the sol containing the precursors is prepared first and then poured onto the porous structures. The metal ions can be directly deposited into the pores using a physical vapor deposition technique. Once the template pores are

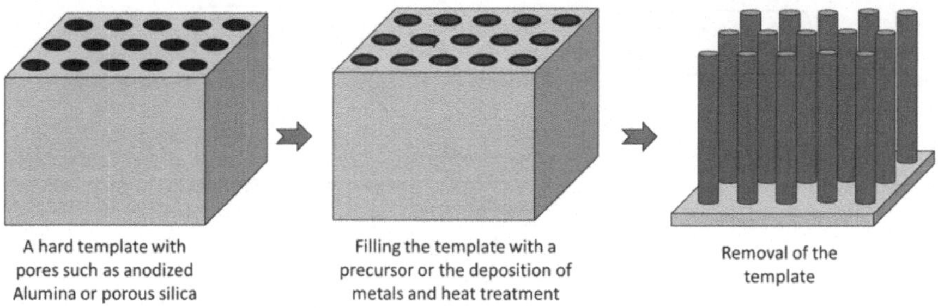

FIGURE 2.14 Template-assisted synthesis of nanomaterials.

filled with the intended materials, the templates are heated at a high temperature to remove the precursor's volatile components and the nanostructures' densification. The typical template-assisted synthesis of nanomaterials is shown in Figure 2.14.

The reactant itself can be a template, e.g., $SrTiO_3$ nanocubes are prepared on the surface of flower-like layered hierarchical titanate spheres as a template, which is the source of Ti. These spheres are prepared by mixing tetrabutyl titanate in a solvent mixture of dimethylformamide and isopropyl alcohol, the solvents, and hydrothermal treatment at 200 °C. In a second hydrothermal treatment, the spheres are heated at 150–180 °C for 10 hours in an aqueous solution containing NaOH, CTAB, and $SrCl_2$ (Kuang & Yang 2013).

Eggshell membranes can be used as porous templates for the fabrication of nanostructures. The membranes are separated from the eggshell by removing the $CaCO_3$ using 1M HNO_3 solution for 5 minutes. The membranes are washed, dried, and immersed in the precursor solution. The membranes saturated with precursor ions are rinsed with de-ionized water and then dried at 95 °C. The membrane/metal salt composite is calcined at 1,000 °C (Dong et al. 2011).

In the soft-template method, polymers, micelles, and surfactants are used as templates, which calcination can remove. The surfactants can be cationic, anionic, or non-ionic surfactants. Three-dimensionally ordered macroporous (3-DOM) structured oxides are synthesized using monodisperse polymethyl methacrylate (PMMA) or polystyrene (Xu et al. 2014) Microsphere colloid crystal template (CCT). In this technique, PMMA microspheres are synthesized using emulsifier-free emulsion polymerization, and the obtained microspheres are then centrifuged to form a CCT, as shown in Figure 2.15 a&b. Then, the precursor salt solution in ethylene glycol/methanol mixture is added to the CCT to fill the gaps between the spheres, which will later form a hard-inorganic precursor framework upon drying. When the CCT/precursor composites are calcined, the polymer degrades, and the precursor will be converted to the respective oxide with a 3DOM structure, as shown in Figure 2.15c.

Templates of biological origin are also used to synthesize nanostructured materials. One such example is the use of a genetically modified M13 virus for the production of nanowires. To prepare the nanowires, the precursor salts are first dissolved in ethylene glycol and then mixed with the virus solution, to which NaOH is added and heated at 80 °C for 4 hours. The virus has an affinity for ethylene glycol bound to the precursor

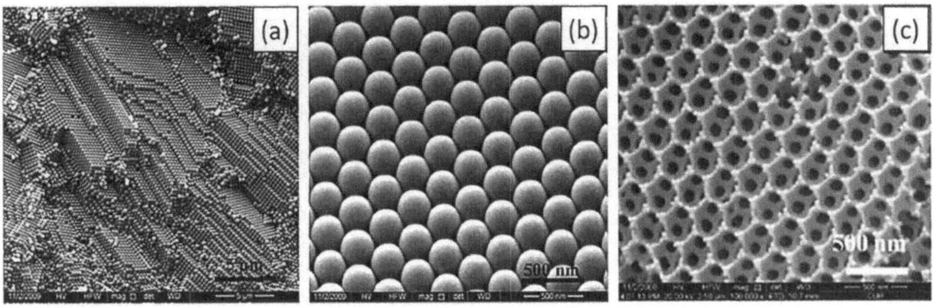

FIGURE 2.15 (a) Low, (b) high magnification images of colloid crystal template (CCT), and (c) Three-dimensionally ordered macroporous (3DOM) structure. (Reproduced with permission of the Elsevier, Xu et al. 2010)

salts at a pH of 10. After incubation for one day, the viruses bound to the precursors are separated and calcined at 600 °C (Nuraje et al. 2012).

2.2.8 Microwave Synthesis

The microwave synthesis is a widely used technique for the bulk and rapid synthesis of organic and inorganic nanomaterials. During microwave synthesis, microwave radiation is applied to initiate the chemical reactions. Even though microwaves contain electric and magnetic fields, only electric fields contribute to the heating process. In such reactions, microwaves employ high-frequency electric fields to heat the materials through dipolar polarization or dipole rotation and ionic conduction. The dipole rotation occurs in the case of polar molecules, during which the molecules try to align themselves with the alternating electric field of the microwave. The rapid change in the electric field applied by the microwave and the attempts of polar molecules to align themselves to the electric field create a rotational motion. This rotational motion of the molecules generates heat by molecular interactions. In the case of materials containing free ions, the ions are rapidly oscillated by aligning themselves with the microwaves with a rapidly alternating electric field. During such rotation of the molecules, the energy is lost in the form of heat through dielectric loss.

Microwave heating of a material is based on its characteristic to absorb electromagnetic energy from the microwave and transform that into heat. Based on the interaction of the materials with the microwave, the materials can be classified into three groups. They are reflectors, transmitters, and microwave absorbers. A microwave reflector is a material that can reflect the microwave, and microwaves do not effectively heat it. The best example of reflectors is metals. Microwave transmitters transmit the microwaves without absorbing, losing, or generating heat. Teflon, quartz, plastics, and some ceramics are a few examples of transmitters, and they are used as containers for chemical synthesis. Microwave absorbers absorb energy from the microwaves and transform it into heat energy instantly; they are high-loss materials.

Therefore, chemical reactions can be performed with the assistance of microwaves. Sol-gel, coprecipitation, microemulsion, hydrothermal, combustion, complexing,

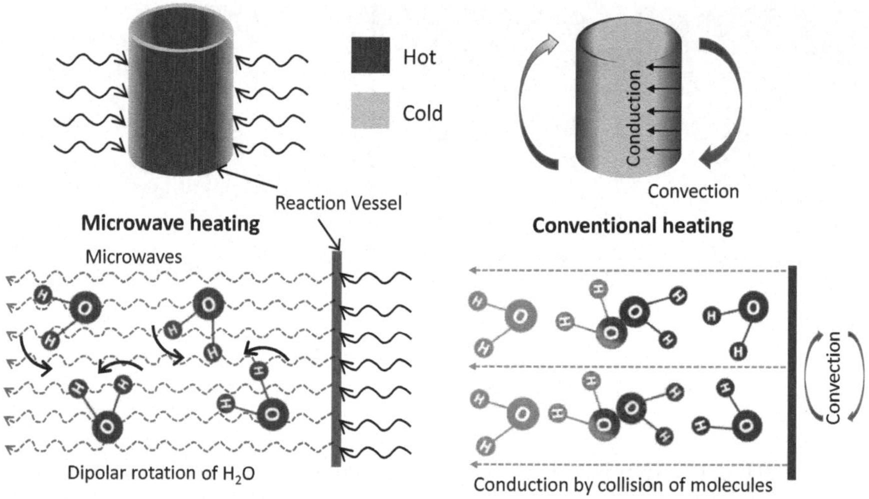

FIGURE 2.16 Comparison of (a) microwave heating and (b) conventional heating of solutions.

etc., are some examples of nanoparticle synthesis techniques that can be assisted by microwave heating. When used for the above chemical reactions, polar solvents, such as DMF, acetonitrile, CH_2Cl_2, ethanol, and H_2O, can be used. Non-polar solvents such as toluene, carbon tetrachloride, diethyl ether, and benzene are microwave-inactive. In a microwave-assisted reaction, heating occurs through the direct interaction of microwave energy with the molecules, such as solvents, reagents, and catalysts present in the recipe, ultimately resulting in the simultaneous heating of the entire reaction mixture. In the typical hotplates, the reaction progresses by conductive heat transfer from the walls of the reaction vessel to the interior through the bombardment of molecules. The mechanism of microwave and the conventional heating of water molecules is compared in Figure 2.16. Therefore, the reaction time for microwave-assisted synthesis is more rapid than that of traditional synthesis using heaters; however, in conventional synthesis, the reaction occurs irrespective of the polar nature of the reactants. The microwave-assisted heating process significantly reduces the time required for traditional heating. For instance, hydrothermal processing of complex oxides requires 24–72 hours of heating in a conventional method, whereas when microwaves are used, the reaction time is reduced to 10 to 30 minutes. If the constituent reactants do not absorb microwaves, microwave-absorbing solid materials like carbon can be used as the local heating element.

2.2.9 Freeze Drying

There are three steps in a typical freeze-drying process: preparation of the precursor solution and primary and secondary drying. Secondary drying involves freezing at temperatures and pressures below the triple point of the precursors to enable sublimation, i.e., without undergoing a liquid state. In the typical synthesis process of

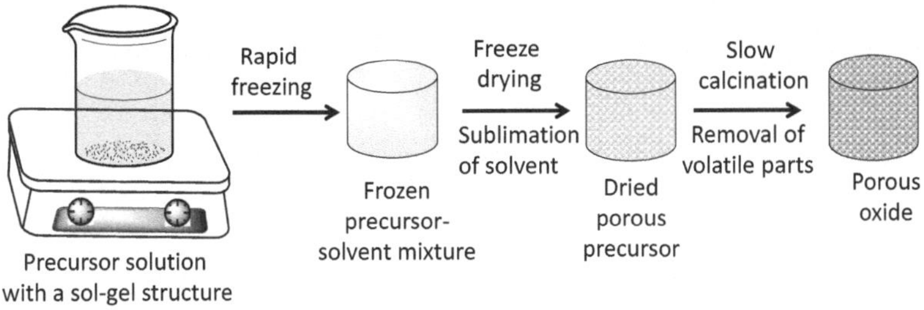

FIGURE 2.17 Steps in freeze-drying process.

oxides, the precursors are first dissolved in a suitable solvent, such as water. The solution containing the precursors is frozen using liquid nitrogen fast enough to retain the chemical homogeneity. The obtained frozen precursor mixture is then freeze-dried in freeze-dried under a high vacuum, during which the unstable solvent part of the precursor mixture is dehydrated or sublimed without undergoing a liquid phase. The obtained porous mixture is then calcined at a high temperature at a slow heating rate to decompose the precursor materials to get the intended highly porous oxide nanomaterials, Figure 2.17.

A sol-gel network-type precursor solution can result in highly porous structures during freeze drying. Therefore, EDTA-citric acid sol containing the nitrate metal salt is an ideal starting material for freeze drying. A polymer can also have a similar effect and is more advantageous than the EDTA-citric acid method since metal salts other than nitrates can be used as the metal source in the polymer-based technique. Freeze spraying is also adopted for the fabrication of ceramic nanomaterials. In freeze spraying, the metal salts are dissolved in a suitable solvent, and then they are sprayed into liquid nitrogen to form precursor particles and freeze-dried. These particles are then calcined at high temperatures to obtain the resulting oxide.

2.2.10 Chemical Vapor Deposition

In chemical vapor deposition (CVD), thin films are deposited from the precursors in the vapor phase onto a substrate. The chemicals decomposed into the intended material on the heated substrate. The deposition process is often thermally driven; however, photo- and plasma-assisted methods can also be used. Since the deposition process is achieved through chemical interaction and the availability of a wide choice of starting materials, this technique is more versatile than PVDs. CVD is a non-equilibrium process, and the control of the precursor vapor or solution determines the phase and morphology of the deposited films or nanostructures. The CVD grows nanomaterials over a large area with high purity and reproducibility with adjustable morphology, composition, tribology, and chemical properties. The significant experimental parameters of the CVD process are the nature of the substrate, the temperature of the substrate and composition of the reaction vapor, the pressure of gas flows, the stability of the precursor during the transport, the efficiency of the precursor to be absorbed and decomposed on the substrate, etc. Accurately controlling the above

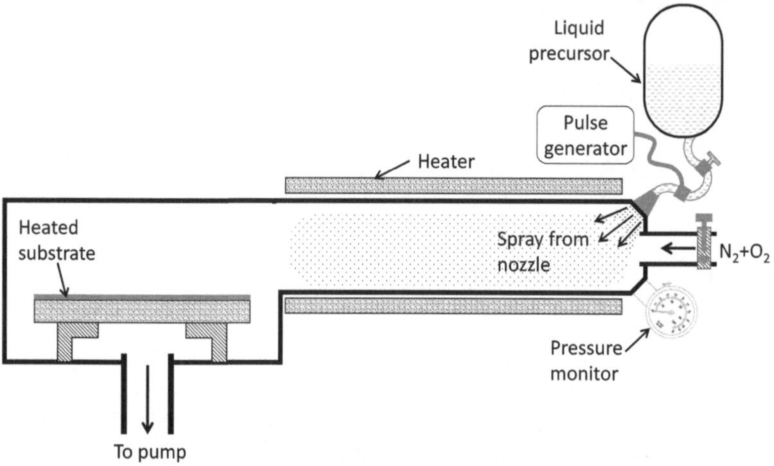

FIGURE 2.18 Direct liquid injection CVD.

parameters in a CVD process can yield products in the form of thin films, powder, or single crystals.

The CVD process synthesizes carbon nanotubes, buckyballs, and metallic and ceramic films. In the typical process, a chemical reaction of gaseous or vapor phase reactants occurs on a heated substrate. A carbon-rich precursor gas with a fuel and a catalyst is used to synthesize carbon allotropes. CVD processes are classified depending on the vaporization of the reactants and their transport to the substrate. Aerosol-assisted CVD (AACVD) and direct liquid injection CVD (DLICVD) are widely used for fabricating complex oxide films. In AACVD, precursors are transported to the substrate using a liquid/gas aerosol. In DLICVD, the precursors are dissolved in a suitable solvent, and the precursor solutions are injected into the vaporization chamber through injectors, similar to the fuel injectors in the automotive. The precursor vapors are then transported to the substrate and deposited as films of required composition. The interaction of precursor vapors with oxygen can lead to the deposition of oxide layers. In that case, annealing/calcination in the presence of oxygen is unnecessary to form oxide layers or particles. The construction of the DLICVD chamber is shown in Figure 2.18.

If the AACVD or DLICVD process uses metal-organic materials as the reactants, it is commonly called metal-organic chemical vapor deposition (MOCVD). In MOCVD, the metal-organic precursor is either evaporated to a vapor phase or dissolved into a suitable solvent and atomized before it is carried to the CVD chamber employing an inert/reactive carrier gas (Argon, N_2/O_2, N_2, etc.). Since the deposition of the films takes place by the interaction of decomposed reactants at a high temperature, both AACVD and DLICVD are known as thermal CVD processes.

The typical CVD process of ceramic materials has the following steps: (1) dissolution of precursors in a stoichiometric ratio to a suitable solvent and atomization or vaporization of target materials with the required composition; (2) transport of the vaporized reactants to the heated substrate; (3) reaction of the target atoms and the reactive gas (such as oxygen, nitrogen, or methane) during the transport; (4) deposition of the vapor phase reactants on the substrate. An inert carrier gas is used to dilute the reactants in the vapor phase and to carry them to the substrate. The growth rate of ceramic nanostructures in a CVD process is much higher than in a reactive PVD process. MOCVD is the most widely used CVD process for the fabrication of oxides.

2.2.11 Molten-Salt Synthesis

If the solubility of the precursor salts is limited in solvents, then the molten salt can be used as a reaction medium. Here, molten salts act as solvents, provided the salt is soluble in water, and the final product is not. Despite the solubility in water, salt must be stable at high temperatures. During the process, salts with a low melting point are added to the precursors and heated above the melting point of the salt, and the molten salts dissolve the precursors. Once the precursors are entirely dissolved in the molten salt, the salt can be cooled down slowly to room temperature. The slow cooling of the melt allows the crystallization of the final product. As the reactant materials are entirely dissolved in the molten salt, solid particles nucleate homogeneously in the liquid phase during cooling. Therefore, by controlling the nucleation sites, one can adjust the size of the crystals. In molten-salt synthesis, the reaction rate is increased at a lower temperature than in solid-state processing. Using molten salt as the solvent, the particle size, shape, and level of aggregation can be controlled. The diffusion rate in solid-state reactions is prolonged, but in the molten-salt method, the mass transport is faster because, in the liquid phase, the mass transport is through convection and diffusion. As many salts, by nature, dissolve in water, molten-salt synthesis (MSS) has the advantage of easy product isolation.

A single crystal with a large size can be obtained by reducing the number of nuclei formed during the cooling of molten salt. Then, the water-insoluble final product is separated from the water-soluble salt flux using excess water. Many nuclei must be present during cooling to prepare particles using molten-salt synthesis. The surfaces of the reactants are the heterogeneous nucleation sites. In such cases, reactants must be partially soluble in the molten salts; in other words, complete solubility for all reactants is undesirable. When insoluble reactants are used as raw materials, the process is called topochemical molten-salt synthesis (TMSS). In TMSS, the morphology of the final product is identical to that of the insoluble raw materials. Therefore, the starting materials act as a template, as shown in Figure 2.19. The starting reactants are transformed into oxide materials by exchanging, deleting, or inserting individual atoms. The cooling rate determines the shape and size of the powder particles, especially during single-crystal growth.

Formerly, a small quantity of molten salts was used as an additive in solid-state reactions to improve the reaction rate. When molten salt is used as the solvent, a large amount of salt (80 to 120 wt% of the reactants) is used to control the morphology and size of the final product. Chlorides and sulfates are often used as salt baths. The eutectic mixtures of the salts are often used to lower the melting temperature of the salts. The salt mixture 0.5 NaCl–0.5 KCl (eutectic composition) with a melting point of 650 °C and 0.635 Li_2SO_4–0.365 Na_2SO_4 with a melting point of 594 °C is the most

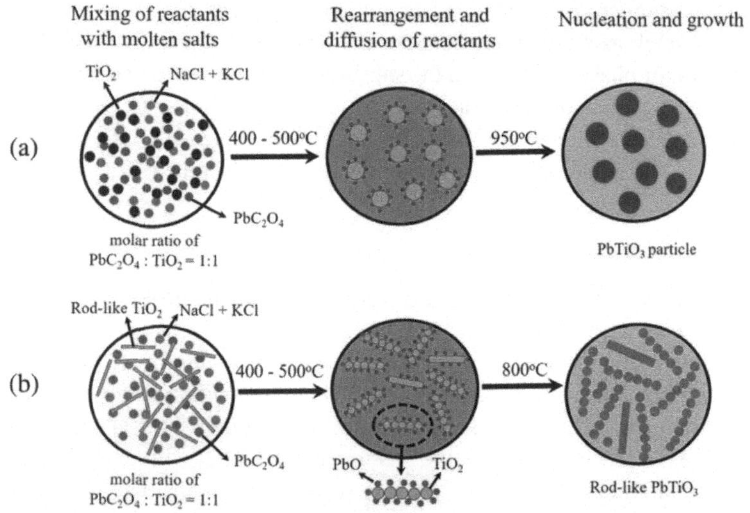

FIGURE 2.19 Steps in the molten-salt synthesis. (Ji et al. 2019)

commonly used salt with low melting temperature and Na_2SO_4–K_2SO_4 mixture is often used for high temperature, which has a melting point of 823 °C.

2.2.12 Electrochemical Synthesis

Electrodeposition, or electroplating, is commonly used to deposit metals into a conducting surface from an ionic salt solution of the intended metal. Electrodeposition is generally adopted for fabricating thin films on metallic surfaces to induce corrosion protection, abrasion resistance, aesthetics, and reduce contact resistance. Electroplating can be extended for the deposition of multiple metallic elements as well. In an electroplating setup, the working electrode, electrolyte, power source or a potentiostat, and the counter electrode form a closed circuit. Sometimes, a reference electrode is also used as a potential for the potentiostat; however, the current flows only between the working and counter electrodes. The working electrode is the conducting material being plated, and the counter electrode is an inert material such as platinum or graphite. The flow of electrons occurs through the ionic electrolyte through the movement of metallic cations in the electrolyte. In other words, as the working electrode is connected to a negative potential, the positively charged metallic ions in the electrolyte move toward the working electrode or cathode to adhere to the surface evenly to form a thin film on the surface. The process continues until all the cations are deposited on the cathode's surface, and the deposition rate is majorly affected by the applied current and cell potential. A schematic of the electrodeposition process is shown in Figure 2.20.

Electrochemical deposition can typically be classified as galvanostatic synthesis and potentiostatic synthesis. In galvanostatic experiments, the working electrode's potential is selected concerning a reference electrode such that the current density

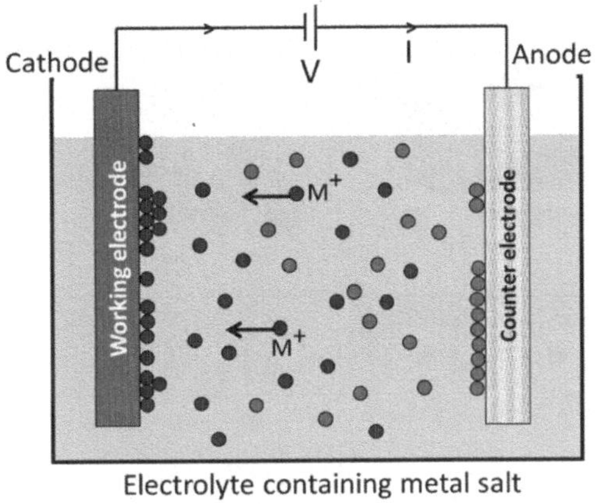

FIGURE 2.20 Schematic of an electrodeposition setup.

is constant. Still, the cell potential drifts as the reactant activity decreases, resulting in different products. In a potentiostatic synthesis, the voltage across the working electrode-counter electrode pair is adjusted to maintain a constant potential difference between the working and reference electrodes in a three-electrode assembly. In this case, the cell current usually decays rapidly as the reaction proceeds; due to low diffusion rates of the reactant molecules from the bulk to the electrode surface, a decrease in the activity of the reactant occurs. However, the reaction is likely to yield a pure single-phase product selected by the applied potential.

2.2.13 Electrospinning

Electrospinning is analogous to electrostatic spraying. In both electrospinning and electrostatic spraying, a charged liquid is deposited on a grounded substrate. Low-molecular-weight solutions are used in electrostatic spraying, but a high molecular weight polymeric material is used in electrospinning. The polymeric fibers are deposited on a collector plate during the electrospinning process. Electrospinning is a bottom-up approach where the polymerized precursors at the molecular level in an appropriate solvent are spun as fibers. At present, electrospinning is the leading technique used in the fabrication of organic, metallic, carbon, and inorganic nanofibers as compared to the other methods, such as vapor phase transport, laser ablation, arc discharge, template-assisted, lithography, phase separation, and hydro/solvothermal methods. Electrospinning is unique for the fabrication of nanofibers because of its simple construction, large-scale productivity, versatility, high aspect ratio and uniformity of the nanofibers, the possibility of automation, large-scale production, low cost, and it does not depend on the operator skills.

An electrospinning unit has three significant components they are 1) a syringe pump, which feeds the polymer solution or melts with the help of a syringe or any other means also known as spinneret; 2) a high-voltage DC supply; and 3) a grounded collector plate, on which the nanofibers are collected. The construction of a simple

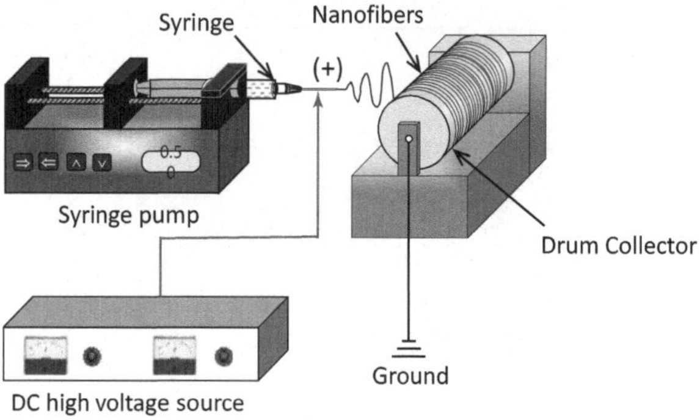

FIGURE 2.21 Schematic of an electrospinning unit.

electrospinning unit is shown in Figure 2.21; the purpose of the spinneret is to deliver the solution at a controlled rate. The high-voltage unit must be able to supply a DC voltage ranging from 1 to 100 kV. The geometry of the collector plate also varies to meet the product's requirements. A static collector plate is used for collecting randomly oriented fibers. However, a rotating drum collector can align the nanofibers in one direction.

As we apply high DC voltage during the electrospinning, the solution droplet at the spinneret is positively charged, and the grounded collector plate attracts the positive charges. If the applied voltage is sufficiently high, this electrostatic charge interaction overcomes the polymer droplet's surface tension, resulting in nanosized jets traveling toward the collector. The solvent evaporates from the spinneret to the collector plate, and the polymeric fibers will be collected on the collector. Electrospun nanofibers are post-processed to fabricate carbon and ceramic nanofibers. The electrospun nanofibers of polyacrylonitrile are heated at ~500 °C under high vacuum for the carbon nanofibers. To fabricate ceramic and metallic nanofibers, precursor solutions containing a polymer and the metallic salts of the respective metal/ceramic are electrospun to get xerogel (precursor fibers without any solvent) fibers. These xerogel fibers are then heated above the degradation temperature of the polymer to get the respective oxide ceramic fibers. The heating at high temperatures is called calcination. During the calcination, the polymer and the volatile components of the metallic salts are removed to form the final oxide structure. One has to heat the metal oxide fibers in a reduced atmosphere to obtain the metallic fibers.

Many organic nanofibers, including protein, biodegradable, and non-biodegradable polymers, are successfully fabricated using electrospinning. Almost any ceramic oxide nanofibers can be fabricated using electrospinning, provided the metallic salt of the respective ceramic must be available. Simple to complex ceramic oxide nanofibers are successfully fabricated using the electrospinning process.

Organic or inorganic metal salts that result in the formation of oxide as they are heated are used to fabricate ceramic oxide fibers. Salts, namely acetates, isopropoxides, formates, citrates, nitrates, chlorides, etc., are added to the polymer solution for ceramic oxide fiber processing. Nitrates and chlorides are widely used metal salts because of their availability, and thermally stable salts such as sulfates are generally avoided because of the difficulty in removing the anions. The homogeneity of the spinnable solutions near the molecular level is achieved by continuous mixing, and the movement of metal ions is expected to be prevented in the polymer sol. Highly volatile solvents, in which the metal salt is sparingly soluble, are avoided for the electrospinning process as the sudden evaporation of the solvent can crystallize the metal salt and ultimately destroy the homogeneity of the precursor fibers. Maintaining an acidic pH may reduce the crystallization tendency of metal salts; otherwise, a solvent with high solubility and boiling point is preferred. Such precautions avoid the rapid gelation of the fibers during the electrospinning process, which otherwise results in large fibers.

During the calcination of ceramic fibers, a low heating rate is often used, succeeded by a certain dwell period. A slow heating rate ensures the removal of organic components without destroying the nano fibrillar morphological features and, in the end, products. It also avoids the disintegration of ceramic fibers. The pyrolysis temperature of the polymer base used determines the minimum calcination temperature. Thermogravimetric analysis (TGA) is widely used to identify the degradation point of the polymer; it is also important to note that the calcination temperature should be well above the oxidation threshold of the metal salt. The calcination of the composite samples is performed at different high temperatures and dwell periods because the calcination temperature and the dwell period have a significant role in controlling the grain size of the fibers, which in turn reflects in the performance of the final products. Sintering for an extended period is recommended if a dense rather than a nanoporous material is desired. The high surface area of the particles leads to rapid densification due to the significant grain growth rate at high temperatures.

2.2.14 Spin Coating

Spin coating deposits nanoscale thin films on glass or single-crystal substrates. Spin coating is the easiest way of depositing a thin film. The thin films of organic and inorganic materials can be deposited using the spin coating. However, the inorganic coatings are obtained by the post-processing of the substrate carrying the precursor coating at a high temperature. The precursor coating must be deposited based on the final composition of the layer. The spin coating utilizes the centrifugal force exerted on the precursor solution. Furthermore, the precursor solution is obtained by dissolving the precursors in a suitable solvent. A spin coater is used to deposit films utilizing the spin coating technique. A spin coater has speed control and a vacuum system for holding the substrate on the table. A droplet of the precursor solution is placed at the substrate's center to deposit a thin film, and the spin coater's table that holds the substrate is rotated at the required speed. As the substrate rotates, the solution spreads uniformly on the substrate while spinning off the excess solution, forming a continuous thin layer. During the process, the solvent evaporates, and thin films with the preferred thickness are formed on the substrates. The spin coating technique is more straightforward than CVD and PVD techniques; however, the thin oxide films obtained through the spin coating are polycrystalline rather than single crystals in CVD or PVD. Unlike PVD or CVD, the selection of materials for spin coating is abundant. The central controlling factors of spin coating are rotation speed

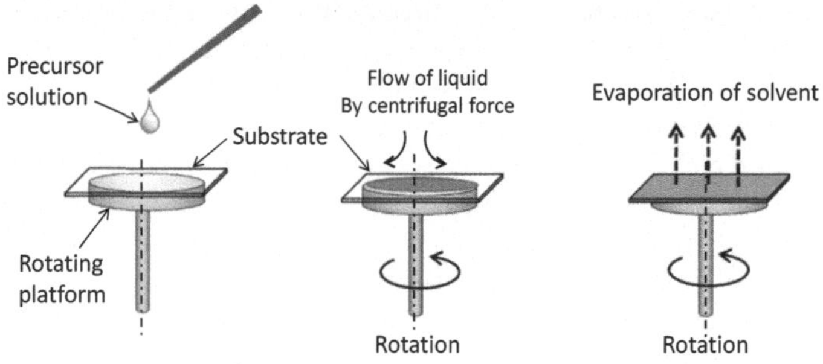

FIGURE 2.22 Steps in the spin coating process.

and the viscosity of the solution. However, factors such as time, the solvent's volatility, and the substrate's wettability play a vital role in determining the quality of the films. The step-by-step process of a spin coating technique is shown in Figure 2.22.

2.2.15 Langmuir Blodgett Film Technique

Langmuir Blodgett (LB) technique is used for depositing films with thicknesses ranging from a monolayer of molecules to multiple layers of molecules. Typically, in the LB technique, a monomolecular layer of an amphiphilic precursor substance is developed on the water surface. It is then transferred to a solid substrate; the precursor molecules are arranged on the air-water interface. Initially, the precursor solution containing the amphiphilic molecules is added slowly to a stagnant water surface. The molecules are arranged so the hydrophilic part is oriented towards the water. The monolayer on the water surface is then compressed using a movable barrier, as shown in Figure 2.23a, to form a continuous layer. The continuous monolayer film is transferred to the required solid substrate by immersing the flat substrate in water and then extracted with the film adsorbed on the surface, Figure 2.23b. The transferring process must be prolonged to retain the film's continuity. By repeating the process, one can obtain multi-layered films. The layer of molecules on the water's surface is termed a Langmuir monolayer, and the film deposited on the substrate is called Langmuir–Blodgett film. This technique is commonly used for depositing layers of biomolecules. The method can be extended for inorganic layers by selecting the appropriate precursor and post-processing obtained LB films.

2.2.16 Spray Pyrolysis

The idea of spray pyrolysis is similar to spray painting. In spray pyrolysis, the spray is the precursor solution of the nanostructured material, and the surface to be painted is heated. Spray pyrolysis is a liquid-phase process, as the precursor is in the form of a solution. At first, the precursor is made into the nanosized mist with a nebulizer's help, and the mist is directed towards a hot substrate kept in a furnace. As the precursor mist is heated at a high temperature, it decomposes to form the final thin film structure. Therefore, one has to select the precursors and the solvent to produce the

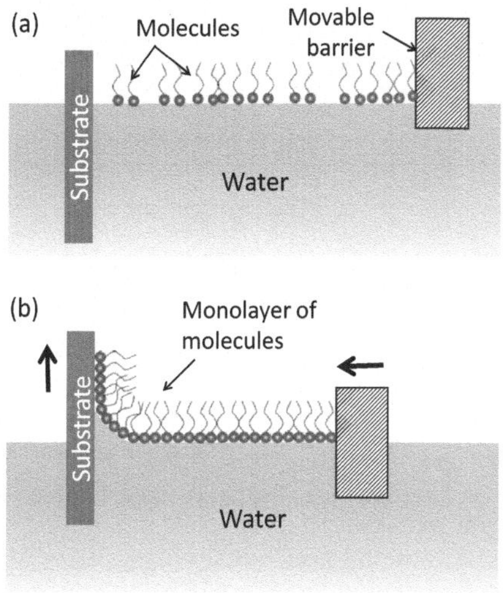

FIGURE 2.23 Langmuir Blodgett film technique (a) moving the loose precursor molecules to a continuous film using a movable barrier (b) transferring Langmuir monolayer to a substrate by dip coating.

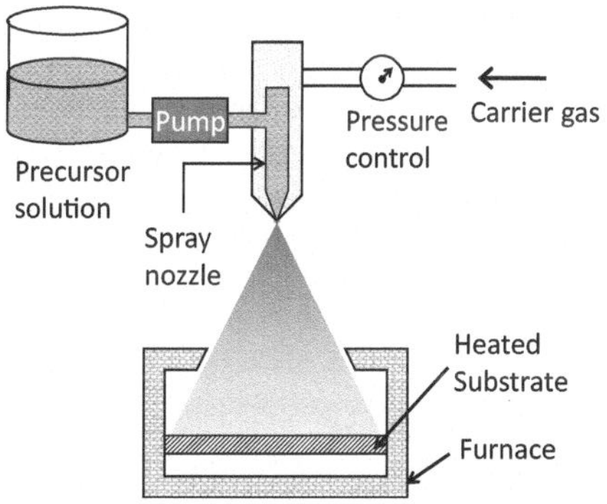

FIGURE 2.24 Spray pyrolysis.

desired products upon heating. The factors controlling the final products are the substrate temperature, carrier gas flow rate, nozzle-to-substrate distance, and the solution content and concentration. The substrate temperature plays the most significant role among the above parameters as the drying, decomposition, crystallization, and grain growth depend on temperature.

The typical spray pyrolysis process is shown in Figure 2.24. The substrate/surface is heated by an electric heater whose temperature can be accurately controlled. At the spraying nozzle, the precursor solution is atomized with the help of compressed carrier gas, which is air in the case of oxide films. The pressure of the carrier gas is controlled by a pressure regulating valve, and if needed, the carrier gas is supplied in

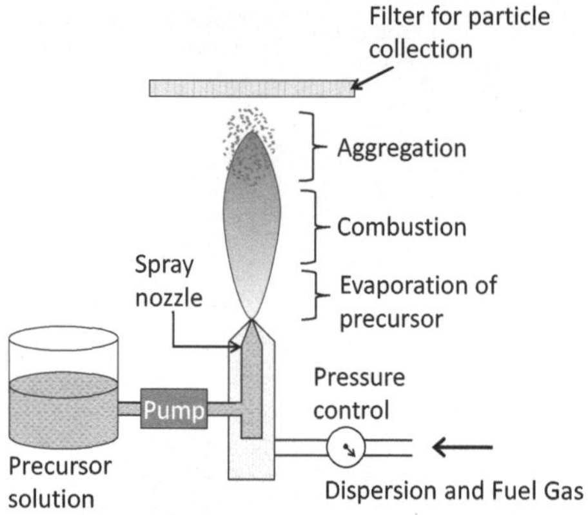

FIGURE 2.25 Flame spray pyrolysis.

pulse mode. By controlling the various parameters, nanostructures with thin film and particle-like architectures can be obtained.

The spray pyrolysis technique used to fabricate nanoparticles is known as flame spray pyrolysis or flame hydrolysis, as shown in Figure 2.25. During the fabrication, the precursor solution of metal salts is atomized using an ultrasonic atomizer or a nozzle fed to an oxy-hydrogen flame. Due to the high drying rates and high temperatures in the flame region, the precursor solution will be instantly converted to nanosized particles. This method is suitable for gaseous, liquid, and solid precursors. Such a technique has a high production rate and yields particles with high purity and uniform size distribution. This technique is highly suitable for inorganic nanomaterials.

REFERENCES

Christen HM, Eres G. 2008. Recent advances in pulsed-laser deposition of complex oxides. J Phys: Condens Matter. 20(26):264005. https://doi.org/10.1088/0953-8984/20/26/264005

Dong D, Wu Y, Zhang X, Yao J, Huang Y, Li D, Li C-Z, Wang H. 2011. Eggshell membrane-templated synthesis of highly crystalline perovskite ceramics for solid oxide fuel cells. J Mater Chem. 21(4):1028–1032. https://doi.org/10.1039/C0JM03061A

Habermeier H-U. 2007. Thin films of perovskite-type complex oxides. Materials Today. 10(10):34–43. https://doi.org/10.1016/S1369-7021(07)70243-2

Ji Q, Xue P, Wu H, Pei Z, Zhu X. 2019. Structural characterizations and dielectric properties of sphere- and rod-like PbTiO3 powders synthesized via molten salt synthesis. Nanoscale Res Lett. 14(1):62. https://doi.org/10.1186/s11671-019-2899-9

Kuang Q, Yang S. 2013. Template synthesis of single-crystal-like porous SrTiO3 nanocube assemblies and their enhanced photocatalytic hydrogen evolution. ACS Appl Mater Interfaces. 5(9):3683–3690. https://doi.org/10.1021/am400254n

Liu S, Tan X, Li K, Hughes R. 2002. Synthesis of strontium cerates-based perovskite ceramics via water-soluble complex precursor routes. Ceramics International. 28(3):327–335. https://doi.org/10.1016/S0272-8842(01)00098-0

Livage J, Henry M, Sanchez C. 1988. Sol-gel chemistry of transition metal oxides. Prog Solid State Chem. 18(4):259–341. https://doi.org/10.1016/0079-6786(88)90005-2

Miotello A, Kelly R. 1999. Laser-induced phase explosion: new physical problems when a condensed phase approaches the thermodynamic critical temperature. Appl Phys A. 69(1):S67–S73. https://doi.org/10.1007/s003399900296

Nuraje N, Dang X , Qi J, Allen MA, Lei Y, Belcher AM. 2012. Biotemplated synthesis of perovskite nanomaterials for solar energy conversion. Adv. Mater. 24(21):2885–2889. https://doi.org/10.1002/adma.201200114

Raza A, Hassan JZ, Ikram M, Ali S, Farooq U, Khan Q, Maqbool M. 2021. Advances in liquid-phase and intercalation exfoliations of transition metal dichalcogenides to produce 2D framework. Adv Mater Interfaces. 8(14):2002205. https://doi.org/10.1002/admi.202002205

Specchia S, Civera A, Saracco G. 2004. In situ combustion synthesis of perovskite catalysts for efficient and clean methane premixed metal burners. Chem Eng Sci. 59(22):5091–5098. https://doi.org/10.1016/j.ces.2004.08.028

Xu J, Liu J, Zhao Z, Zheng J, Zhang G, Duan A, Jiang G. 2010. Three-dimensionally ordered macroporous $LaCo_xFe_{1-x}O_3$ perovskite-type complex oxide catalysts for diesel soot combustion. Catalysis Today. 153(3):136–142. https://doi.org/10.1016/j.cattod.2010.01.063

Xu J-J, Wang Z-L, Xu D, Meng F-Z, Zhang X-B. 2014. 3D ordered macroporous $LaFeO_3$ as efficient electrocatalyst for Li–O_2 batteries with enhanced rate capability and cyclic performance. Energy Environ Sci. 7(7):2213–2219. https://doi.org/10.1039/C3EE42934B

3 Characterization of Nanomaterials

Nanomaterials exhibit properties different from the bulk materials. Their unique properties depend on their high surface area, size, shape, crystal structure, elemental composition, and other physical properties. The reactivity at the nanoscale is close to atomic or molecular level interaction. Therefore, in characterizing nanomaterials, we need techniques that are different from those of bulk materials. Often, these techniques are complementary, and one can confirm the properties obtained from one characterization with another. Since many synthesis techniques are available to produce a wide range of nanomaterials, we need precise characterization protocols for reliable and reproducible results. We need multiple characterization tools to confirm the result obtained by one characterization tool with the results obtained from another. Moreover, different characterization techniques are used to analyze nanomaterials of different origins, and the information revealed is different for different applications.

There are many characterization tools available to the researchers to under the properties of the nanomaterials. Each characterization technique is unique and suitable for understanding one particular property. However, during the analysis of nanomaterials, many characterization tools are combined to conclude the suitability of that nanomaterial for the intended application. This chapter covers the most important characterization tools commonly used to analyze nanostructured materials. We can generally classify them as morphological, compositional, structural, electronic, optical, and thermal characterization tools. This classification is based on their capability of the characterization technique; however, such a classification overlaps in many instances. For instance, the same characterization tool can be used for morphology, structure, and electronic properties. The morphological characteristics can be studied using electron microscopic techniques, but the attachments with the electron microscopes can analyze the composition, structure, and electronic properties. The typical characterization technique and the information retrieved are listed in Table 3.1.

3.1 MORPHOLOGICAL CHARACTERISTICS

The morphological characterization of synthesized nanomaterials is essential before it can be used for any application. Nanomaterials exist in different sizes and shapes

DOI: 10.1201/9781003317272-3

TABLE 3.1
Characterization and the Information Retrieved

Characterization Tool	Information Retrieved
Scanning Electron Microscope	Size, morphology, compositional images
Transmission Electron Microscope	Size, morphology, defects, crystal structure
X-Ray Diffraction	Crystal structure, impurities, crystallite size
Atomic Force Microscope, AFM	Size, morphology, 3D images of the samples, mechanical properties, magnetic properties
Brunauer–Emmett–Teller Surface Area, BET	Surface area analysis
Fourier Transform Infrared Spectroscopy FTIR	Molecular interaction, Surface modification, ligand binding
XAS (EXAFS, XANES)	X-ray absorption coefficient (element-specific)—chemical state of species, interatomic distances, Debye-Waller factors, also for non-crystalline NPs
SAXS	Particle size, size distribution, growth kinetics
XPS	Electronic structure, elemental composition, oxidation states, ligand binding (surface-sensitive)
Dynamic Light Scattering	Particle size, size distribution, surface charges
Thermogravimetric Analysis, TGA	Mass and composition of stabilizers
UV-Vis Spectroscopy	Size, concentration, electronic structure
Photoluminescence Spectroscopy	Luminescence characteristics
ICP-MS	Elemental composition, size, size distribution, NP concentration
Electron Diffraction	Crystal structure, lattice parameters, study order-disorder transformation, long-range order parameters

based on the crystal structure, composition, and synthesis method. Based on the shape of the nanomaterials, such as particles, tubes, wires, flowers, cubes, etc., they behave differently in their respective applications. Controlling the morphology can alter the surface area of the nanomaterials and, as a result, their functionality. The nanomaterials can be fine-tuned based on their morphology. Therefore, morphological characterization is performed on a newly synthesized nanomaterial. High-resolution imaging at the atomic scale can reveal the presence of defects, impurities, different phases, and their morphology. The most common nuclear scale and nanoscale imaging techniques are scanning electron microscope (SEM), transmission electron microscope (TEM), atomic force microscope (AFM), and scanning tunneling microscope (STM). One can study the nanomaterials' structural and electronic characteristics and chemical composition using electron microscopy. AFM and STM allow the manipulation of the atomic scale.

3.1.1 Electron Microscopy (SEM)

Ordinary optical microscopes are unsuitable for analyzing nanomaterials because their resolution is limited to one-half the wavelength of visible light. The wavelength of light is 400–700 nm; therefore, the resolution of the optical microscope is ~200 nm. In electron microscopes, high-energy electron beams are used as the illumination source, which allows higher resolution than optical microscopes, with the highest resolution of about 0.1 nm. When the electron beam interacts with the material, the depth of penetration of the electron beam can reach up to 2 μm, depending on the accelerating voltage applied to the electron gun.

In electron microscopes, a focused beam of high-energy electrons is used as an illuminating source. These accelerated electrons can interact with the atoms of the materials differently, and the signals obtained are shown in Figure 3.1. The origin of these signals depends on the electron beam's penetration depth. Suppose the electron beams penetrate the sample without interaction or undergo elastic or inelastic scattering. Both elastically and inelastically scattered electrons are used for imaging and other information from the sample using TEM. Backscattered electrons (BSEs) are electrons in the primary electron beam elastically reflected by the nuclei of atoms. They are high-energy electrons. The secondary electrons (SE) are the electrons from the sample's atoms, knocked off by the high-energy primary electron beam. Both SE and BSE are used for imaging in SEM. SE are produced from a small volume and in small quantities from the surface; therefore, they are commonly used for imaging the morphology of the surface. BSE is produced in large quantities and from a large volume and is suitable for compositional imaging. The production of BSE is a function of atomic weight. If the atomic weight is high, the BSEs emitted will be

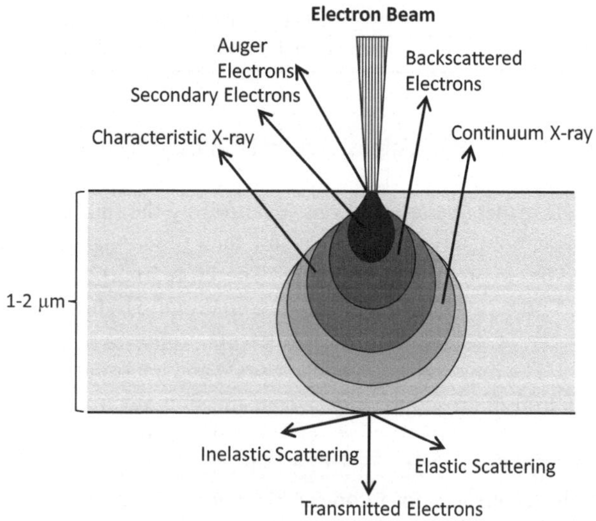

FIGURE 3.1 Signals produced by the interaction of electrons with materials and the depth of penetration.

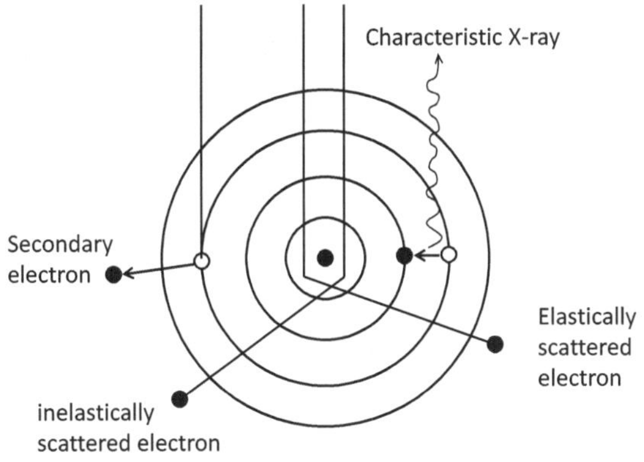

FIGURE 3.2 Signals analyzed in an SEM.

more significant, and the resulting image will be bright. Imagine the sample contains regions with different compositions; then, BSE imaging can identify those regions.

The high-energy primary electrons can eject one of the atom's inner shell electrons (K, L, or M) to create a hole. The hole will be filled by an electron from a higher-energy level, releasing energy during its transition from the high to the lower energy state, as shown in Figure 3.2. This X-ray is analyzed for the sample's composition using energy-dispersive and wavelength-dispersive X-ray spectroscopy.

Auger electrons are generated when the primary electron beam knocks off an electron from the inner shell, and the vacancy created is filled with an electron from a higher-energy shell. The energy released during this transition knocks another electron in the higher-energy shell called the Auger electron. Analyzing Auger electrons can reveal the sample's composition and chemical bonding state. Cathodoluminescence (CL) is a characteristic of certain materials that emit photons when illuminated by electrons. This light emission can also be used for imaging in an SEM, provided the CL detector is attached to it.

3.1.1.1 Scanning Electron Microscope

In a scanning electron microscope (SEM), a high-energy electron beam is focused on the solid sample surface to produce a variety of signals at the surface of solid specimens. These signals can be used for imaging and identifying chemical composition and crystal structure. SEM is used to scan a small area on the surface, and the scanning is done line by line, also known as a raster scan. The image obtained from SEM is on greyscale, and the variations in the grey are used to display the differences in the morphology of the sample surface. Since the conductive materials generate SE and BSE electrons effectively, the non-conducting samples must be coated with a thin layer of conducting materials, such as gold, before imaging those samples. In conventional SEMs, magnification from 20X to 200,000X is possible with a spatial resolution of 1 to 10 nm. The SEM supports the analyses at selected points or locations on the sample surface for analyzing the composition and crystal structure. The construction of an SEM is shown in Figure 3.3.

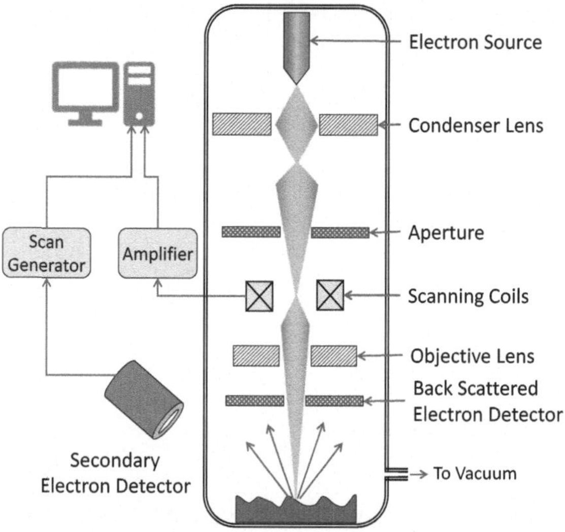

FIGURE 3.3 Construction of SEM.

The major components of an SEM are:

Electron Gun: There are two types of electron guns used in SEM: a tungsten filament for thermionic emission and a field emission gun. An electron gun acts as the electron source.

Electron Lenses: Unlike glass lenses in optical systems, electromagnetic lenses focus and control the electron beam to achieve the required resolution.

Specimen Chamber: The specimen chamber is designed to load and unload the sample for examination. The sample is kept under a high vacuum and SEM instrumentation column to avoid any interaction of electrons with air molecules, which otherwise degrade the quality of the images.

Scanning Coils: The scanning coils help monitor the sample in a raster manner, so they control the motion of the electron beam.

Detectors: There are different types of sensors attached to an SEM. These detectors are placed above or beside close enough to collect the signals produced from the sample after electron bombardment. The standard SEM detectors are the SE, BSE, EDS, and WDS. SE generates surface features, and BSE is used to create compositional features.

Image Formation: The signals detected by the detectors are used to develop the image. The image is created based on the signal strength received by the detector as the electron beam scans the surface line by line.

3.1.1.2 Transmission Electron Microscope

In TEM, an electron beam is transmitted through thin specimens and projected to a fluorescent screen to form an image. The thickness of the specimen must be less

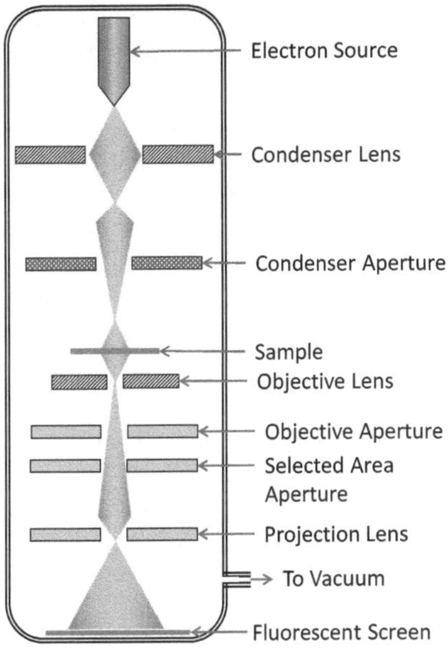

FIGURE 3.4 Construction of a TEM.

than 100 nm to make it transparent to the electron beam. The image obtained from a transmission electron microscope results from the electron-sample interaction as the electron beam passes through the sample. The magnified image is then projected to a fluorescent screen, a photographic film, or a scintillator screen.

The construction of TEM is schematically shown in Figure 3.4. In TEMs, a high-voltage electron beam is used to image the sample. The electron beam from the electron gun placed at the top of a TEM travels through the column of the microscope. To avoid any scattering of the electron beam, the column of the TEM is maintained under high vacuum. The electron beam is focused to a fine point using the electromagnetic lens. The fine electron beam is transmitted through the thin specimen as the electron enters the specimen and scatters elastically or inelastically with the specimen. The signals from the scattered electrons form the final image on the fluorescent screen at the microscope's bottom. Electrons interact with the sample through elastic or inelastic scattering, and the signals are based on the above types of scatterings. The image obtained by TEM is based on the thickness or density variation of the sample.

The preparation of thin samples is vital in TEM imaging, since the samples must be transparent to the electron beam. The solid samples of soft materials, such as biological materials and polymers, can be made into thin sections using ultramicrotome. For metallic materials, abrasion followed by electrochemical etching is generally used. The image nanomaterials, nanomaterials are loaded onto carbon film-supported copper grids. The copper grids are available in different mesh sizes. The nanomaterials are dispersed in a suitable medium, usually ethanol, and a drop of the mixture is placed on the grid to load the nanomaterials into the copper grids. The grid is then allowed to dry at room temperature before loading into the TEM instrument.

TEM is operated in different modes, and the modes are selected by changing the apertures appropriately corresponding to the technique. Some of the standard modes are listed below;

Bright-Field Imaging: Bright-field imaging uses transmitted electrons to obtain the image. Here, the sample appears light, and the background appears dark.

Dark-Field Imaging: The scattered electrons do the imaging here, while an aperture blocks the transmitted electrons. In dark-field images, the sample appears dark, and the background seems light, suitable for atomic-resolution images.

Selected Area Diffraction: By inserting a "selected area diffraction aperture," one can limit the region of interest, and the image forms a set of diffuse rings or an array of dots. These rings or dots correspond to the diffraction of electron beams by the samples. Therefore, diffused rings or arrays correspond to the crystal structure of the samples.

X-Ray Spectroscopy: Similar to EDX and WDX analysis in SEM, compositional analysis is possible using TEM by detecting the energy and wavelength of the X-rays emitted from the sample.

Electron Energy-Loss Spectroscopy (EELS): EELS measures the ionization of the sample by the electron beam. Here, the energy loss of the inelastically scattered electrons is measured. EELS spectra reveal information on core-shell transitions, plasmon resonance, valence-to-conduction band transitions in semiconductors, and vibrational modes (phonons).

3.1.2 Atomic Force Microscope (AFM)

AFM uses a probe to scan the surface to get the information from the surface, which is like a blind man's touch. The probe is arranged as a mechanical cantilever with a diamond tip of atomic-scale resolution. AFM gathers the information from the surface by feeling the surface with the probe. The surface of the nanostructured materials can be studied using AFM.

AFM can be used for surface imaging, force measurement, and manipulation at atomic scale. The surface or topographic imaging provides a 3D structure of the sample surface. The force between the sample and the probe can be used to estimate mechanical properties such as Young's modulus, hardness, etc. The properties or surface characteristics of the sample can be manipulated using the probe by controlling the probe accurately.

An AFM measures the sample and probe interaction through atomic forces. The quality of the measured data in AFM depends on the cantilever and the sharp tip at its end. The cantilever scans the sample surface; as the cantilever tip interacts with the sample surface, the cantilever is deflected, and the deflection is measured and combined to get the image of the entire surface. The major components of an AFM are the cantilever and the tip, laser source, photodiode, piezoelectric crystal, and feedback system, as shown in Figure 3.5.

The cantilever is made of silicon or silicon nitride single crystals. The tip is attached to the free end of the cantilever. The tip is sharp, with a nose diameter of a few nanometers. The standard material used for the fabrication of tips is diamond.

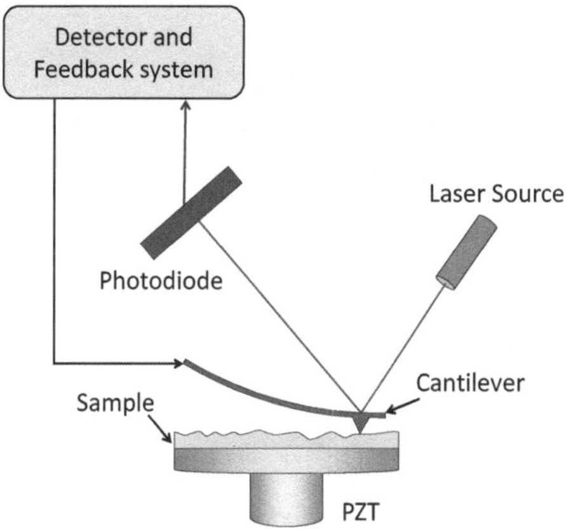

FIGURE 3.5 Working of an AFM.

The interaction of the tip and the sample causes the cantilever to deflect, and the laser beam focused on the back of the cantilever and photodiode arrangement detects this deflection. The feedback system and a piezoelectric crystal to which the cantilever is attached maintain a constant force between the tip and the sample surface.

Any adjustment to maintain a constant height difference between the tip and sample by keeping a constant deflection of the cantilever is translated to a 3D surface map.

There are three modes through which an AFM can be operated:

Contact Mode: In this mode, the sample and the tip are in constant contact and are used for flat and rigid surfaces.

Tapping Mode (Non-Contact Mode): The tip oscillates near the resonance frequency with the help of a piezoelectric crystal and makes contact with the surface intermittently. This mode is useful for soft and delicate samples to prevent any damage to the sample.

Non-Contact Mode: The cantilever tip does not contact the sample surface but oscillates near the resonant frequency. The interaction of the sample and the cantilever decreases the cantilever's resonance frequency. The feedback loop maintains a constant oscillation by adjusting the tip-to-sample distance. The feedback loop records the input to keep the tip-to-sample distance constant, and the data is used to construct the topographic image of the sample surface.

3.1.3 Scanning Tunneling Microscopy (STM)

Like AFM, STM creates a 3D image of the surface science at the atomic level. The operation of STM is based on quantum tunneling. Under a high vacuum, if a conducting tip is placed very close to another conductive surface, electrons flow to the latter without touching it, provided a voltage must be applied between them. The current flowing between the tip and the conductive surface is called tunneling

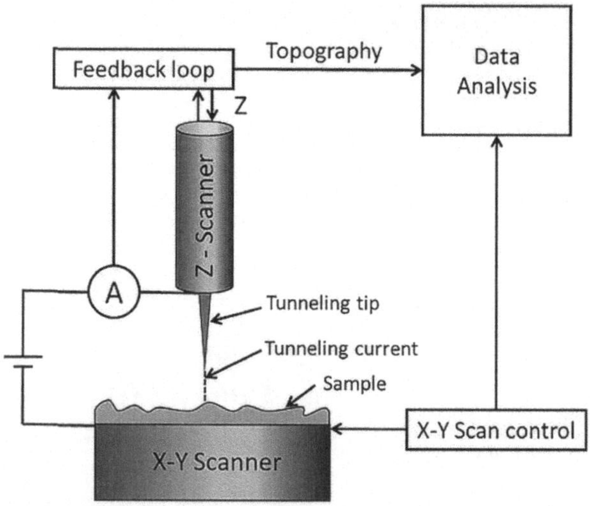

FIGURE 3.6 Working of an STM.

current, and it is a function of the distance between the conductor and the surface. As the current-carrying tip scans the surface line by line, the tunneling current varies depending on the surface texture. Any change in the tunneling current is recorded, and a topographical map is generated using an atomic-scale resolution.

In an STM, a conductive made of tungsten or platinum-iridium is used, and the tips are sharpened to a single atom at the apex of the tip to obtain high resolution. The tip can precisely move along X, Y, and Z directions. A feedback loop maintains a constant tunneling current between the tip and the sample surface, see Figure 3.6. To do so, the feedback loop positions the tip in response to changes in the tunneling current such that the distance between the tip and the sample surface is constant, and the variation in the tunneling current is plotted as the surface's topography. The samples must be conductive or semi-conductive with a flat and clean surface, making the preparation extremely complex.

STM is commonly used to study the atomic structure of surfaces, understand the arrangement of atoms, determine the presence of any defects, and determine the distribution of electronic states. The manipulation of individual atoms is possible using STM, which allows the construction of nanostructures with tailored properties. The surface modification of nanomaterials and its interaction with other mediums can be studied using STM, which will provide more insight into catalysis, adsorption, etc.

3.1.4 Dynamic Light Scattering

Dynamic light scattering (DLS) works on the Brownian motion principle of particles when dispersed in a liquid/solvent. Brownian motion represents the random motion of the particles in all directions when suspended in a liquid. The particles collide with solvent molecules, and energy transfer from them to the particles makes them move randomly. The collision affects smaller particles than large particles, resulting in them moving more quickly than large particles. One can deduce the hydrodynamic diameter by considering factors such as the solvent's viscosity and the system's temperature, affecting the speed of the Brownian motion of the particles.

The Stokes-Einstein equation relates the speed and the particle size in pure Brownian motion, in which the speed of the particle is given by the translational diffusion coefficient *D*. The diffusion coefficient "*D*" is a function of the solvent's viscosity and the system's temperature.

$$D = \frac{k_B T}{6\pi\eta R_H} \tag{3.1}$$

*Where D is the t*ranslational diffusion coefficient in m²/s, k_B is the Boltzmann constant in m^2kg/Ks^2, *T* is the Temperature in K, h is the Viscosity Ns/m^2, and R_H is the hydrodynamic radius in meters.

Hydrodynamic diameter is the effective diameter of the particle when dispersed in a liquid. When a particle is in liquid, a thin electric dipole layer forms around it, affecting the particles' motion. The diameter of the particle, including the electric dipole layer, is called hydrodynamic diameter. In DLS, the onset of sedimentation is the upper size limit, and the signal-to-noise ratio is the lower limit.

The schematic of a DLS measuring system is shown in Figure 3.7. A monochromatic laser illuminates the sample placed in the sample contained in a cuvette. Due to the suspended nanoparticles in the sample, the laser is scattered in all directions. The intensity of the scattered laser is detected by the detector placed at a certain angle over time. The variation in the intensity of the scattered light is due to the movement of the particles. This fluctuation in intensity depends on the particle size. The fluctuation in the intensity is faster for smaller particles and slower for large particles. The diffusion coefficient is determined by correlating this fluctuation, and the particle size is calculated from the Stokes-Einstein equation.

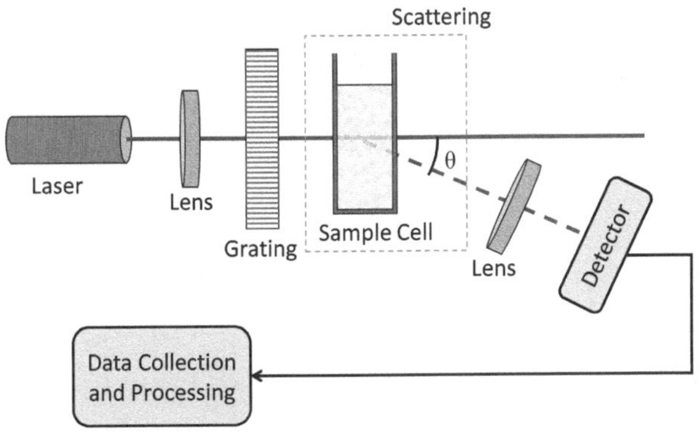

FIGURE 3.7 Working of a DLS.

3.2 STRUCTURAL CHARACTERIZATION

3.2.1 X-Ray Diffraction

X-ray diffraction (XRD) is used to characterize the nanomaterials, including crystalline structure, nature of the phase, impurities, lattice parameters, and crystalline grain size. Powder XRD is simply the most common technique for characterizing inorganic nanomaterials. XRD is considered as the fingerprint of inorganic materials. It is an example of non-destructive testing. As the monochromatic X-rays from the X-ray source incident on the powder sample, the X-rays interact with the regular array of atoms in the crystal lattice, as shown in Figure 3.8. The electrons present in the atoms can repulse the X-rays and scatter them elastically. These scattered X-rays result from constructive or destructive interference depending on the angle of incidence. Moreover, for constructive interference, the path difference between the beam of X-rays after scattering must be equal to the beam's wavelength or its multiples of whole numbers "nλ" (n = 1, 2, 3, 4……… ..).

Figure 3.8 shows two monochromatic X-rays with wavelength "λ" incident on the sample at an angle θ. The X-ray beams are scattered by the first and second layers of atoms arranged in a specific (h k l) plane. Then, for a constructive interference between the X-ray-1 and X-ray-2, the path difference must equal nλ.

Then

$$AB + BC = n\lambda : (\mathrm{AB} = \mathrm{BC}) \tag{3.2}$$

If the spacing between diffracting planes of atoms is *d*, then from triangle ABC,

$$AB = BC = d\sin\theta : \text{and } n\lambda = 2AB\, or\, 2BC \tag{3.3}$$

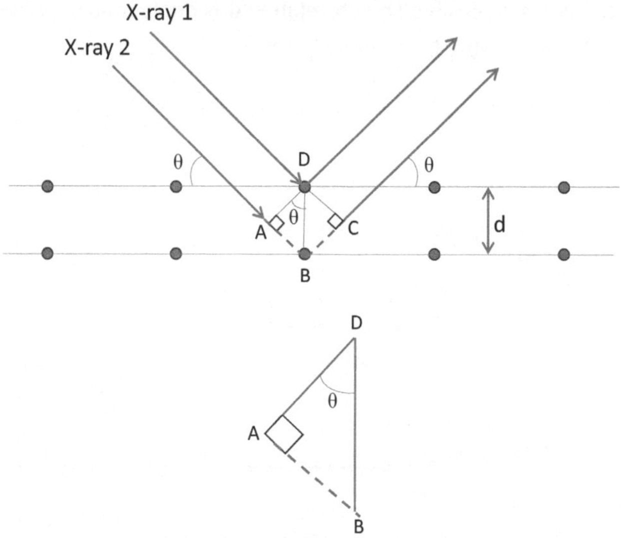

FIGURE 3.8 Diffraction of monochromatic X-rays.

Therefore,

$$n\lambda = 2d\sin\theta \tag{3.4}$$

This (Equation 3.4) relationship is known as Bragg's law.

Where n is the order of diffraction, d is the spacing between diffracting planes, θ is the angle of incidence, and λ is the beam wavelength.

In XRD, the X-rays are generated in a cathode ray tube. First, the electrons are produced by heating a suitable filament (generally tungsten) to produce electrons. These electrons are then accelerated toward a metallic target (Eg. Cu, Al, etc.) by applying a potential difference. As the electrons bombard the target metal, the inner shell electrons will be excited to higher-energy states. These excited electrons can return to the ground state, emitting characteristic X-rays. The characteristic X-rays are directed to the sample, which rotates constantly. A detector rotating with the sample records the intensity of X-rays falling on it, as shown in Figure 3.9. When the incident angle of the X-rays satisfies Bragg's law, one can observe a sudden increase in the intensity. The detector records the X-ray signal, converts it to a count rate, and then displays it on a computer monitor. Once all the peaks and the angles corresponding to different planes are obtained, the peak position is compared with the position and intensity of the peaks using the reference patterns in the International Centre for Diffraction Data (ICDD) database. The nanomaterials likely exhibit broad XRD peaks as the *d*-spacing at the surface is larger than the bulk materials, as the atoms on the surface of the nanoparticles lack any periodic arrangement. Thus, the nanomaterials exhibit both amorphous and crystalline nature. Additionally, the lattice strains on the surface of nanoparticles are remarkably high due to the large surface area of the particles.

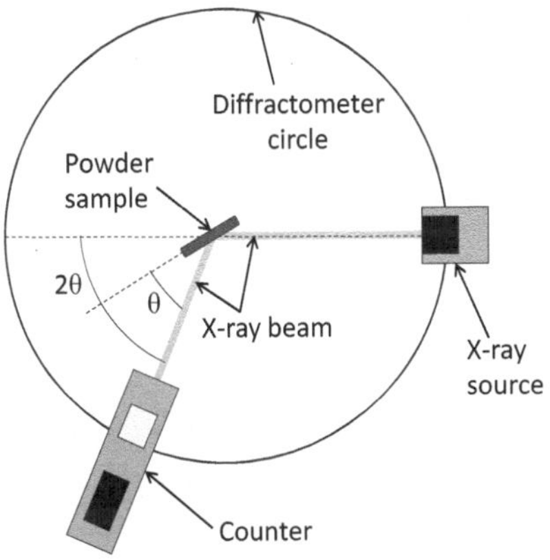

FIGURE 3.9 Components of an XRD machine.

The crystallite size can be determined using the Scherrer equation by measuring the broadening of the sample's most intense peak of an XRD pattern. The Scherrer equation is given by;

$$D = \frac{K\lambda}{\beta cos\theta} \quad (3.5)$$

Here, λ is the X-ray wavelength, θ is the diffraction angle, *and β is the total wave half-width maximum,* meaning the width at half height of the maximum intensity peak. K is a constant with a value close to 1.

Usually, the crystallite size obtained from the XRD analysis is smaller than the actual size of the nanomaterials obtained from a microscopic technique. This can be due to the nanoparticles' amorphous regions or the nanomaterials' polycrystalline nature. XRD analysis can distinguish the morphology of the nanostructures. The XRD pattern of single-crystal nanostructures grown along a particular axis will have a sharp, intense peak along the plane perpendicular to that axis. Moreover, the nanostructures with different morphology, *e.g.,* cubes, plates, and rods, change the relative intensities of the peaks compared to the bulk material's XRD pattern.

Applications of XRD

- A helpful tool for designing new materials.
- Qualitative and quantitative phase analysis.
- Analysis of phase changes under temperature, humidity pressure, etc.
- Analysis of crystallite size (diameter), crystal orientation, residual stress, and purity.

3.2.2 Small-Angle X-Ray Scattering (SAXS)

SAXS is another X-ray-based analysis mainly used for the structural characterization of nanomaterials and biomaterials. It measures the intensity of the X-rays scattered by the sample. The measurements are made between small angles between 0 and 5°. SAXS can detect the structural features in nanomaterial with a size <100 nm. Wide-angle X-ray scattering (WAXS) is a similar technique used to analyze the scattered X-rays at a wider angle, and it is used to analyze materials at bulk scale.

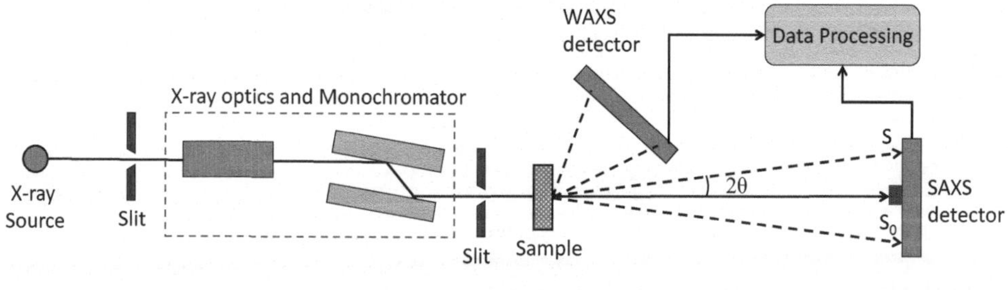

FIGURE 3.10 Schematic of SAXS construction.

During SAXS measurement, a collimated X-ray beam is transmitted through the sample with a thickness of ~1 mm. A monochromatic narrow X-ray beam is essential to record weak scattering by the sample, as in Figure 3.11. The sample can be in the form of a solution, suspension, or solid material. However, the samples should be in such a way that X-rays must penetrate and interact with them. The elastically scattered beam of X-rays is detected using a 2D area detector in 360°. The angle between the scattered beams and the direct/transmitted beam is determined, referred to as the scattering angle "2θ". The magnitude of the scattering vector (q) is a function of scattering angle, given as,

$$q = 4\pi \sin\theta/\lambda \tag{3.6}$$

Where λ is the wavelength of the X-rays (λ = 0.154 nm for the Cu X-rays).

The scattering angle or the vector (q) is plotted against the intensity of the scattered X-rays. The significant peaks observed at lower angles determine the nanostructures' overall size, shape, morphology, and orientation. Additionally, SAXS provides insight into the nanomaterials' porosity, surface area, degree of crystallinity, structural order, size distribution, etc.

3.3 SPECTROSCOPIC TECHNIQUES

Nanomaterials exhibit unique properties associated with their structure. Establishing the relationship between the structure and the properties of the nanomaterials requires specialized spectroscopic techniques. Many spectroscopic techniques are commonly used for characterizing nanomaterials' structural, optical, magnetic, and electronic behaviors. A combination of spectroscopic techniques is generally required to understand the nanomaterials fully. In any spectroscopy, the absorption, reflection, and scattering of electromagnetic waves of different energy by materials at the molecular level are analyzed to deduce the properties of the materials. Electromagnetic waves can excite the molecules from the ground to the excited state. Table 3.2 lists the spectroscopic technique, the type of electromagnetic wave used, and the type of excitation and extracted information.

3.3.1 X-Ray Absorption Spectroscopy (XANES)

As the name suggests, X-ray absorption spectroscopy (XAS) is the X-ray absorption coefficient of a material in terms of energy. XAS is used to determine the characteristic absorption edges of electrons of each element in the atom, which is called an X-ray absorption near-edge structure (XANES). The XANES spectra reveal the electronic structure and symmetry of the metal site. Extended X-ray absorption fine structure (EXAFS) is a subdivision of XAS, which determines a selected atom's chemical state and local atomic structure. EXAFS can identify the chemical state of atoms even at deficient concentrations. EXAFS measures the absorption of X-rays by atoms above the absorption edge. XANES and EXAFS are complementary techniques for analyzing nanostructures, as they can provide structural information of nanomaterials with atomic-scale resolution. These techniques can identify the defects and position of dopant atoms in the lattice of the nanostructure.

TABLE 3.2
List of Different Spectroscopic Techniques and the Information Retrieved

Technique	Electromagnetic Wave	Interaction	Main Information Derived
X-ray Photoelectron Spectroscopy (XPS)	X-ray	Emission of photoelectrons	Electronic structure, elemental composition, oxidation states, ligand binding (surface-sensitive)
Fourier Transform Infrared Spectroscopy (FTIR)	Infrared light	Vibration of molecules	Surface composition, ligand binding
Nuclear Magnetic Resonance (NMR) Spectroscopy	Radio waves	Interaction with nuclei	Ligand density and arrangement, electronic core structure, atomic composition, influence of ligands on NP shape, NP size
UV-Vis Spectroscopy	UV-visible light	Electron excitation	Electronic transitions, plasmonic resonance, optical properties, size, and concentration
Raman Spectroscopy	Infrared light	Scattering of light by the molecules	Structural information
Fluorescent (PL) Spectroscopy	UV-visible light	Electron excitation	Optical properties—relation to structure features such as defects, size, composition.
Energy-Dispersive Spectroscopy (EDS) And Wavelength-Dispersive Spectroscopy (WDS)	Electron beam	Characteristic X-ray emission	Oxidation state, symmetry, surface spins, magnetic ordering of Fe atoms, magnetic anisotropy energy, thermal unblocking, distinguish between iron oxides.
Electron Energy-Loss Spectroscopy (EELS-STEM)	Electron beam	Inelastic scattering of electrons	Type and quantity of atoms present, chemical state of atoms, collective interactions of atoms with neighbors, bulk plasmon resonance.

In XAS, the X-ray from the source passes through a monochromator and a grating, ensuring that a single X-ray wavelength reaches the sample. A collimating system is used to focus the beam on a small area. The powder samples are pelletized to get a uniform shape. The liquid samples are filled in the pans to get a uniform shape. As the X-ray strikes the sample, the sample will absorb some of the X-rays, and the remaining X-rays reach the X-ray detector. The X-rays absorbed by the sample can

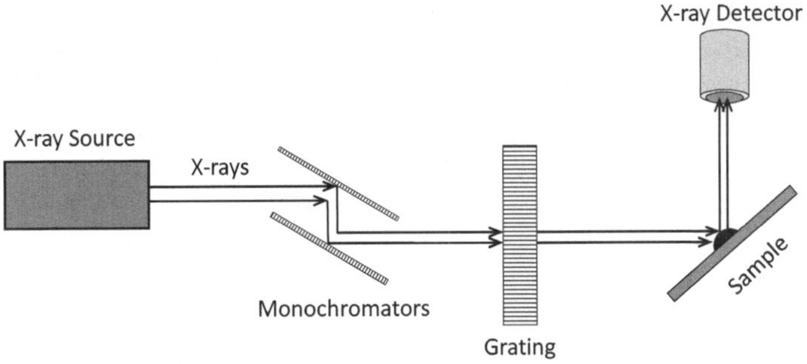

FIGURE 3.11 Schematic of XANES construction.

be determined by subtracting the intensity of X-rays collected by the detector from the intensity of the incident beam of X-rays. X-rays can be detected using a photo-multiplier tube or semiconductor diode. The working XANES is schematically shown in Figure 3.10.

3.3.2 UV-Visible Spectroscopy

UV-Vis spectroscopy uses a UV-visible spectrum to excite the materials. During the analysis, the instrument measures the wavelengths of UV or visible light absorbed or scattered by the sample concerning a reference or blank sample. The amount of light absorbed or scattered varies from material to material. Light with different wavelengths carries a specific amount of energy with it. The material absorbs a particular amount of energy to excite electrons from the ground state to the high-energy state, and this absorption of light is detected by UV-Vis spectroscopy. Electrons in different energy levels require a range of energy to excite the electrons from the corresponding ground state to the excited state. As a result, different materials absorb light waves of various wavelengths. It measures the intensity of light reflected from a sample and compares it to the intensity reflected from a reference material. NPs have optical properties sensitive to size, shape, concentration, agglomeration state, and refractive index near the NP surface. UV-Vis spectroscopy is a vital tool to identify, characterize, and investigate these materials and evaluate the stability of NP colloidal solutions.

The main components of a UV-Vis spectrometer are light, a monochromator, a sample compartment, and a detector. The light source can emit steady light over a wide range of wavelengths. A xenon lamp is the most common light source, as it can produce an extensive range of light with different wavelengths in both the UV and visible spectra. However, some instruments scan UV and visible ranges using two light sources. A tungsten or halogen lamp is used for visible light, and a deuterium lamp is used for UV light. The switching of light sources takes place between 300–350 nm.

During the analysis of the samples, the operator needs to select the scanning range, and the instrument should allow a single wavelength of light to pass through the sample at a time. It is possible with the help of monochromators.

The selected wavelength can pass through the sample placed in the sample compartment. At the same time, the chosen wavelength passes through a reference sample, often called a "blank sample." A cuvette filled with the solvent used for the sample is placed as the reference. One has to select the transparent cuvette to the selected wavelength range. The plastic or glass cuvettes can be used for visible regions and quartz for UV wavelengths.

Once the light passes through the sample and the reference, a detector compares the intensity of light passed through the sample and the reference. The difference will be determined as the % of absorption. The detector converts the light to an electronic signal. The detector output is a plot of absorbance or transmittance along the y-axis as a function of wavelength along the x-axis. According to Beer–Lambert's law, the absorbance (A) can be written as follows.

$$A = log_{10}\left(\frac{I_o}{I}\right) \tag{3.7}$$

Here I_0 is the intensity of light before passing through the sample, and I is the intensity of light after passing through the sample. The ratio of I *to* I_0 is known as transmittance (*T*),

Then,

$$A = log_{10}\left(\frac{1}{T}\right) = -log_{10}(T) \tag{3.8}$$

Ultraviolet-visible (UV-Vis) spectroscopy is a powerful tool for analyzing nanomaterials. This technique determines the amount of UV-visible light absorbed by the molecules or nanoparticles at different wavelengths, indicating electronic transitions.

The major components of a UV-Vis spectroscopy are:

Light Source: It uses a source that can emit a broad spectrum of light. Since the samples are illuminated with UV and visible lights, a tungsten filament lamp and a deuterium lamp are used as visible light and UV sources, respectively. The light source is the start of the operation of UV-Vis spectroscopy.

Monochromator: The light from the source is passed through a monochromator to separate the light into individual wavelengths. The UV-visible spectroscopy measures the absorption at specific wavelengths.

Beam Splitter: A beam splitter splits the light from the monochromator into two parts: a transmitted beam reaches the reference cell, and the reflected beam reaches the sample cell. Both the reference and the sample cells are made of a transparent material. The sample can be a liquid, gas, or solid, and the light must pass through it. Some of the light is absorbed by the molecules in the sample.

Detector: the light leaving the sample cell then reaches a detector, where the intensity of the light is compared to that of the incident beam at individual wavelengths to measure the absorption.

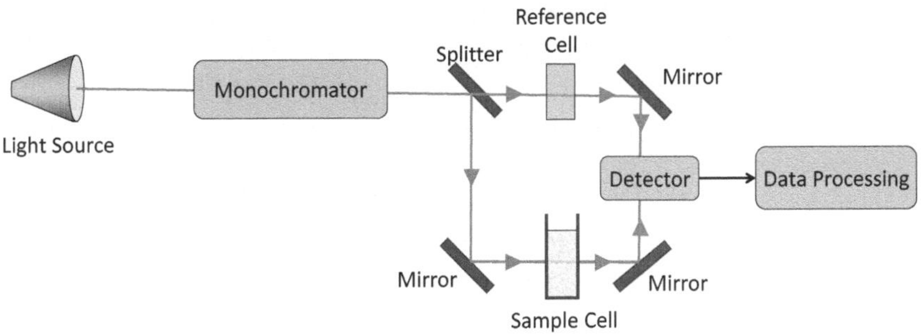

FIGURE 3.12 Working of UV spectroscopy.

Data Analysis: The data analysis part processes the signals from the detector to generate a spectrum. Usually, the spectrum is a plot of absorbance (or transmittance) versus wavelength. A representative diagram of UV-visible spectroscopy analysis is shown in Figure 3.12.

One can obtain various information, such as electronic transitions occurring in the sample from the UV-Vis spectrum. The peaks on the spectrum correspond to the excitation of the molecules composing the nanoparticles from a ground state and an excited state by absorbing light. Additionally, UV absorbance can be correlated to the concentration of the nanomaterials, size and shape, surface functionality, surface plasmon resonance, etc.

The optical properties of materials can be analyzed using UV-Vis spectroscopy. It is helpful to determine the optical properties of nanomaterials and their suitability in different applications. UV-Vis spectroscopy is essential for the fabrication of photovoltaic devices based on nanomaterials. The optical band gap of materials can be determined using this technique, and that property is necessary to select appropriate materials for solar energy conversion. The UV-Vis techniques are helpful for the photocatalytic properties of various nanomaterials. Using UV-Vis, the degradation of organic molecules in the presence of nanomaterials accurately. The surface plasmonic effect of nanomaterials is evident in UV-Vis analysis. Therefore, this technique is reliable for finding the size of metallic nanoparticles.

3.3.3 Fourier Transform Infrared Spectroscopy (FTIR)

Fourier Transform Infrared (FTIR) spectroscopy is used to analyze the vibration and rotation of molecules when excited by infrared radiation. The wavelength of the infrared spectrum is higher than the UV and visible spectrum. The IR light has sufficient energy to excite the bonds between atoms, and the bonds between atoms are different; therefore, they absorb light with other energies or wavenumbers ($1/\lambda$). The covalent bonds absorb IR light between wavenumbers 400–4,000 cm^{-1}. The bonds between atoms are like springs, which can vibrate in different modes and absorb the corresponding IR wavenumber. These bonds can stretch, scissor, wag, twist, or rock,

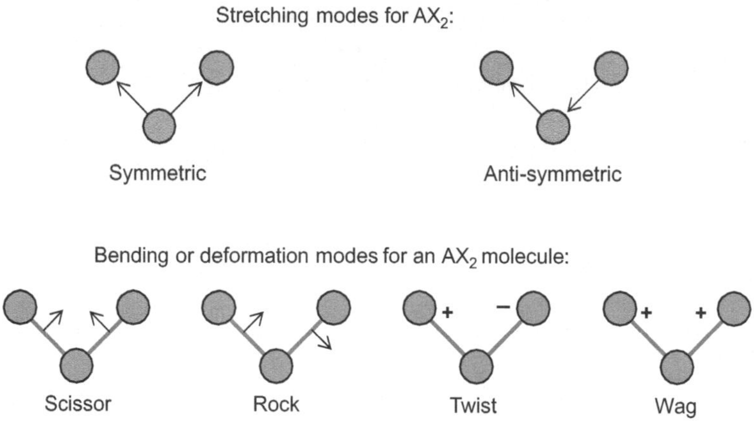

FIGURE 3.13 Vibration modes of molecules under IR excitation. (Reproduced with permission of the Elsevier, Harris et al., 2010)

as shown in Figure 3.13. An FTIR instrument produces an infrared spectrum, and the instrument will detect the infrared light absorbed by the material.

By analyzing the molecular vibrations, one can deduce the composition, functional groups, molecular structure, and chemical compositions of organic and inorganic nanostructures. FTIR is essential for analyzing the interaction of nanostructured materials with organic molecules. Therefore, the FTIR technique is relevant in nanocomposites, nanomaterials sensors, drug delivery, etc., and we will see these topics and many others in detail in the upcoming chapters. FTIR is also important in studying the surface modification of nanomaterials and the absorption of molecules to the nanomaterials surface. FTIR can be used to analyze solid, liquid, and gas samples. An FTIR spectrum collects the position of bands concerning the strength and nature of bonds, as well as the functional groups attached, thus providing information on molecular structures and their interactions.

Figure 3.14 shows the construction of an FTIR spectrometer. An FTIR instrument has an interferometer composing an IR source, a collimator, a beam splitter, two mirrors, a laser, and a detector.

The broad spectrum of IR radiation from the source passes through the collimator and reaches the beam splitter, which splits the beam into two parts. One part is transmitted and strikes a moving mirror, while the other is reflected and strikes a fixed mirror. The moving mirror moves back and forth by a calibrating laser source. These two beams are reflected by the mirrors combined at the beam splitter and sent to the sample. When the two reflected beams are combined at the beam splitter, the moving mirror's position is adjusted to form an interference pattern. The combined beam from the beam splitter is sent to the sample compartment where the sample is placed. The beam is transmitted through the sample and reaches the detector. The interference pattern carries information about the intensity of the infrared radiation at different wavelengths after the sample absorbs it. The interferograms are subjected to Fourier transform to obtain the spectrum corresponding to the absorption of infrared

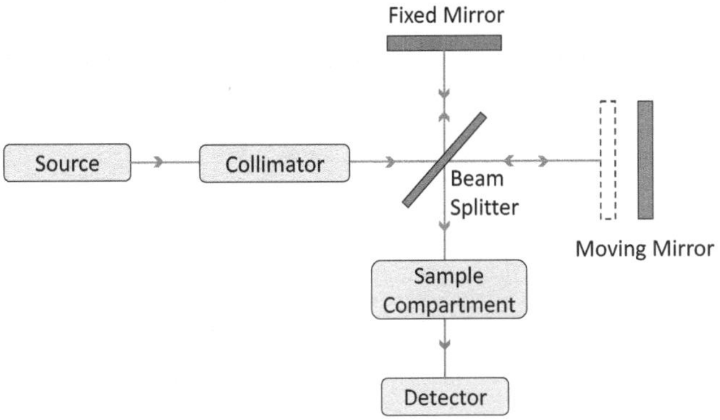

FIGURE 3.14 Working of an IR spectrometer.

radiation by the sample at different frequencies. The obtained FTIR spectrum is then analyzed to identify characteristic peaks corresponding to other functional groups in the sample. A spectral database is often used to match the experimental data and to determine the unknown sample.

FTIR is commonly operated in three modes: transmission, reflection, attenuated total reflection (ATR), and KBr pellet method. The operating modes are selected based on the sample's nature and the absorbance frequency range. The transmission mode is suitable for the analysis of transparent polymer films. Reflectance and ATR modes are commonly used for nontransparent samples, and the spectrum is recorded in the range of 4,000–500 cm^{-1}. In the case of ceramic nanostructures, the absorption is below 500 cm^{-1}. Therefore, the KBr pellet method is used, in which the samples are mixed with KBr and pressed to form a transparent pellet.

A representative FTIR spectra is shown in Figure 3.15. The absorbance peak corresponding to the different functional groups can be classified into four regions: single bond stretching, triple bond, double bond, and fingerprint regions; the most relevant absorbance peaks can help to identify the compound that appears in the fingerprint region, which is below 1,500 cm^{-1}. The vibration modes of different functional groups are listed in Table 3.3.

3.3.4 Raman Spectroscopy

Raman spectroscopy is an optical spectrometer that utilizes a monochromatic laser light with wavelengths ranging from 400–1,100 nm to illuminate and analyze the inelastically scattered light from a sample. As the intense laser light interacts with the sample, there are two possibilities: elastic and inelastic scattering. The central part of the laser beam is elastically scattered, meaning no change in the wavelength, but the direction may change. However, a small fraction of the laser beam undergoes an inelastic scattering, which means a change in the wavelength and direction. During inelastic scattering, the laser beam interacts with the sample, and the energy of the laser beam changes, and it is emitted as waves of different wavelengths or colors. The inelastic interaction of materials with solid laser light is called the Raman effect, and

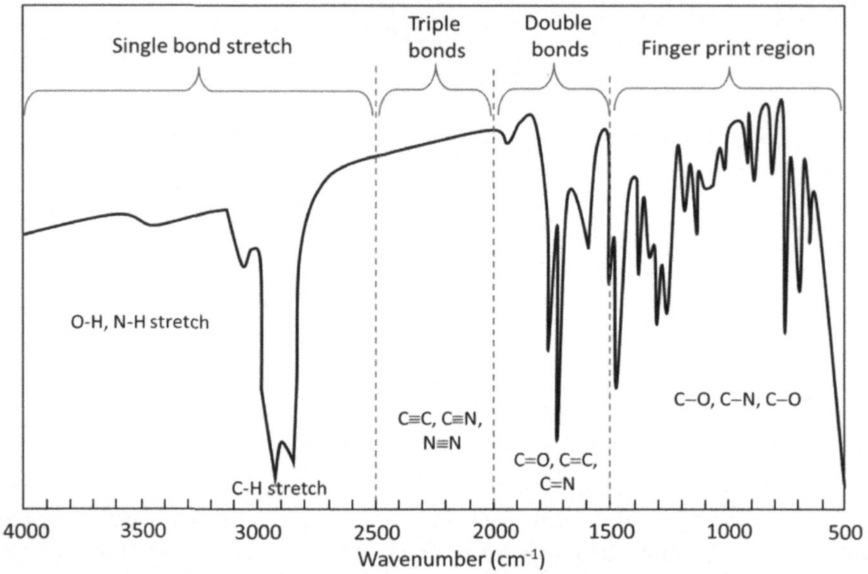

FIGURE 3.15 A representative FTIR spectra.

TABLE 3.3
Vibrational Modes of Functional Groups and the Corresponding Wavenumber Absorbed

Vibrational Modes	Frequency (cm^{-1})
O–H hydroxyl group	3570–3200 (broad)
N–H stretch	3400–3380, 3345–3325
Aromatic C–H stretch	3130–3070
C–H stretch	2970–2950/2880–2860
S–H stretch	2600–2550
C=C alkenyl stretch	1680–1620
N–H bend	1650–1590
C=O	1610–1550/1420–1300
C–H bend	1470–1430/1380–1370
P=O stretch	1350–1250
Si–O–C stretch	1100–1080
Si–O–Si stretch	1095–1075/1055–1020
C–N stretch	1090–1020
C–O stretch	~1050

the change in energy is called the Raman shift. The inelastic interaction of laser light can either increase (anti-Stokes lines) or decrease (Stokes lines) the energy of the incident laser light. The dominant part is Stokes lines, as energy is more likely to be lost. The wavelength of the inelastically scattered light depends on the bonds present

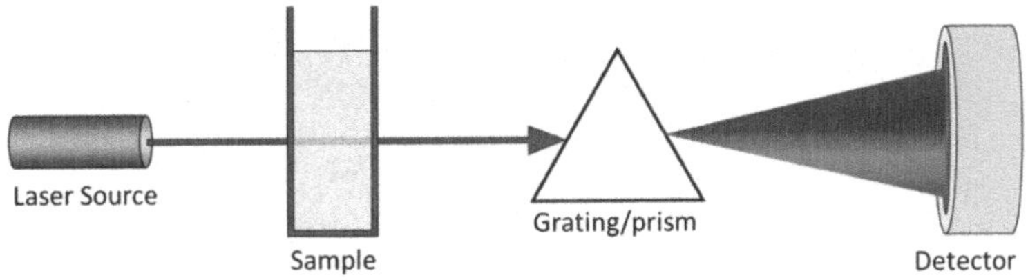

FIGURE 3.16 Simple construction of a Raman spectrometer.

in the material, and the analysis of the spectrum can reveal the composition of the materials under study.

The construction of a simple Raman spectrometer is shown in Figure 3.16. The monochromatic light from the laser source interacts with a sample; most scattered light has the same energy (wavelength) as the incident light, known as elastic scattering or Rayleigh scattering. The inelastically scattered light is separated with the help of the detector based on the intensity of the light received at the detector compared to the intensity of the incident light.

Raman spectroscopy is employed to study nanostructured materials' composition, shape, structure, crystallinity, defects, and orientation. Any factor that modifies the vibration or rotation of the molecules can change the peak position in the Raman spectra compared to the pure bulk materials.

Surface-Enhanced Raman Spectroscopy (SERS)

Surface-enhanced Raman spectroscopy is a modified version of the Raman spectroscopy. In SERS, the sample is placed on or near a roughened noble-metal substrate so that localized surface plasmons excrete the Raman scattering signal. SERS is helpful for the trace analysis and detection of molecules adsorbed on the surface of nanomaterials. Thus, SERS is sensitive to understanding the surface properties of nanoparticles.

3.3.5 X-Ray Photoelectron Spectroscopy (XPS)

X-ray Photoelectron Spectroscopy (XPS) is a widely used technique for the surface characterization of nanostructured materials. XPS is a surface-sensitive technique known as electron spectroscopy for chemical analysis (ESCA). XPS can determine the elements in the nanostructured materials, their oxidation state, and their electronic structure. This technique is also helpful in analyzing the surface modification of the nanomaterials and the interaction of nanomaterials with other chemical species. XPS works on the principle of the photoelectric effect. When a monochromatic X-ray of sufficient energy is made to interact with the materials, the X-ray photons can eject electrons, called photoelectrons. This interaction takes place with the electrons in the outermost atomic layers of the material, making it a surface-sensitive measuring technique. The kinetic energy of the emitted photoelectrons corresponds to the binding energy of the electrons in the respective atomic orbitals. By measuring the kinetic energy of the emitted photoelectrons, one can quantitatively determine the elements present in the sample and their chemical states.

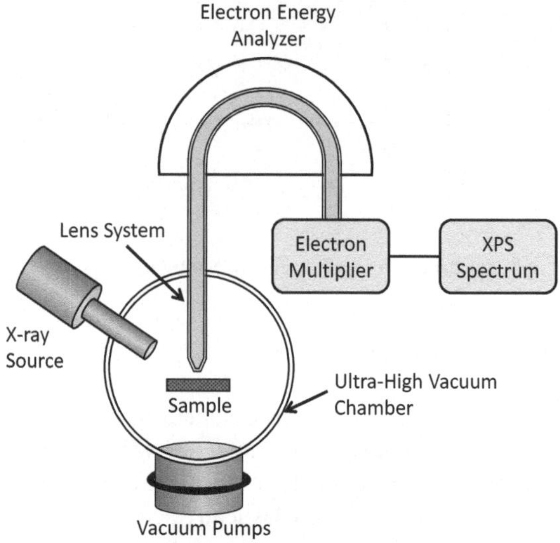

FIGURE 3.17 Construction of an XPS spectrometer.

In an XPS spectrometer, the major components are an X-ray source, a sample chamber, and an electron analyzer. The monochromatic X-ray of specific energy from the X-ray source is directed to the sample in a high vacuum chamber. The vacuum is maintained to prevent any interactions of X-rays and the generated photoelectrons with different molecules in the air, otherwise producing errors during the XPS measurement. The kinetic energy of the emitted photoelectrons from the sample is then analyzed using a hemispherical electron energy analyzer. The signals from the analyzer are sent to a detector to obtain the intensity of the photoelectrons at different energies. The binding energy and the corresponding intensity are finally received as a spectrum. The comprehensive scan or survey spectrum reveals all the elements present in the material, and high-resolution spectra reveal the oxidation states of the individual components. Figure 3.17 shows the construction of an XPS spectrometer.

XPS can identify the surface chemistry and stoichiometric elemental composition of the nanomaterials at the surface. XPS is useful in determining the presence of defects and impurities in the nanostructured materials. Additionally, depth profiling allows the analysis of thin films and coatings and the interfaces at the nanoscale. XPS is used in conjunction with other analytical techniques to study nanostructures.

3.3.6 Photoluminescence Spectroscopy

Photoluminescence spectroscopy studies light emitted by a material upon excited by photons. During the excitation by light (photons), the electrons absorb the energy of the photons. They are excited from the equilibrium or ground state to a higher-energy level or excited state. The excited electrons return to the ground state by releasing the extra energy in either UV or visible light, known as a radiative emission, or non-radiatively, through crystal defects. The radiative light emission is known as photoluminescence, and the wavelength emitted depends on the energy of the excitation light and energy difference in the discreet electronic states of the materials under study. The radiative process determines the quantum light emitted. The energy (or wavelength) and intensity of the emitted photons provide valuable information about the material's electronic structure, composition, and physical properties.

Photoluminescence can be classified into three different types they are;

Resonant Radiation: The absorbed and emitted photons have the same energy or wavelength. The process takes place within ten nanoseconds after the excitation.

Fluorescence: the material undergoes excitation and emission of a lower energy photon as the electrons return to the ground state in a short lifetime of about 10^{-8} to 10^{-4} s.

Phosphorescence is the radiative release of energy through electronic transition with different spin states, with a lifespan ranging from 10^{-4} to 10^{-2} s.

The device used for recording the photoluminescence spectra is a spectrofluorometer, also known as a spectrofluorophotometer. A typical spectrofluorophototmeter has an excitation source that provides the photons to excite the sample to induce photoluminescence. The light source must be able to emit light in a wide wavelength range, starting from UV to near IR. Xenon lamps are the most used excitation source, which can cover a wide spectral range and are valid for samples that require excitation at multiple wavelengths. LEDs are another excitation source with a broad wavelength emission. Lasers are used when monochromatic and high-intensity emission is required. An optical system consisting of lenses, mirrors, and sometimes optical fibers directs the light from the excitation source to the sample and collects the emitted light efficiently. This system includes monochromators to separate specific excitation wavelengths and filters to prevent the excitation wavelength from reaching the detector. The light from the detector reaches the sample and places it in the sample chamber. Separate sample holders analyze solid, liquid, and gas samples. The light emitted from the sample reaches the detector, and the detector converts this light signal to an electrical signal. There are two detectors, one for the signal and the other for the reference. A photomultiplier tube and a photon-counting module serve as a signal detector which counts the photon falling on it. The reference detector is a silicon photodiode that monitors the light from the xenon lamp for wavelength correction and time-dependent output of the lamp. The photon count measured at the detector is converted to an electrical signal, which will then be converted to a spectrum with the help of software. Figure 3.18 shows the working of a photoluminescence spectrophotometer.

To do the measurement, the sample is fixed to the sample holder placed in the sample chamber, and the excitation source is adjusted to obtain the required wavelength and directed to the sample. The light emitted by the sample reaches the detector through a monochromator and filters. The detector and the associated software system process the data to a photoluminescence spectrum, showing intensity as a function of wavelength.

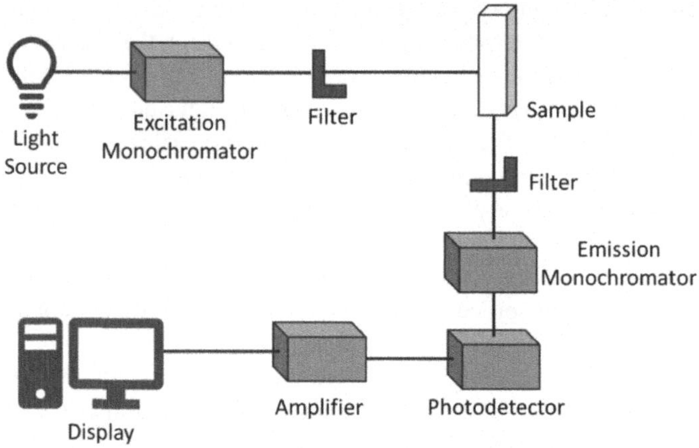

FIGURE 3.18 Components of a photoluminescence spectrophotometer.

Photoluminescence spectroscopy is used to study fluorescent nanomaterials, such as carbon quantum dots, semiconductor quantum dots, metal nanoclusters, phosphors, etc. Photoluminescence spectroscopy helps identify the materials for optical imaging, optoelectronics, and sensing. It can provide information on the band gap of the materials, defects present, and the effect of quantum confinement. Photoluminescence spectroscopy is an essential tool for photovoltaic research. It is the primary instrument used for characterizing the materials used in solar cells, such as perovskites and silicon. By understanding the photoluminescence properties, one can optimize the design and efficiency of solar cells. It is also helpful in analyzing various pigments, paints, inks, and coatings.

3.4 COMPOSITIONAL ANALYSIS

3.4.1 Energy-Dispersive X-Ray Spectroscopy (EDS) and Wavelength-Dispersive X-Ray Spectroscopy (WDS)

Energy-dispersive X-ray spectroscopy (EDS) and wavelength-dispersive X-ray spectroscopy (WDS) analyze the characteristic X-ray produced by the primary electrons to identify the constituent elements and chemical composition. The EDS and WDS detectors are used as attachments to a TEM or SEM to extract information on the localized composition of the sample observed under the microscope. The elemental maps of the sample surface can be retrieved with the help of these detectors. EDS measures the energy of the characteristic X-rays to deduce the elements present in the sample. Still, WDS analyzes the wavelength of the characteristic X-rays to identify the elements present in the sample. The WDS analysis is more accurate and provides more precise measurements than the EDS analysis, and Table 3.4 presents the significant difference between EDS and WDS. EDS and WDS detectors are attached to an electron microprobe and placed in the microprobe, as shown in Figure 3.19.

TABLE 3.4
Comparison of EDS and WDS

	EDS	WDS
Mechanism	Analyzing the energy of X-rays emitted by the sample	Analyzes the wavelength of the X-rays emitted by the sample
Data Collection	Collects all the X-rays simultaneously	Collects one wavelength at a time
X-Ray Separation	No separation	Separation by X-ray diffraction
Elemental Analysis	Qualitative	Both qualitative and quantitative
Time for Measurement	Less	More
Detection Limit	Lower	Higher
Sensitivity	Poor	Good
Resolution	Lower	Higher

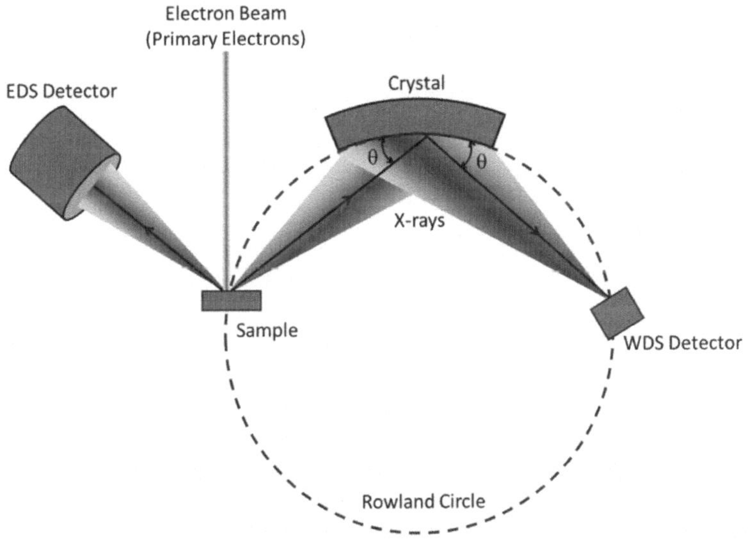

FIGURE 3.19 Arrangement of EDS and WDS detectors.

Energy-Dispersive X-Ray Spectroscopy (EDS)
The EDS detector analyzes the characteristic X-rays emitted by the sample under an intense electron beam. The detector converts the energy of X-rays to a proportional electrical signal, usually into a voltage signal. The X-ray is converted into a charge by ionizing atoms in the semiconductor crystal. Then, the ionized charge is converted into the voltage signal by the field-effect transistor (FET) preamplifier. Finally, this voltage signal is converted to the EDS spectrum.

Wavelength-Dispersive X-Ray Spectroscopy (WDS)

During WDS analysis, the characteristic X-rays are generated from the sample, and X-rays are selected based on their wavelength using analytical crystals with specific lattice spacings. The analytical crystals are placed so that the characteristic X-ray always reaches the analytical crystal at a particular angle of incident θ. The X-rays that satisfy Bragg's condition are diffracted, and only a single wavelength reaches the detector. By changing the position of the crystal, one can change the wavelength of the X-rays to reflect it so that the same crystal can be used for analyzing multiple elements present in the sample. The analytical crystal's geometry also determines the X-rays' reflected wavelengths. Usually, the analytical crystal is designed to a radius of 2R, where R is the radius of the focusing circle, the Rowland circle in Figure 3.19. Analytical crystals are single-crystal materials with a known lattice spacing, and many analytical crystals with different lattice spacing and spectrometers are used to cover all the elemental wavelengths of interest.

For a given sample, intensities of characteristic X-rays of each element of interest are "counted" in the detector, and the count rates are compared to those of standard samples containing known values of the elements. The X-ray intensity received at the detector is corrected based on atomic number (Z), absorption (A), and fluorescence (F) with the assistance of software to get the actual value. The spectra are obtained as count vs energy of the X-ray emitted.

3.5 SURFACE AREA

3.5.1 Brunauer–Emmett–Teller (BET) Surface Area Analysis

The *Brunauer–Emmett–Teller* (BET) technique is named after Brunauer, Emmett, and Teller, who developed the theory. A BET surface area analyzer is commonly used to find the surface area of porous materials, and it is beneficial for finding the surface area of the nanomaterials. This technique provides information on the surface structure and the area as these surface characteristics determine the catalytic activity, adsorption, sensitivity, etc. BET surface area works on the principle of physical adsorption of a gas on a solid surface. BET is a relatively accurate, rapid, and simple method for analyzing nanomaterials.

In BET surface area analysis, an inert gas, usually nitrogen, adsorbs on the material's surface under test. At room temperature, this adsorption is poor; therefore, the sample is cooled to cryogenic temperatures using liquid nitrogen. This temperature is kept constant for the sample, and the pressure of the adsorbing gas increases gradually. As the pressure gradually increases, the adsorption of molecules to the surface also increases. Finally, the surface will be covered by a thin layer of inert gas. The number of gas molecules that form the thin layer can be determined from the volume adsorbed. The amount of adsorbed gas molecules is plotted against the temperature and pressure, known as the adsorption isotherm. From this, one can calculate the sample's surface area using the appropriate mathematical model, the BET equation.

During the BET analysis, the sample is placed in a container that can be sealed to maintain a controlled environment to measure the adsorbed gases. The gas dosing

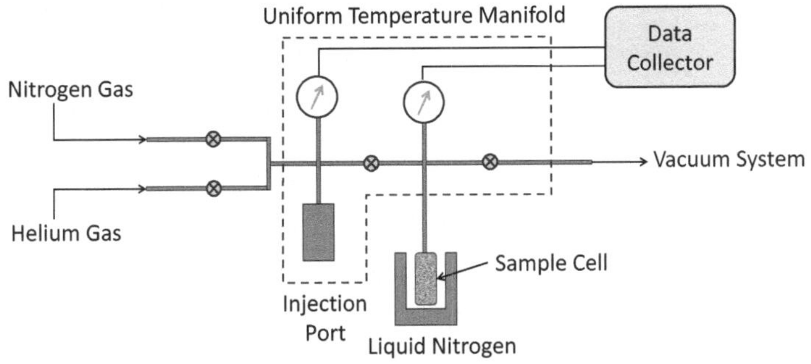

FIGURE 3.20 Components of a BET surface area analyzer.

system allows the adsorbing gas to be absorbed at a controlled rate and pressure in the sample container. Susceptible pressure and temperature sensors are used to monitor the pressure and temperature precisely before subjecting the sample to the adsorption of inert gas. The sample is degassed to remove any adsorbed gases at a degassing station. A vacuum system is used at the degassing station. A data acquisition and control system automatically controls, monitors, and records the various parameters during the measurement, as shown in Figure 3.20. Finally, the software analyzes the recorded data, generates adsorption isotherms, and then fits the data to the BET equation to estimate the specific surface area.

3.6 THERMAL CHARACTERISTICS

3.6.1 Thermogravimetric Analysis (TGA)

Thermogravimetric analysis (TGA) measures the thermal stability of the materials. The technique measures weight loss with gradual temperature rise. The weight loss is attributed to the vaporization and the decomposition of various components at different temperatures. TGA is useful for determining the thermal stability of the nanocomposites, identifying and quantifying the organic ligands attached to nanoparticles after surface modification, adsorption of gases by the metal-organic frameworks, organic phases present in nanostructures, etc. TGA can be coupled to an FTIR to identify the components released during the heating process.

The components of the TGA apparatus are shown in Figure 3.21. In TGA, a small sample is placed in the sample pan of the microbalance, which is kept in a furnace. The sample pan is tiny and usually made of aluminum (for low temperature), Al_2O_3 (for high temperature), etc. A reference pan is also paced in another rod of the microbalance. The temperature slowly increases to the set value at a predetermined heating rate. The atmosphere for the analysis can be changed according to our requirements: nitrogen, air, carbon dioxide, etc. The temperature of the sample pan and the reference pan are noted, and the difference is plotted as a derivative thermogram. The mass change with the temperature rise is constantly monitored with the help of microbalance to plot a graph of mass versus temperature. As the temperature increases, the sample may lose weight due to decomposition, oxidation, dehydration, or other processes. TGA is a

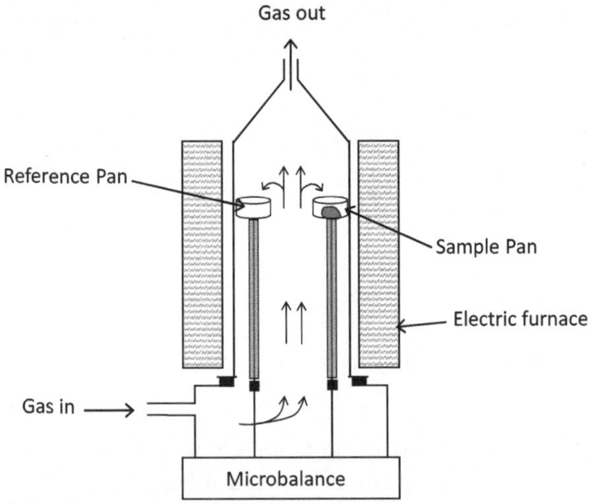

FIGURE 3.21 Components of TGA.

valuable tool in materials science, polymers, pharmaceuticals, and many other fields to obtain the thermal stability of various formulations.

REFERENCE

Harris, D., Machavaram, V.R., Fernando, G.F., 2010. 14 – Process monitoring and damage detection using optical fibre sensors, in: Goodship, V. (Ed.), Management, Recycling and Reuse of Waste Composites, Woodhead Publishing Series in Composites Science and Engineering. Woodhead Publishing, pp. 369–424. https://doi.org/10.1533/9781845697662.4.369

4 Micro and Nano Machining

4.1 INTRODUCTION

The growing demand for materials with enhanced properties such as high strength, improved hardness, superior corrosion resistance, elevated temperature tolerance, and lightweight characteristics has spurred significant research efforts among materials scientists and researchers. These materials find applications across diverse fields including defense, aerospace, power generation, bio-implants, as well as in conventional sectors like automotive and electronics. To meet these demands, researchers are actively exploring novel and advanced materials. However, conventional machining processes such as milling, turning, shaping, and drilling are often inadequate for machining these advanced materials efficiently. These conventional methods not only result in prolonged production times but also yield suboptimal surface finishes due to excessive tool wear when applied to hard materials. In response, there has been a surge in the development of "Advanced Manufacturing Processes" capable of machining hard materials effectively. Various advanced machining methods have emerged in recent years, each employing unique techniques for material removal. These methods include Chemical Machining, Electrochemical Machining, Photochemical Blanking, Electrochemical Blanking, Electrochemical Honing, Electrical Discharge Machining, Ion Beam Machining, Electron Beam Machining, Laser Beam Machining, and physical vapor techniques like Sputtering and Epitaxy. These advanced manufacturing processes leverage different energy sources for material removal, enabling them to reduce raw materials to the desired size and shape efficiently. It's crucial to categorize these techniques based on their material removal mechanisms. Moreover, the efficiency and applicability of each process vary, underscoring the importance of thoroughly studying machining parameters and process capabilities before selection for a specific application.

4.2 CLASSIFICATIONS OF ADVANCED MACHINING PROCESSES

Advanced machining processes are classified based on several parameters that influence their outcomes, primarily by the energy involved in the machining operation. Mechanical, chemical, thermal, and electrical energies are utilized to machine raw materials, leading to the development of finished products. Table 4.1 provides a basic

DOI: 10.1201/9781003317272-4

TABLE 4.1
Classification of Modern Machining Processes

Type of Energy	Mechanism of Metal Removal	Transfer Media	Energy Source	Processes
Mechanical	Plastic shear	Physical contact	Cutting tool	Tool-based machining (TBM)
	Brittle fracture/ shear	Hig h velocity particles	Pneumatic/ hydraulic pressure	Abrasive jet machining (AJM), ultrasonic machining (USM), water jet machining (WJM)
	Physical sputtering	Ion stream in vacuum	Kinetic energy of ion	Ion beam machining (IBM)
Chemical	Corrosive reaction	Reactive environment	Corrosive agent	Ch emical machining (CM)
Electrochemical	Ion dissolution	Electrolyte	High current	Electrochemical machining (ECM), electrochemical grinding (ECG)
Thermoelectric	Fusion	Hot gases	Ionized material	Plasma arc machining (PAM)
	Melting/ablation	Electrons	High voltage	Electrical discharge machining (EDM)
	Melting/ablation	Radiation	Amplified light	Laser beam machining (LBM)
	Melting/ablation	Electron stream	Kinetic energy of electron	Electron beam machining (EBM)

classification of advanced machining processes based on the energy sources utilized. This chapter delves into various advanced machining processes utilizing chemical, thermal, electrothermal, electrical, and mechanical energies for machining operations.

Chemical machining (CM) employs chemical energy, while electrochemical machining (ECM) utilizes electrochemical energy. ECM operates through anodic dissolution, whereas CM relies on the corrosive actions of chemicals. ECM typically boasts a higher material removal rate (MRR) compared to CM, making it widely used for machining conductive materials. ECM offers advantages such as no tool wear, minimal heat damage, and absence of burrs during the process.

Thermal/heat energy from sources like sparks, photon beams, electron beams, and ionized gases can also remove materials. Electrical Discharge Machining (EDM) uses sparks, Laser Beam Machining (LBM) utilizes condensed photons, Electron Beam Machining (EBM) employs focused electron beams, and Plasma Beam Machining

(PBM) utilizes ionized gases, typically argon, for material removal. EDM is prevalent for machining hard and conductive materials, while EBM, PBM, and LBM processes offer faster Material Removal Rates. For applications demanding minimal thermal damage to specimens, Plasma Beam Machining (PBM) is considered, though it has less accuracy compared to other processes. Selection of advanced machining processes depends on various parameters including machining parameters, material type, desired output, and process cost. Despite advancements, these processes have limitations in achieving accuracy in shape and surface finish. Scientists are exploring hybrid machining processes, combining different advanced manufacturing techniques to overcome these limitations. Electric Discharge Grinding is one such hybrid process, which combines electric and abrasive/chemical sources to improve efficiency. Other hybrid processes include electrochemical discharge machining and electrochemical grinding.

Developing machining processes for micro and nano applications is crucial for improving component accuracy. Analyzing advanced machining processes is essential for enhancing their efficiencies and ensuring products meet desired standards. The efficiency of a machine depends not only on the shape and size of the product but also on internal cracks due to stresses and surface damages. Thus, the development of new hybrid machining processes aims to produce products with high accuracy, especially in the nano regime. Table 4.1 outlines various advanced manufacturing processes, their mechanisms, energy sources, and types of tools used for manufacturing. In this chapter, the authors explore recent advancements in machining operations, particularly micro and nano machining, based on material removal mechanisms and various operational parameters. They provide detailed explanations of different machining operations, their construction, merits, demerits, and latest improvements (Machinability Data Center, 1980).

4.3 CHEMICAL MACHINING (CM) PROCESS

Chemical Machining (CM) stands as one of the oldest machining processes, enduring through time due to its efficacy in material removal. It involves the removal of materials from a specimen through chemical etching, employing highly reactive chemical electrolytic solutions, which can be either acidic or alkaline. This technique traces back to ancient Egypt, where copper was etched using citric acid around 2300 BC. Chemical machining encompasses various types such as chemical milling, engraving, blanking, and etching. The appeal of chemical machining lies in its ability to machine intricate designs with precision, leveraging chemical actions within a controlled environment. In contemporary applications, CM processes are found to be extensively used in producing small pocket holes and intricate geometric shapes without inducing thermal damage to the surface. A protective layer, known as a maskant layer, shields areas where material removal is not intended. Automobile, electronics, and aerospace industries prominently employ chemical machining processes. Furthermore, this method is widely used in fabricating components for micro electromechanical systems (MEMS). CM processes demonstrate adeptness in creating multiple micro or nanosized holes or cavities, contributing to the development of storage devices in microelectronic applications (Bard and Faulkner, 2001; Bhattacharyya, 2015; Koryta et al., 1993; Taniguchi, 1983).

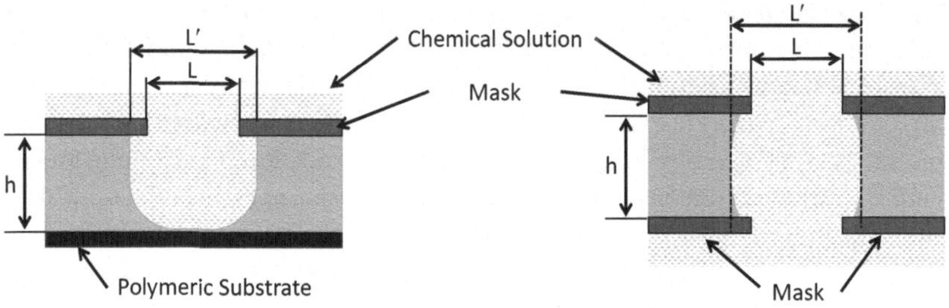

FIGURE 4.1 Pictorial representation of undercut in one and dual sided chemical milling operations.

4.3.1 Working Principle of the Chemical Milling Process

Chemical etching serves as the fundamental method for removing materials from metal specimens. Various chemical etchants such as HCl, H_2SO_4, HF, and $FeCl_3$ find widespread use, catering to metals like Fe, Al, and Cu. These chemicals react with metals upon immersion, facilitating material removal through chemical reactions. Prior to commencing machining operations, specimens undergo thorough cleaning with acetone and drying to eliminate impurities. Following cleaning, the specimen is coated with a chemically inert material, termed maskant, to prevent chemical interaction with the metal surface. Consequently, coating is applied according to the desired design, where material removal should not occur, leaving the uncoated area susceptible to material removal. Thus, material removal is a controlled process confined to the exposed area. Material removal rates vary depending on the metal specimen and chemical used, typically ranging from 0.0023 to 0.01 mm/min. However, a notable drawback of the chemical milling process is the removal of materials not only in the downward direction but also sideways beneath the maskant layer, resulting in what is termed an undercut. The extent of this undercut, known as the etch factor, can be calculated for all chemical machining components, as illustrated in Figure 4.1.

$$Etch\,Factor(E) = \frac{Depth\,of\,Cut(h)}{Under\,cut(x = L' - L)} \tag{4.1}$$

The degree of undercut in an etched specimen depends on several factors, including the type and concentration of etchants, the material of the specimen, and the depth of cut. Control over the specimen's undercut can be achieved through effective masking techniques and by adjusting variables, such as the immersion time of the specimen in the etchant solution. Formation of gas bubbles at the surface of the specimen can impede the material removal rate. Therefore, it's essential to manage the formation of bubbles at the interface of the specimen and etchant to ensure proper material removal. Although the chemical milling process is time-consuming, it offers the capability to produce high-quality pocket holes and intricate shapes. With proper control and optimization of parameters, chemical milling proves to be a valuable method for achieving precision machining outcomes (Benedict, 1987). The precision of the

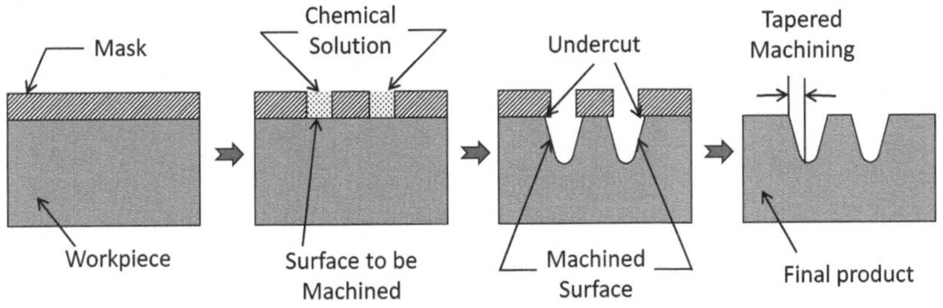

FIGURE 4.2 Schematic representation of materials removal in chemical milling process.

chemical milling process relies on various process parameters, including the type of maskant utilized, the technique employed for maskant coating, the interaction method between the specimen and etchant, and the type of etchant used. A schematic illustration depicting the material removal process is presented in Figure 4.2

4.3.2 Steps Involved in Chemical Milling Process

The chemical milling process consists of the following steps:

1. Cleaning of specimen material.
2. Coating of chemically inert mask.
3. Removal of unwanted mask.
4. Etching of the specimen.
5. Removal of mask from the specimen.

Cleaning of Specimen Material: The specimen should be cleaned prior to machining operation. The surface of the specimen should be free from dust, oil or any other foreign substance that can reduce the surface finish of the component. The clean surface also induces the better adhesion of mask with the specimen surface. The specimens are cleaned using acetone or toluene so as to remove the impurities from the surface of the specimen.

Coating of Chemically Inert Mask: The mask material is coated over the specimen surface. The mask will resist the erosion of specimen during chemical reaction. These mask materials are not water soluble, it can be removed after etching operation, if treated by certain alcohols such as acetone.

Removal of Unwanted Mask: In this step, the mask resist materials are removed as per the requirement such as design, size, and shape of the component using toluene.

Etching of the Specimen: In this step, the material removal due to chemical actions takes place. Due to the exposure in etchant, the areas without mask undergo chemical reaction and material removal takes place. After the etching process the specimen is cleaned using distilled water to remove the etchant from specimen surfaces.

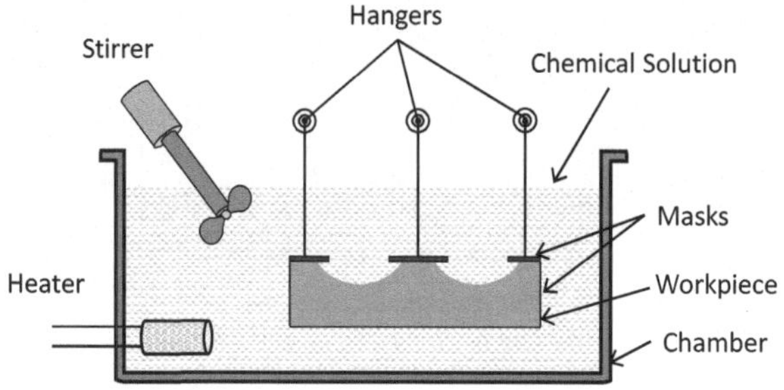

FIGURE 4.3 Experimental setup of chemical milling process.

Removal of Mask From the Specimen: the final step of chemical milling process is the removal of mask from the specimen surface. The specimen is cleaned using acetone or toluene so as to remove the mask from the specimen (Bähre et al., 2013). The schematic representation of chemical milling process is shown in Figure 4.3.

4.4 CHEMICAL BLANKING PROCESS

The chemical blanking process is one of the unconventional machining process utilized for the production of complex shapes in thin metal plates and foils using chemical reactants. The chemical blanking is used to produce slots and holes in thin sheets and foils. In the chemical blanking process, the masks are utilized for blanking operation. There are different types of maskings available such as photoresist, cut and peel methods. The chemical blanking over comes the disadvantages of mechanical blanking, such as vibrations and surface distortion. The chemical blanking process consists of four steps:

1. Cleaning of specimen material.
2. Coating of chemically inert mask.
3. Removal of unwanted mask.
4. Etching of the specimen.
5. Removal of mask from the specimen.

As discussed earlier in chemical milling section.

4.5 PHOTOCHEMICAL MACHINING (PCM) PROCESS

The photochemical machining process (PCM), also referred to as photochemical milling, is a method extensively employed for crafting thin sheet metal components. This process employs chemical etchants to selectively remove materials from designated areas through corrosive action. PCM has gained prominence, particularly in the electronic industry, for manufacturing printed circuit boards due to its precision and cost-effectiveness (Allen, 2004). The photochemical machining process serves as an alternative to operations like stamping, punching, and water jet machining for manufacturing thin parts. It boasts excellent dimensional tolerances and produces clean edges without burrs or sharp imperfections. This method is applicable to a wide range of metals and alloys. Photochemical machining shares similarities with photolithography, utilizing photo engraving techniques to create patterns on thin metal sheets. The process begins with the development of the pattern in photographic film, comprising two negative image film sheets representing the desired pattern. Metal sheets are then precisely cut to size and cleaned using acetone to remove dirt and impurities. Subsequently, the metal specimen undergoes spin coating with a masking material, onto which a photoresist material is applied—either positive or negative. The specimen is positioned between the two phototool sheets, creating a vacuum to ensure proper contact. Exposure to UV radiation hardens the areas of the photoresist corresponding to the design in the phototool. Next, the metal undergoes etching using alkaline or acid solutions to develop the desired pattern. The etchant solutions interact with the unmasked areas of the metal, removing metal particles. Upon completion of the etching process, the maskants are removed from the finished metal plates. A visual representation of the steps involved in the photochemical machining process is depicted in Figure 4.4. The applications of photochemical milling extend to the manufacture of micro sensors, micro channels, fuel cells, and semiconductor devices (Allen, 2004).

4.6 MERITS AND DEMERITS OF CHEMICAL MACHINING PROCESS

The chemical machining process (CM) offers several advantages:

- The process does not alter the physical properties of the specimen.
- It does not induce internal stresses in the material.
- Burr formation is avoided.
- The setup costs are low.
- Skilled labor is not required.
- It yields a good surface finish.
- Mass production.

While chemical machining processes offer numerous advantages, they also present several disadvantages:

- They are unable to produce sharp edges.
- Thick materials cannot be effectively machined.
- Due to the use of chemicals, there are potential hazards to both workers and the environment.
- Accuracy is relatively minimal compared to other advanced machining processes.

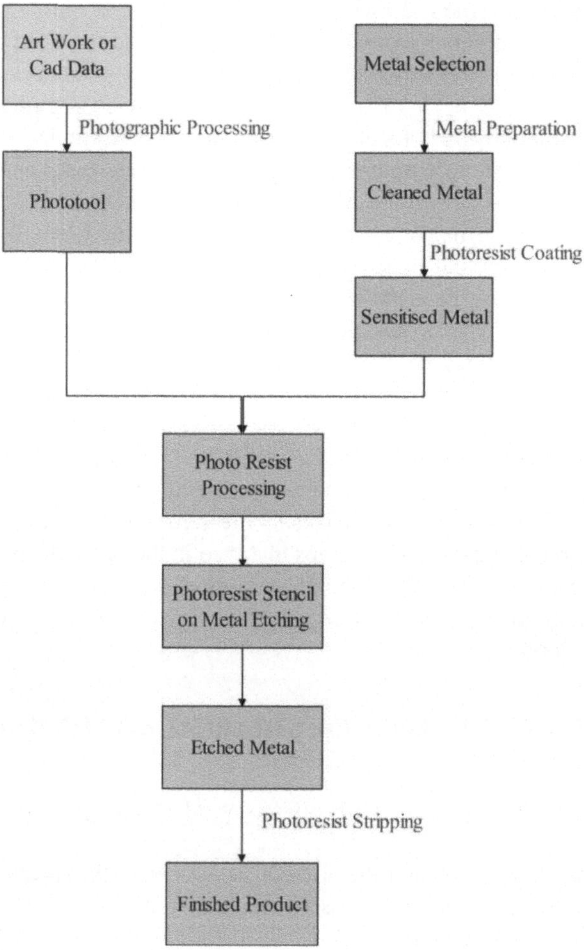

FIGURE 4.4 Steps involved in photo chemical machining.

4.7 APPLICATIONS OF CHEMICAL MACHINING PROCESS

The chemical machining process finds applications across a spectrum of component sizes, from micro to large. Some of its applications include:

- Development of instrument panels, IC circuits, and other semiconductor devices.
- Grinding or reducing the thickness of hard metals such as titanium, nickel, and various other metals and alloys (Black and Kohser, 2008).
- This process is used to drill holes, engraving on metals, and to develop contours, pockets on metal specimens (Fadaei Tehrani and Imanian, 2004).
- This process is employed for manufacturing rocket and missile components.

FIGURE 4.5 Applications of chemical machining process. (Reprinted with permission from Allen, 2004. Copyright © 2004, Elsevier.)

- It finds application in producing thin metal sheets with precise narrow cuts, commonly utilized in the aerospace industry.
- PCM processes are extensively employed in various electronic and electro-mechanical applications.

Due to the increasing demand for the production of thin and intricate parts at an economical rate, the chemical machining process has become a crucial manufacturing technique for developing complex shapes. For instance, it is used in fabricating lead frames for IC circuits, head assemblies for hard disks, and micro and nano-etched gaskets used in mobile phones, as illustrated in Figure 4.5.

4.8 ELECTRO CHEMICAL MACHINING PROCESS

Electrochemical Machining (ECM), commonly referred to as ECM, stands as an advanced machining technique capable of fabricating products that are challenging to produce using conventional metalworking methods. ECM is distinguished for its exceptional accuracy and is particularly well-suited for machining conductive metals. The process operates on the principle of electrolysis, where the specimen serves as the anode and the tool as the cathode, immersed in an electrolytic solution. Material removal from the specimen occurs upon the application of electric current through the electrolytic solution. ECM boasts significant potential due to its diverse range of applications and is regarded as one of the premier machining processes in advanced manufacturing industries.

In 1833, Michael Faraday introduced two fundamental principles of electrolysis, marking a pivotal advancement toward modern manufacturing techniques like electrodeposition and electro-dissolution. Subsequently, during the 1950s and 1960s, notable progress occurred in electrochemical machining. This process was increasingly recognized for its capabilities in shaping, milling, deburring, and finishing operations, particularly in aerospace applications (Wilson, 1971). Initially, ECM was primarily employed for the fabrication of large components that posed challenges for conventional machining processes. However, the rapid development of various types of electrochemical machining machines, particularly by countries like the US, Japan, and Germany, has propelled this process to achieve remarkable success. This success is attributed to its accuracy, quality of output, cost-effectiveness, and manufacturing flexibility. Only after the late 1990s, ECM was utilized in various other applications such as automobile industries, medical industries, electronic industries and also by aerospace industries (Chryssolouris et al., 1984; Taniguchi, 1983). Now a days, ECMs are widely used for micro nano engineering applications, such as microfluidics and nano fluidics (Bard and Faulkner, 2001).

4.8.1 Basics of Electrolysis

Electrolysis operates on the principle that when an electric field is applied between two conductive electrodes submerged in a conductive electrolytic solution, metal removal occurs from the anode, as illustrated in Figure 4.6. In this setup, two electrodes composed of copper are immersed in a $CuSO_4$ electrolytic solution, with the electrodes connected to a DC power supply.

An electrolytic cell is established when the electricity passes through the $CuSO_4$ electrolytic solution. As the result of this current flow between two conductive electrodes, a chemical reaction takes place at the interface of the electrolytic solution and the electrodes. The reactions are both anodic and cathodic in nature. The cations which are positively charged will move towards cathode and anions which are negatively charged move towards anode (Pandilov, 2018).

The electrons always tend to move out to maintain the anodic and cathodic reactions, because of this the dissolution of copper from anode to cathode to takes place in an electrolytic cell.

$$Cu = Cu^{++} + 2e \tag{4.2}$$

Neutralization of cation by the action of electrons takes places at the cathode as follows:

$$Cu^{++} + 2e = \mathrm{Cu} \tag{4.3}$$

$$2H^{+} + 2e = \mathrm{H2} \tag{4.4}$$

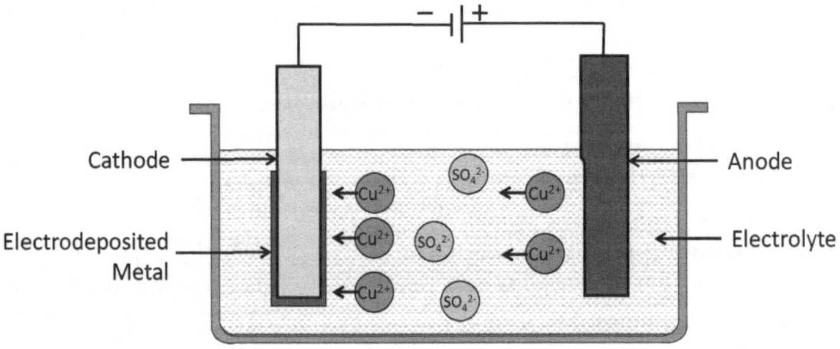

FIGURE 4.6 Schematic representation of electrolysis process.

The neutral nature of the electrolytic solution is the important factor which is to be maintained during an electrolysis process. Hence, the percentage of reaction should be same on both anode and cathode. So in the electrolysis, which involves copper sulfate electrolytic solution and copper electrodes, the dissolution of copper from anode to cathode takes place. As a result, positively charged Cu ions migrate from the anode and deposit onto the cathode. Consequently, at the conclusion of the electrolysis process, the weight loss of the anode equals the weight gained by the cathode. This electrolysis process is principally governed by Faraday's laws, which are as follows (Taniguchi, 1983). According to the first law, the mass of the substance deposited onto the electrode during electrolysis is directly proportional to the amount of electricity transferred at the electrode. The quantity of electricity is measured in coulombs, thus referring to electric charge rather than electrical current. The second law states that for a given amount of electric charge, the mass of the altered electrode material is directly proportional to the equivalent weight of the electrode material. The equivalent weight of a material is determined by its molar mass divided by a constant integer based on the reaction experienced by the electrode material. Therefore, based on these two governing laws, it is feasible to calculate the quantity of metal removed or coated during the electrolysis process (Yuan, 2021).

4.8.2 Working Principle of Electro Chemical Machining (ECM) Process

Electroplating and ECM both rely on electrolysis principles. However, in the ECM process, this principle is employed to remove metal from the specimen. In this process, the specimen is connected to the positive terminal while the tool is connected to the negative terminal in the presence of a conducting electrolytic solution. Anodic dissolution occurs at the specimen, leading to the removal of materials. Figure 4.7 illustrates the material removal process in electrochemical machining (ECM). As described earlier, iron ions dissolve from the anodic workpiece and react with hydroxyl ions in the electrolyte before precipitating onto the cathodic tool, forming metal hydroxide. This metal hydroxide settles at the bottom surface of the machining cell due to gravity. Consequently, as current passes through the machining cell, more metal ions dissolve into the electrolyte, resulting in the removal of metal from the anodic workpiece. When a potential difference is applied to the electrodes, various electrochemical and chemical reactions occur in the electrolytic cell. The reaction with the lowest oxidation potential occurs at the cathode, while the reaction with the highest oxidation potential occurs first at the anode. Several factors, such as the metal's nature, the electrolyte type, current density, temperature, and others, influence the oxidation potential and determine the type of reactions that will occur (Bard and Faulkner, 2001).

The reactions occurring at the cathode and anode involving the anodic workpiece and electrolyte solution can be described as follows:

1. *Anodic Reactions:*
   At the anode, two potential reactions can occur:
   a. Generation of oxygen or halogen gas.
   b. Dissolution of metal ions.

The reactions responsible for oxygen evolution can be described as follows:
In acidic electrolyte

$$2H_2O \rightarrow O_2 + 2H^+ + 2e^- \tag{4.5}$$

In alkaline electrolyte

$$OH^- \rightarrow 2H_2O + O_2 + 4e^- \tag{4.6}$$

During water hydrolysis, oxygen is released, and hydrogen ions are generated, causing localized increases in electrolyte acidity. The reaction governing metal dissolution can be represented as:

$$M \rightarrow M^+ + e^- \tag{4.7}$$

$$M^+ + OH \rightarrow MOH \tag{4.8}$$

2. *Cathodic Reactions:*
   At the cathode, two potential reactions occur:
   a. Formation of gas bubbles (hydrogen).
   b. Neutralization of positively charged metal ions.
   In acidic electrolyte

$$2H^+ + 2e^- \rightarrow H_2 \text{ (Evolution of hydrogen gas)} \tag{4.9}$$

In alkaline electrolyte

$$2H_2O + 2e^- \rightarrow 2OH^- + H_2 \text{ (Evolution of hydrogen gas)} \tag{4.10}$$

Sometimes, metal ions can also reach the cathode and undergo deposition through neutralization:
In acidic electrolyte

$$M^+ + e^- \rightarrow M \text{ (metal)} \tag{4.11}$$

For example, when iron undergoes anodic dissolution using NaCl electrolyte solution, the following reactions occur:

$$\text{At the anode: } Fe \rightarrow Fe^{++} + 2e \tag{4.12}$$

$$\text{At the cathode: } 2H_2O \rightarrow H_2 + 2(OH)^- \tag{4.13}$$

The overall electrochemical reaction can be represented as:

$$Fe^{++} + 2(OH)^{--} \rightarrow Fe(OH)_2 \tag{4.14}$$

$$4Fe(OH)_2 + 2H_2O + O_2 \rightarrow 4Fe(OH)_3 \tag{4.15}$$

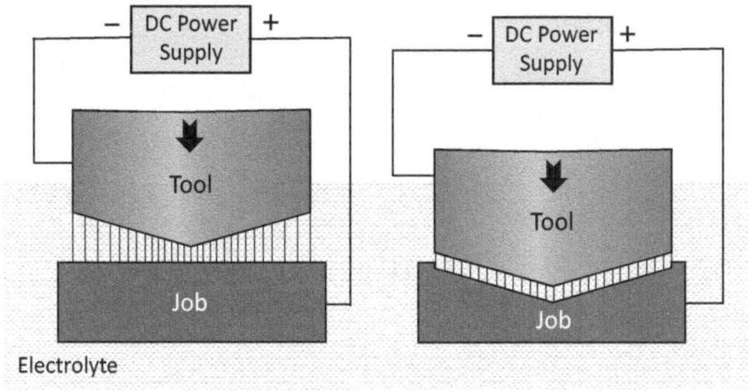

FIGURE 4.7 Schematic representation of ECM machining using shaped tool (a) Before machining (b) After machining process.

A conductive metal can be selected as a tool during machining.

Based on the described electrochemical reactions, several relevant observations can be made regarding ECM:

1. Material removal occurs at the anode workpiece, while hydrogen gas bubbles are produced, and ferrous hydroxide ($Fe(OH)_2$) and ferric hydroxide ($Fe(OH)_3$) precipitate in the electrolyte, forming sludge.
2. The rate of metal dissolution from the anode depends on factors such as the metal's atomic weight (M), its valency (Z), the current (I) passing through, and the duration of time (t) for which the current flows. Thus, the dissolution or machining rate is unaffected by the metal's hardness or other characteristics (e.g., physical or mechanical properties) during machining. conductive metal can be chosen as a tool during machining.
3. Only hydrogen gas is evolved at the cathode tool, and the shape of the tool remains unchanged during machining. Consequently, any soft electrically conductive metal can be chosen as the tool during the ECM process (Rajurkar et al., 2013).

In Figure 4.7, the machining process in ECM is depicted, where a shaped tool is employed to work on a workpiece. The gap between the tool and the workpiece is filled with electrolyte. As current flows, the anode undergoes dissolution. It is observed that the dissolution is more pronounced in the smallest gap between the tool and the workpiece, where a higher current density is present, as illustrated in Figure 4.7. As the tool moves closer to the workpiece, the work surface gradually takes on the shape of the tool. Over time, when the system reaches a steady-state condition, the gap becomes uniform, and the profile of the tool's surface is replicated on the workpiece, as shown in Figure 4.7.

4.8.3 Components of Electro Chemical Machining Machine

An industrial Electro Chemical Machining (ECM) equipment consists of four essential systems or components, as shown in Figure 4.8 (Li et al., 2016; Rajurkar et al., 2013, 1999; Saravanan et al., 2021):

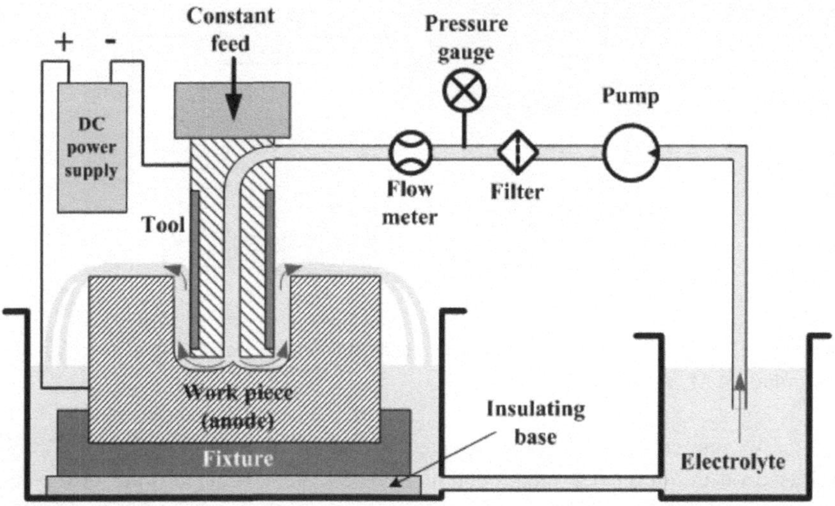

FIGURE 4.8 Representation of different components involved in ECM process. (Reprinted with permission from Keerthivasan et al. (2021). Copyright © 2021, Elsevier.)

DC Power Supply: The speed of machining in ECM (electrochemical machining) correlates directly with the electric current density. To attain rapid machining rates, ECM commonly employs a high direct current surpassing 1,000 A. For accurate machining with minimal clearance between the tool and the workpiece, a low voltage is necessary to avoid short circuits. Voltage levels in ECM typically vary between 5 and 25 V. A portion of this electrical power is harnessed by the control system for operation.

Electrolyte Circulation System: Efficient removal of the electrochemical reaction by-products from the space between the workpiece and the tool is essential. The buildup of these by-products can hamper process efficiency and impede machining speed. Hence, a swift electrolyte flow is vital, typically falling within the range of 300 to 3,000 m/min (1,000–10,000 ft/min). The inlet pressure of the electrolyte should ideally range between 0.15 to 3 MPa. The electrolyte system incorporates a sturdy pump, and continuous filtration is employed to trap the deposited reaction by-products (referred to as sludge).

Control System: The control system oversees the electrical aspects of the process (voltage and current), regulates the tool feed speed, and monitors parameters of the electrolyte circulation system (such as inlet and outlet pressure and electrolyte temperature). Its primary role is to guarantee the stable and efficient operation of the ECM equipment.

Mechanical System: The mechanical system encompasses various components such as the table, frame, work enclosure (to prevent electrolyte spillage), and the work head, which houses the tool. The tools, known as electrodes, are integral parts of this mechanical setup. Maintaining a consistent voltage level is critical in ECM, achieved by the control system through uniform tool movement at a speed corresponding to the desired linear machining rate. The ECM process operates under steady conditions with a fixed gap typically ranging from 0.1 to 0.4 mm (0.004–0.016 inches). Ensuring a secure fixation of the workpiece is paramount, facilitated by fixtures, tables, and frames, ensuring stable system operation with a constant gap. Conventional machining equipment, including CNC machines, can be adapted for ECM processes (Kozak et al., 2004; Leese and Ivanov, 2016).

4.8.4 Parameters Which Influence Metal Removal Rate, Overcut and Surface Finish in ECM Process

- Concentration
- Flow rate of electrolyte
- Temperature
- Applied voltage
- Feed rate
- Current density
- Type of electrolytes

4.8.5 Applications of ECM Process

Electrochemical Machining (ECM) finds its primary application in situations where conventional machining techniques are not suitable (Rajurkar et al., 2013b). In theory, ECM can be employed for machining all types of electrically conductive metals and alloys. The typical applications of ECM include:

1. *Machining Hard and Heat-Resistant Materials:* ECM is unaffected by the hardness of the workpiece, thus making it a preferred method for machining hard materials.
2. *Machining Complex Cavities and Irregular Holes:* ECM excels in crafting intricate cavities in forging dies and drilling irregularly shaped holes, which are often challenging to achieve with conventional methods.
3. *Machining Complex Profiles:* ECM is particularly adept at shaping complex profiles like turbine wheels, turbine blades, and jet blades.
4. *Die Sinking:* ECM provides an alternative to cavity-type Electric Discharge Machining (EDM) for die sinking applications.
5. *Thin-Walled Part Fabrication:* ECM imposes no surface stress on the workpiece, enabling the machining of thin-walled shapes, even from brittle and easily deformable materials.
6. *Electrolytic Grinding:* ECM facilitates grinding of hard and brittle materials by utilizing a conductive rotating wheel embedded with non-conductive hard particles, submerged in an electrolyte.
7. *Deburring:* ECM can smooth rough corners or edges, effectively removing burrs from workpieces.

In summary, electrochemical machining offers significant advantages in various applications, overcoming limitations posed by traditional machining techniques.

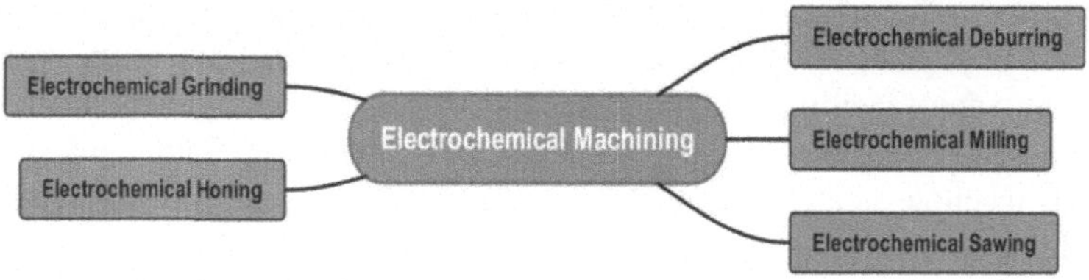

FIGURE 4.9 Classification of electrochemical machining process.

4.8.6 Different Types of Electro Chemical Milling Processes

Electrochemical Machining (ECM) stands out for its exceptional flexibility and versatility in machining, owing to its unique material removal mechanism. By precisely controlling anodic dissolution, ECM enables the production of products with diverse shapes, sizes, and surface finishes. This method involves the coordination of multiple parameters, providing the opportunity to adjust these factors to suit various applications.

ECM demonstrates remarkable efficacy across a spectrum of tasks, including shaping, turning, milling, sawing, drilling, deburring, smoothing, and finishing, among others. Furthermore, it can be seamlessly integrated with other advanced machining techniques to develop innovative hybrid machining approaches that leverage the strengths of each constituent process. The potential of anodic dissolution can also be effectively utilized for micro-machining applications. In Figure 4.9, the fundamental mechanisms of ECM and its versatile applications are illustrated, showcasing its significant potential.

4.8.6.1 Electrochemical Deburring

Burrs are thin, typically triangular ridges that form along the edges of workpieces during various manufacturing processes, such as machining, sheet material shearing, trimming, forging, and casting. The removal of burrs in drilling alone accounts for over 40% of total machining time, reducing production efficiency and increasing costs (Choi and Kim, 1998). Achieving a successful deburring process within a manufacturing system that is highly efficient and fully automated poses a significant challenge. Manual methods are commonly used, but removing complex internal burrs, which are difficult to access, presents a challenge.

However, electrochemical deburring, also referred to as electrochemical machining (ECM), utilizes the same principle of metal removal mechanism as ECM to effectively eliminate metal burrs produced during conventional tool-based machining in finished items. In the electrochemical deburring process, the flow rate, electrolyte pressure, and current density are comparatively lower than in the standard ECM process. The cathode tool in electrochemical deburring typically remains stationary during operation. Illustrated in Figure 4.10 is a schematic of an electrochemical deburring system, where the tool is fully insulated except for the section adjacent to the burrs. The electrolyte is introduced into the deburring process by flowing through the gap between the stationary tool and the burrs. The positioning of the deburring tool and

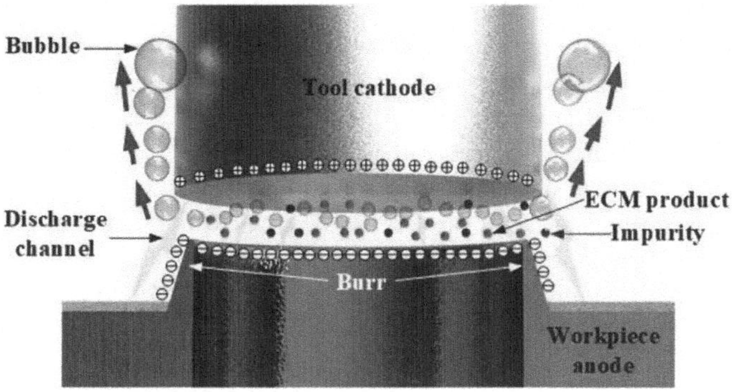

FIGURE 4.10 Schematic representation of electrochemical deburring process. (He et al., 2022)

the electrolyte flow path is contingent upon the geometry of the burrs, encompassing their height, width, shape, and location. The equipment utilized in electrochemical deburring boasts a simple construction due to the absence of tool and workpiece movement, although a pump is required to supply the electrolyte into the machining zone (Kadam and Mitra, 2021).

In electrochemical deburring, the tool is positioned in a way that promotes preferential anodic dissolution on the burrs by increasing localized current density. Considerable care is taken to strategically position the tool in order to concentrate the metal dissolution process specifically at the base of the burrs. Consequently, the orientation of the burrs becomes a critical factor in the design of electrochemical deburring tools. The design of the electrolyte flow path is also crucial to effectively flush out the free burrs after their removal from the gap between the tool and the workpiece surface, as failure to do so may cause a short circuit during the electrochemical deburring operation (Sarkar et al., 2004). Important parameters considered in electrochemical deburring include low applied voltage (5–20 V), low machining current (100–1,000 A), low electrolyte pressure (60–300 kPa), and a low flow rate of 3 to 10 L/min. The duration of the machining process is a crucial factor that can be determined beforehand by considering the characteristics of the burrs and utilizing mathematical formulations discussed earlier in the dynamics of ECM. Depending on the specific parameter settings employed during the deburring operation, the removal of burrs can be achieved within a relatively short timeframe, typically ranging from approximately 5 to 30 seconds. This efficient process yields a high-quality surface finish in the range of 0.08 to 1.2 μm (He et al., 2022).

4.8.6.2 Electrochemical Milling Process

In recent years, electrochemical machining (ECM) has garnered substantial attention in the industrial sector due to its array of advantages over other non-traditional machining methods. However, the process of designing and fabricating intricate-shaped tools in ECM can be time-consuming and resource-intensive, resulting in heightened costs for final products. Moreover, larger projected areas of the tool lead to increased power consumption, particularly when operating at high currents, notably with pulsed DC, which can be both costly and impractical. Furthermore, utilizing a constant DC power supply in ECM may result in inaccuracies in the machined profile due to inefficient sludge removal.

To tackle these issues, electrochemical milling (EC milling) has emerged as a solution. EC milling operates on a principle akin to conventional ECM, with a machining strategy comparable to traditional end milling processes. It allows for the milling of relatively larger areas using simpler geometric tools, eliminating the necessity for designing intricate 3D-shaped tools as in conventional ECM. In EC milling, achieving the same level of current density requires less current, facilitating the use of pulsed DC power supplies compared to die sinking ECM. Consequently, this leads to enhanced machining precision, reduced power consumption, and decreased production costs for final products. By incorporating tool rotation and implementing flushing through the tool, sludge removal can be notably improved, further enhancing machining accuracy.

For the realization of 3D features in a workpiece, a cylindrical tool tip with a basic shape can be effectively employed. This is achieved by controlling the scanning movement of the tool electrode along a predetermined tool path within an electrolytic cell, as illustrated in Figure 4.11. Ensuring the precision of 3D shape generation in EC milling necessitates accurately generating tool paths, considering factors such as stray current and microsparks. A tool with a simple shape is guided along the desired path, and the machining process is executed using the bottom tip of the tool. At the end of each cycle, a small downward tool feed is applied. This process continues until the desired shape and depth of the 3D feature are achieved. This scanning approach enables the generation of complex 3D shapes with a high aspect ratio. It is important

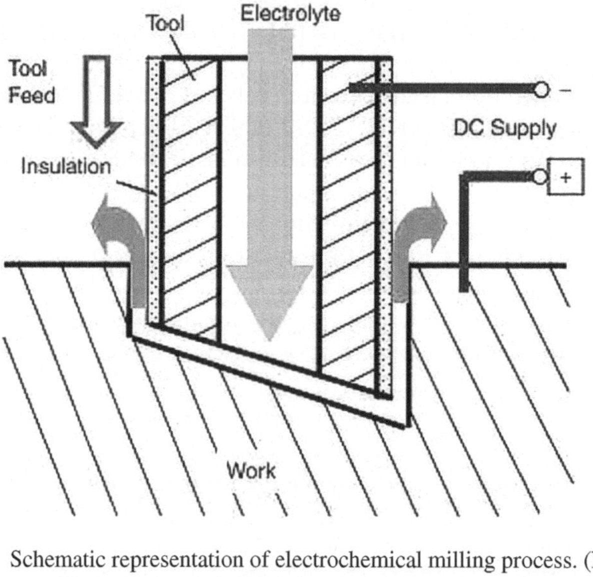

FIGURE 4.11 Schematic representation of electrochemical milling process. (Reprinted with permission from Ghosh, 2012. Copyright © 2012, Springer.)

to maintain the machining gap within an optimal and minimal range throughout the process.

4.8.6.3 Electrochemical Sawing

The principle of electrochemical machining (ECM) finds application in sawing, especially for cutting extremely thin and fragile workpieces. Conventional cutting methods often exert substantial mechanical force, posing a risk of breakage or deformation to delicate workpieces. Furthermore, traditional sawing techniques usually demand a sturdy fixture for holding the workpiece, which might not be compatible with thin and fragile materials (Liu et al., 2016).

In EC sawing, the delicate workpiece is separated using anodic dissolution without subjecting it to the adverse effects of cutting forces. In this process, as illustrated in Figure 4.12, a thin circular tool serves as the cathode and rotates over the anode workpiece. Electrolyte is supplied to the cutting zone through a nozzle. The rotation of the tool effectively forces the electrolyte into the cutting area, eliminating the need for high-pressure electrolyte supply in EC sawing. Sludge and gas bubbles are efficiently removed from the cutting area due to the centrifugal effect of the rotating tool. Since no cutting force is generated, the workpiece holding fixture can be simple in nature for EC sawing (Rajurkar et al., 2013).

The tool used in EC sawing is typically disc-shaped with minimal thickness and made of electrically conductive materials such as copper, brass, or graphite. It is crucial to maintain a very low gap between the rotating tool and the workpiece to achieve controlled metal removal. For improved cutting efficiency, a low rotational speed is recommended in EC sawing. The rotating tool is fed forward across the workpiece to achieve parting off in a single-pass operation. Recommended electrolytes for EC sawing include NaCl and $NaNO_3$, depending on the material being cut. The cutting efficiency depends on various parameter settings, such as machining voltage, current, tool rotational speed, and feed rate.

EC sawing offers several advantages over conventional sawing methods. It can be used to cut any conductive material, regardless of its mechanical properties. EC sawing produces no thermal or mechanical damage to the workpiece, and there are no burrs generated during the parting-off process, eliminating the need for post-cutting finishing operations. However, it is essential to protect all machine components from

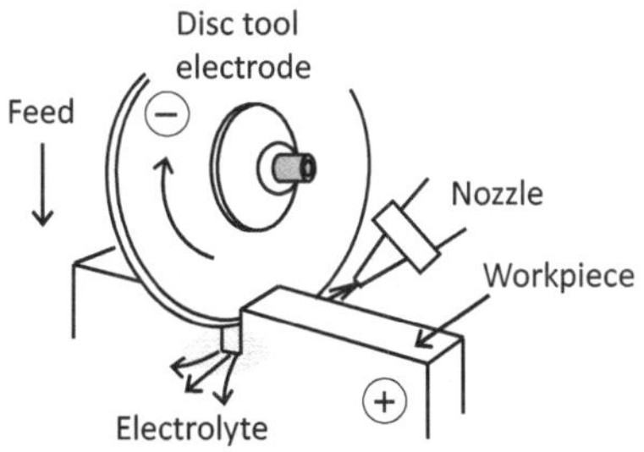

FIGURE 4.12 Schematic Representation of electrochemical sawing process. (Reprinted with permission from Bhattacharyya and Doloi, 2020a, Copyright © 2020, Elsevier.)

the corrosive environment associated with the electrolyte used in EC sawing, which may result in higher maintenance costs.

4.8.6.4 Electrochemical Grinding Process

Electrochemical Grinding (EC Grinding) is a specialized grinding process that utilizes the electrochemical material removal mechanism. It shares similarities with mechanical grinding in terms of achieving a good surface finish. However, instead of using an abrasive bonded grinding wheel, a metallic wheel is employed. In EC Grinding, a small gap is maintained between the metal wheel and the workpiece, which is submerged in an electrolyte solution. Unlike conventional grinding methods, there is no cutting force, heat generation, distortion, or stress development on the workpiece since there is no direct contact between the tool and the workpiece.

Figure 4.13 illustrates a schematic diagram of EC Grinding. The metallic wheel acts as the cathode and rotates over the anodic workpiece, both of which are immersed in an electrolytic chamber. This process is particularly useful for producing flat surfaces with a high-quality surface finish, which necessitates a very small gap between the workpiece and the metallic wheel. The wheel is fed downwards by controlling the Z drive of the setup, and a servo feed system is employed to maintain and control the small gap between the metallic wheel and the workpiece. The workpiece undergoes a horizontal motion beneath the metallic wheel using an X-Y stage, and the feed rate can be controlled using a servo system. The rotational speed of the metallic wheel creates a circulation of the electrolyte solution within the machining gap.

EC Grinding is primarily employed for generating flat surfaces with high precision. However, an abrasive bonded metallic wheel can be used for faster material removal and more efficient machining. In this case, both micro cutting by abrasives and electrochemical machining occur. This process is widely utilized for machining and grinding applications.

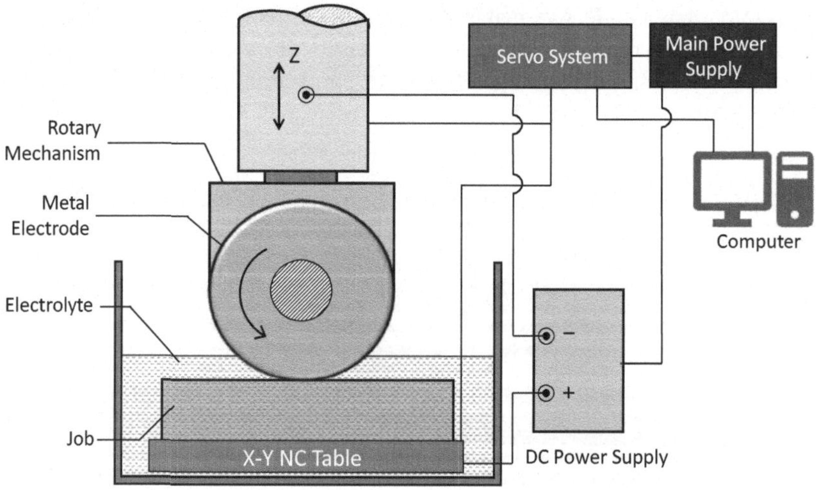

FIGURE 4.13 Schematic representation of electrochemical grinding process.

Working Principle of Electrochemical Grinding Process

In electrochemical grinding (ECG), a grinding wheel (cathode) is used, which is connected to the negative terminal of a DC power supply, similar to a conventional grinding wheel. However, in ECG, the bonding material of the wheel is electrically conductive. The electrolyte is supplied through the interelectrode gap (IEG) between the wheel and the workpiece. The height of the abrasive particles protruding from the bonding material helps maintain a constant IEG because these particles act as spacers. The lifespan of an ECG wheel is approximately ten times longer than that of a conventional wheel due to two main reasons: (i) only 10% of the material removal is attributed to abrasive action, and (ii) the arc of contact between the wheel and workpiece is very short (Xu and Wang, 2021).

In ECG, there are three fundamental modes of material removal, involving both mechanical abrasion and electrochemical dissolution, with varying levels of overcut:

1. Total mechanical material removal occurs during the initial stages of the grinding process.
2. Electrochemical dissolution is combined with mechanical abrasion, resulting in a zero overcut.
3. Electrochemical dissolution is combined with mechanical abrasion, resulting in an overcut greater than zero.
4. Totally electrochemical anodic dissolution occurs when a large interelectrode gap is created (Atkinson and Noble, 1979).

The grinding area between the wheel and workpiece can be divided into three interaction zones: Zone I, Zone II, and Zone III as shown in Figure 4.14

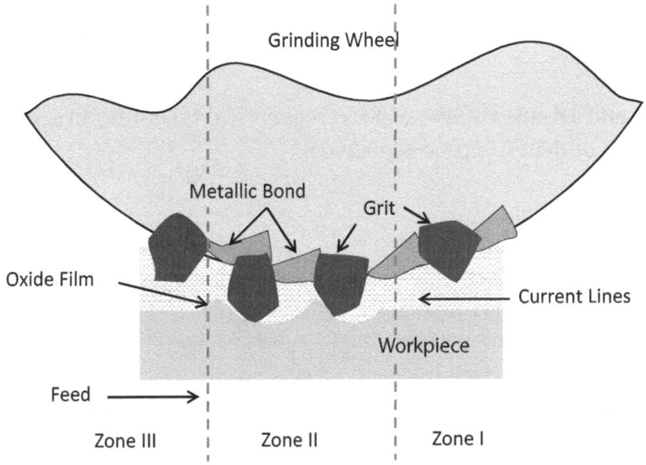

FIGURE 4.14 Represents the different zones of material removal in electrochemical grinding process.

Zone I: As the grinding wheel rotates, fresh electrolyte is drawn into the initial larger interelectrode gap. Material is primarily removed through electrochemical anodic dissolution. As the process continues, the electrolyte becomes contaminated with reaction products, gases, and oxide layers, which decreases the electrical conductivity of the medium and reduce the interelectrode gap. Material removal rate decreases due to reduced current flow. Zone I ends when the wheel comes into contact with the workpiece.

Zone II: In this zone, material is removed from the workpiece surface by the abrasive particles of the ECG grinding wheel, which is in direct contact with the workpiece. The protruding abrasive grains maintain the interelectrode gap to prevent short circuits and remove material as ground chips through mechanical grinding action. After removing the oxide passive layer, the interelectrode gap increases, exposing a fresh surface of the workpiece. The rotation of the wheel creates a cavitation effect, drawing in fresh electrolyte and increasing local electrolyte pressure while suppressing the formation of gas bubbles. As a result, electrochemical anodic dissolution occurs at a higher rate due to increased current flow. Zone II combines mechanical abrasion and partial electrochemical dissolution of metal oxides. Zone II ends when the wheel lifts off the workpiece surface.

Zone III: In this zone, material removal occurs solely through electrochemical anodic dissolution of the workpiece. The interelectrode gap is larger, resulting in less material being removed. The electrolyte pressure is lower in this zone. Zone III serves as the finishing zone, where any scratches or microburrs formed in Zone II are removed through electrochemical dissolution.

The lifespan of the grinding wheel in ECG is approximately 8 to 10 times longer than that of a conventional grinding wheel due to minimal wheel wear, small contact length, and reduced contact time between the wheel and the workpiece. In ECG, about 90–95% of the material is removed through electrochemical dissolution (ECD), while the remaining 5–10% is removed through mechanical grinding (MG) action (Ilhan et al., 1992).

Advantages and Disadvantages of Electrochemical Grinding Process

The advantages of the ECG process include:

1. Good surface quality
2. Elimination of grinding burrs
3. Absence of distortion in thin, fragile, or thermo-sensitive parts
4. Absence of work hardening
5. Production of narrow tolerances
6. Longer grinding wheel lifespan
7. Low induced stress
8. Ability to achieve large depths of cut
9. High material removal rate

However, the ECG process also has some limitations, which are as follows:

1. Higher capital cost compared to conventional machines.
2. Limited to electrically conductive materials.
3. Corrosive nature of the electrolyte used.
4. Requires proper disposal and filtering of the electrolyte.

Application of ECG

The ECG process demonstrates particular effectiveness in the following applications:

1. Machining parts made from difficult-to-cut materials, such as sintered carbides, creep-resisting (Inconel, Nimonic) alloys, titanium alloys, and metallic composites.
2. Similar applications to milling, grinding, cutting off, sawing, and tool and cutter sharpening.
3. Production of tungsten carbide cutting tools, fragile parts, and thin-walled tubes.
4. Grinding of Inconel 718 using a brazed diamond wheel.
5. Grinding of ceramic-metal tungsten plate using a diamond metal-bonded wheel.
6. Machining of metal-bonded diamond composite wheels.
7. Removal of fatigue cracks from steel structures under seawater. ECG has been utilized to create holes, approximately 25 mm in diameter, in steel 12 to 25 mm thick, at the ends of fatigue cracks. This process effectively stops further crack development and allows for the removal of specimens for metallurgical inspection.
8. Production of specimens for metal fatigue and tensile tests.
9. Machining of carbides and various high-strength alloys.

These applications demonstrate the versatility and capability of the ECG process for handling challenging materials and achieving desired machining outcomes.

4.8.6.5 Electrochemical Honing Process

Electrochemical honing combines anodic dissolution and mechanical abrasion to achieve a fine finish on internal surfaces. This process is similar to conventional honing, but the majority of material removal occurs through anodic dissolution. EC honing can be employed to finish the inner surface of electrically conductive materials of varying hardness (Rao et al., 2017).

The schematic diagram in Figure 4.15 illustrates the setup for EC honing. The tool consists of a hollow stainless steel body with non-conductive honing stones positioned around its circumference. These stones, which possess corrosion-resistant properties, are similar to conventional honing stones. A minimum of three stones is typically used, applying equal pressure against the workpiece through spring action. Ideally, the length of the stones should be half the length of the bore to be finished. The length of the main steel body serves as the cathode and should be long enough to cover the entire bore length during reciprocation. An electrolyte is supplied to the

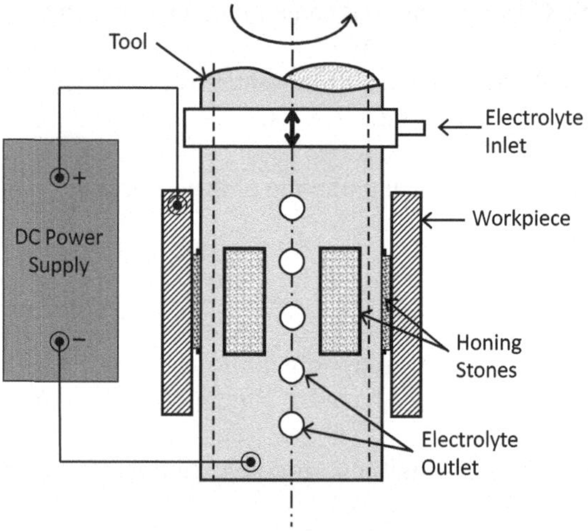

FIGURE 4.15 Schematic representation of electrochemical honing process.

sleeve housed on the main steel body and distributed along the inner wall of the hole through multiple outlets as depicted in Figure 4.15. The main steel body moves back and forth inside the bore while also rotating to ensure uniform metal removal from the entire inner surface (Dubey, 2008).

During EC honing, an electrochemical reaction leads to the formation of metal oxide, which is then removed by the abrasives of the honing stones. As material is removed from the bore through anodic dissolution, the gap between the tools (steel body and the wall) increases. However, the stones expand to maintain continuous contact with the inner wall of the bore due to the spring-loaded mechanism. Recommended electrolytes for EC honing include NaCl and $NaNO_3$ solutions.

Compared to conventional honing, the EC honing process is significantly faster due to the combination of anodic dissolution and mechanical erosion for material removal. It yields an exceptionally smooth surface finish, leaving no micro scratches on the finished bore. Additionally, dimensional tolerance is improved compared to conventional honing. Unlike ECM, EC honing induces compressive residual stress on the work surface through mechanical action, enhancing the fatigue strength of the finished component (Senbel, 2000).

4.9 THERMAL ENERGY BASED MICRO/NANO MACHINING PROCESSES

Thermal energy plays a crucial role in overcoming the limitations of conventional machining processes that rely on cutting tools or abrasives. By directly utilizing thermal energy, engineers can remove material and achieve the desired shape of a workpiece through processes such as melting, vaporization, or ablation. Various forms of thermal energy, including electrical sparks, plasma arcs, laser beams, and electron beams, are utilized by engineers to machine advanced materials, catering to the needs of modern manufacturing industries.

Modern machining processes such as Electrical Discharge Machining (EDM), Plasma Arc Machining (PAM), Laser Beam Machining (LBM), and Electron Beam Machining (EBM) have gained popularity due to their inherent advantages over traditional machining methods. These processes harness different forms of thermal energy, and researchers and scientists are continually working to improve their applications in material removal. The objective is to generate precise shapes with high accuracy and superior surface quality, aligning with the evolving properties of materials and the complex design considerations in modern engineering applications.

As materials become more advanced and engineering applications grow more intricate, the complexity of machining also increases. To address these challenges, engineers are dedicated to understanding the fundamental mechanisms of material removal and achieving precise control over the aforementioned machining processes. However, there are still areas that require further improvements, emphasizing the need for in-depth studies on advanced machining processes that utilize thermal energy for material removal. Therefore, this discussion will delve into the details of various thermal-based advanced machining processes.

4.9.1 Electrical Discharge Machining

Currently, Electrical Discharge Machining (EDM) stands as the most widely used advanced machining process that relies on thermal energy for material removal. It is a popular nonconventional machining technique that utilizes spark discharges to machine electrically conductive materials, irrespective of their mechanical properties. Unlike traditional methods, EDM does not require direct contact between the tool electrode and the workpiece. This feature enables cost-effective machining of difficult-to-machine materials and complex-shaped jobs, providing significant support to the manufacturing industry.

The controlled removal of material from the workpiece through a series of small electrical sparks is known as electrical discharge machining. The origins of EDM can be traced back to 1770, when English chemist Joseph Priestly discovered the erosive effects of electrical discharge. However, it was not until 1943 during World War II that physicist B. Rand N. I. Lazarenko at Moscow University investigated sparking phenomena between electric power contacts. Lazarenko aimed to minimize erosion caused by electrical discharges and subsequently harnessed the destructive properties of electrical sparks for productive purposes. This breakthrough marked the beginning of the development of electrical discharge machining, utilizing controlled discharge conditions to achieve precision machining through melting and vaporization of challenging materials (Dwivedi and Choudhury, 2016).

The early EDM machines, until 1950, utilized the power supply developed by Lazarenko and served as the foundation for subsequent advancements. Since then, EDM technology has rapidly improved through the efforts of various researchers and has become indispensable in manufacturing applications such as die and mold making, prototyping, and micro-machining. Figure 4.16 illustrates the basic setup of EDM, where pulse discharge occurs in a gap filled with an insulating medium, typically a dielectric liquid, between the tool electrode and the workpiece.

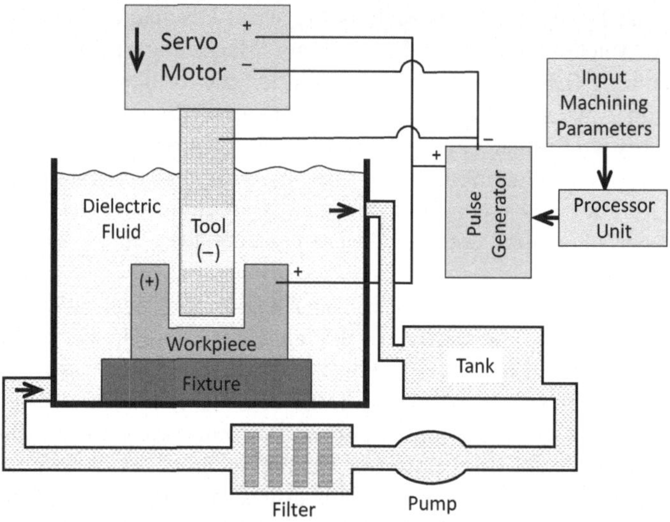

FIGURE 4.16 Schematic representation of electrical discharge machining process. (Abu Qudeiri et al., 2019)

The discharge is initiated by a high voltage, causing the breakdown of the dielectric in the small gap. This electrical discharge generates intense heat in the sparking zone, leading to the melting and evaporation of materials from both the workpiece and the tool electrode. Effective machining requires spark discharges of short duration and high repetition rate. The spark energy can be concentrated in a small area by submerging the discharge in the liquid dielectric. The negative shape of the tool electrode is replicated on the workpiece surface, with an offset equal to the gap size, allowing for precise shape generation. To achieve this, the gap between the tool and the workpiece must be maintained within the range of 10–100 μm.

Discharges should occur at a high frequency, typically in the kHz to MHz range, as the metal removal per discharge is minimal. Each pulse creates a discharge at the location with the shortest gap between the tool and workpiece surface. As a result, materials are evaporated or ejected in the molten phase from that spot, creating a small crater on both the tool electrode and the workpiece surfaces. The molten materials are then cooled and solidified in the dielectric medium, forming spherical debris that is subsequently removed from the gap by the flowing dielectric.

After each pulse, the discharge spot shifts to the next shortest gap, allowing sparks to travel across the workpiece surfaces. This repetitive discharge pattern leads to uniform metal removal, ultimately creating a mirror image of the tool electrode shape on the workpiece surface. Servo control tool feeding arrangements are employed to maintain a small spark gap between the tool and the workpiece.

4.9.1.1 Important Parameters of EDM Process

The parameters governing the EDM process can be categorized into two types: electrical parameters and non-electrical parameters. Electrical parameters include discharge voltage, peak current, pulse duration and interval (pulse-on and pulse-off times), electrode gap, polarity, and pulse waveform. Non-electrical parameters encompass dielectric fluid flushing, workpiece rotation, and electrode rotation. Let's discuss these parameters in detail.

The discharge voltage refers to the average voltage across the spark gap during machining. It directly affects the regulation of the spark gap size and overcut. High-conductivity materials typically require low voltage, while low-conductivity materials benefit from higher voltage. The discharge voltage significantly influences the material removal rate (MRR), tool wear rate (TWR), and machining accuracy. Increased current enhances MRR and TWR but adversely affects accuracy. This characteristic of current led to the development of high wear-resistant electrodes for high-current conditions.

The pulse-on time represents the duration of the discharge application. The amount of energy generated during the pulse-on time directly influences the MRR. Hence, a longer pulse-on time leads to increased MRR by augmenting the discharge energy. The pulse-off time is the duration without any discharge, allowing debris resulting from sparking and erosion to be flushed out from the gap between the workpiece and the electrode. Flushing improves ionization conditions and prevents the formation of an insulating layer. Proper selection of the pulse-off time ensures stable machining. A shorter pulse-off time increases MRR if sufficient debris flushing occurs; otherwise, it may lead to unsuitable conditions during the next pulse-on period. Pulse waveforms can take various forms such as rectangular, trapezoidal, or composite geometric shapes. It has been demonstrated that among the standard forms, the trapezoidal waveform minimizes electrode wear. Recent developments include generators that produce a high-voltage pulse with a narrow duration before the main pulse, facilitating the main pulse.

The polarity in EDM depends on factors like electrode and workpiece materials, current density, and pulse length. One of the electrodes or the workpiece is positively charged while the other is negatively charged. Negative electrode polarity is recommended for high-precision machining when the MRR is high. In wire EDM, the electrode wire typically has a negative polarity to maintain a high machining rate, and since the wire wear continues continuously during arcing, its wear rate is lower. The electrode and workpiece are kept at a small predetermined distance known as the "discharge gap," controlled by the discharge gap servo. The discharge gap is usually maintained between 0.005 mm and 0.1 mm. For finishing or high-precision machining, a relatively low voltage in the range of 50 V to 300 V is required, as excessively high voltage reduces machining precision.

Non-electrical parameters include dielectric fluid flushing, workpiece rotation, and electrode rotation. The dielectric fluid serves multiple purposes such as providing insulation against premature discharging, reducing temperature during machining, and cleaning debris from the machining area. Ideal dielectric fluids possess high dielectric strength, flushing ability, and fast recovery after breakdown. Sink-type EDMs typically employ hydrocarbon- and silicon-based dielectric oils or kerosene with increased flash points. Wire EDM uses deionized water and oil. Recent studies have explored oil-based synthetics to minimize adverse effects on workers and the environment. The dielectric type and flushing method have a notable impact on MRR, TWR, and surface roughness (SR). Improved flushing conditions can be achieved through workpiece and electrode rotation. Electrode rotation enhances flushing, resulting in better SR, higher MRR, and reduced density of surface cracks and recast layers. Explaining the impact of EDM process parameters on response parameters is challenging due to the stochastic nature of the discharge mechanism. Numerous researchers have conducted studies to investigate EDM processes and examine the influence of process parameters on performance measures (Banu and Ali, 2016; Peças and Henriques, 2009; Selvarajan et al., 2018; Singh and Ghosh, 1999; Tarng et al., 1997).

4.9.2 Wire-Electrical Discharge Machining Process

Wire EDM involves the use of a continuously moving metallic wire with a small diameter to cut a workpiece. Erosion in the machining area is caused by discrete sparks that occur between the wire and the workpiece. Similar to sinker type EDM, there is no contact between the wire and the workpiece during the machining process. Both the wire and the workpiece need to be immersed in a dielectric fluid. The wire used in wire EDM is typically thin, with a diameter as small as 0.1 mm, and it is commonly made of materials such as copper, brass, coated steel, or even metals like molybdenum. This process employs a high peak current with a short duration. The machine control unit, equipped with a computer numerical control (CNC) system, facilitates the necessary movement to cut complex shapes. It includes a dedicated microprocessor to maintain the appropriate gap range (25–50 μm) and controls the wire's speed as it passes through the workpiece, resulting in surfaces with high accuracy. Deionized water is frequently used as the dielectric fluid in wire EDM. The wire EDM process finds applications in various industries, including die making, medicine, electronics, and the automotive industry (Das A et al., 2022; Alam et al., 2022; Patel and Maniya, 2018). A schematic diagram of wire EDM is depicted in Figure 4.17.

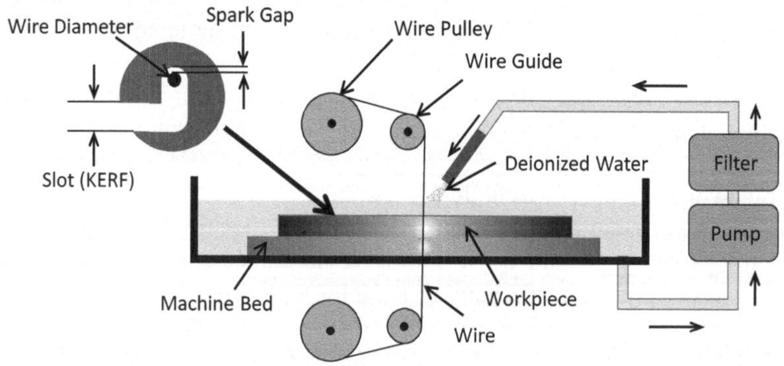

FIGURE 4.17 Schematic representation of wire-electrical discharge machining process. (Abu Qudeiri et al., 2019)

4.9.2.1 Applications of Wire-Cut Electrical Discharge Machining Process

Wire EDM is a versatile process used for cutting a wide range of materials including aluminum, brass, copper, carbides, graphite, steels, and titanium. The choice of wire material depends on the specific application requirements. For faster cutting action, zinc-coated brass wires are utilized, while molybdenum wires are preferred for more precise applications. The applications of wire EDM include:

1. Aerospace industries.
2. Medical applications.
3. Electronics component manufacturing.
4. For manufacturing die, fixtures, punches, gears, cams for industrial applications.
5. Manufacturing micro tools for micro- EDM.

Wire EDM provides precise cutting capabilities and is employed in various industries for its ability to handle diverse materials and meet specific machining requirements.

4.9.3 Plasma Arc Machining (PAM) Process

There are three fundamental states of matter: solid, liquid, and gas.. However, when gases are subjected to extreme temperatures of approximately 5,500°C, a fourth state of matter known as plasma is achieved. In this state, gases undergo partial ionization, resulting in a mixture composed of free electrons, positively charged ions, and neutral atoms. This unique combination is referred to as plasma.

When a gas is heated, the frequency of collisions between its atoms increases. Consequently, the gas becomes ionized as the outer electrons of the atoms are stripped away, resulting in the formation of electrons and ions. The newly generated electrons, in turn, collide with other atoms, boosting their kinetic energy and causing further ionization. As a result, more electrons and ions are produced within the plasma. This characteristic of plasma, containing electrons, enables it to conduct electricity effectively.

As the gas becomes fully ionized, the central part of the plasma reaches temperatures ranging from 11,000°C to 28,000°C. This highly ionized gas can be directed onto a work piece through a high-velocity jet, which leads to the removal of metal by melting. This process, known as plasma arc machining, is utilized in cutting various materials such as stainless steel, nickel alloy steels, cast iron, titanium, copper, and aluminum.

In summary, plasma, the fourth state of matter, arises when gases are intensely heated, causing partial ionization. This ionized gas mixture, consisting of electrons, ions, and neutral atoms, possesses remarkable electrical conductivity. When employed in plasma arc machining, the high-temperature plasma effectively melts and removes metal from a workpiece, making it an essential process for cutting a wide range of materials.

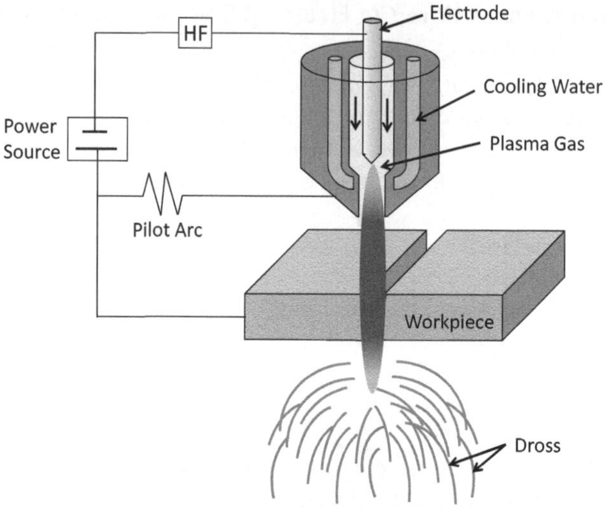

FIGURE 4.18 Schematic representation of plasma arc machining (PAM) process. (Ilii et al., n.d.)

4.9.3.1 Working Principle of Plasma Arc Machining (PAM) Process

The process begins by applying a direct current (DC) power to the circuit, which generates a powerful arc between the cathode (electrode) and the anode (nozzle). A gas, typically hydrogen (H_2) or nitrogen (N_2), is introduced into the chamber. This gas is then heated to an extremely high temperature ranging from 11,000°C to 28,000°C by utilizing the electric arc formed between the electrode and the nozzle (Chamarthi et al., 2013).

At such elevated temperatures, the gases undergo ionization, resulting in the release of a significant amount of thermal energy. The ionized gas, known as plasma, is characterized by its high velocity and temperature. This plasma jet is directed onto the surface of the workpiece through the nozzle.

The intense heat and velocity of the plasma jet cause the metal of the workpiece to melt, while the force of the high-velocity gas stream effectively propels the molten metal away. It's important to note that the heating of the workpiece material is not a result of any chemical reaction but rather the continuous impact of the plasma on the workpiece material as shown in Figure 4.18. Consequently, this machining process can be safely applied to any metal, including those susceptible to chemical reactions (Manjunathan et al., 2023; Nemchinsky and Severance, 2009).

4.9.3.2 Different Modes of Operation in Plasma Arc Machining (PAM) Process

Plasma Arc Machining (PAM) systems can be categorized into two types: Transferred Arc PAM and Non-Transferred Arc PAM.

In the Transferred Arc PAM process, the plasma arc is transferred between the electrode and the workpiece. The positive terminal of the DC power source is connected

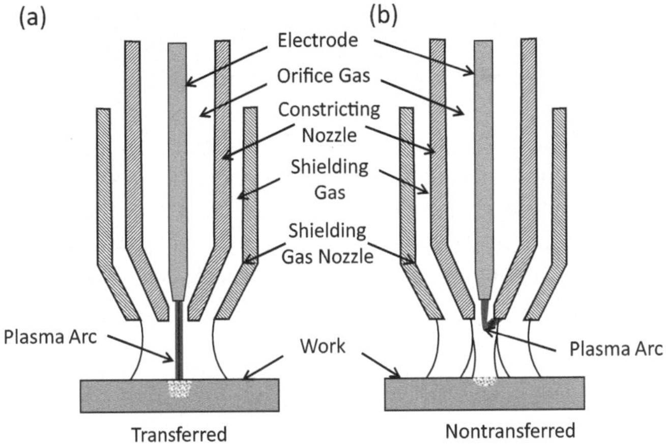

FIGURE 4.19 Different modes of operation in PAM process.

to the workpiece materials, while the negative terminal is connected to the electrode. For this process, the workpiece must be electrically conductive. The Transferred Arc PAM process allows for high power and achieves high electrothermal efficiency, reaching up to 90%. Refer to Figure 4.19 (a) for an illustration of the Transferred Arc PAM process.

In the Non-Transferred Arc PAM process, the arc is maintained between the electrode and the nozzle without being transferred to the workpiece. The negative terminal of the DC power source is connected to the electrode, while the positive terminal is connected to the nozzle. In this case, the nozzle material should be electrically conductive, and this type of PAM can be used with both electrically conducting and non-conducting materials. The gas is ionized by the arc, forming a plasma arc between the electrode and the nozzle. The plasma arc then strikes the workpiece with high velocity. The electro thermal efficiency for the Non-Transferred Arc PAM process is comparatively lower, up to 75%, compared to the Transferred Arc PAM process. Please refer to Figure 4.19 (b) for an illustration of the Non-Transferred Arc PAM system.

4.9.3.3 Advantages of the Plasma Arc Machining (PAM) Process

The advantages of plasma arc machining (PAM) process are as follows:

- PAM can be used to cut both electrically conducting and non-conducting materials, offering versatility in material selection.
- PAM allows for faster cutting compared to oxy-fuel cutting methods.
- It is suitable for cutting thicker materials, enabling efficient processing of various workpiece sizes.
- By utilizing high-power plasma torches, a high material removal rate can be achieved.
- PAM is effective for high-speed profile cutting of engineering materials such as stainless steel, super alloys, aluminum, ceramics, and more.

- This machining process eliminates the need for a cutting tool, simplifying the operation.
- PAM does not require the use of chemicals, minimizing the potential harmful effects of corrosion.
- Automated Plasma cutting can be implemented using CNC (Computer Numerical Control) technology, enhancing precision and repeatability.
- Underwater cutting is feasible with PAM, offering benefits like reduced noise, fume emission, high solubility of removed materials in water, and limited oxidation impact when appropriate plasma forming gas is chosen for the workpiece material.

4.9.3.4 Disadvantages of Plasma Arc Machining (PAM) Process

The disadvantages of plasma arc machining include:

- Cutting thicker materials at high speed requires large power supplies.
- The high heat involved in the process can potentially damage the cut edge of the workpiece.
- Toxic fumes are produced during PAM, necessitating proper ventilation and safety precautions.
- Achieving dimensional accuracy and sharp corners in the cutting process can be challenging.
- Generally, the kerf width (width of the cut) is wider, although the use of High Tolerance Plasma Arc Cutting (HTPAC) systems can achieve narrower kerf widths.
- The cutting edge generated by PAM tends to have a tapering effect, which may require additional finishing processes to achieve the desired geometry.

4.10 LASER BEAM MACHINING

LASER, an acronym for Light Amplification by Stimulated Emission of Radiation, represents a form of light that is coherent and monochromatic. Laser beams possess the remarkable ability to be focused into extremely small spot sizes, typically on the order of a few microns. This characteristic leads to an exceptionally high-energy intensity at the focal point. As a result, lasers find diverse applications as an energy source in numerous areas of manufacturing.

In stark contrast to ordinary torch lights, laser light cannot be concentrated into a tiny spot. Common light sources, such as incandescent light bulbs, emit photons in an incoherent manner, scattering in almost all directions and spanning a wide range of wavelengths. On the other hand, laser beams, comprised of photons, outperform electrons due to their electrically neutral nature and smaller size. Unlike electrons, photons lack volume and can be concentrated into a minuscule space. Additionally, photons do not repel each other like electrons.

The foundation of laser theory was laid by Einstein, who based his work on the concept of stimulated emission. As early as 1900, Planck proposed the theory of quanta, which was widely accepted by 1920. It was established that lasers exhibit particle-like characteristics during interactions with matter and exchange energy in the form of photons.

The first laser, known as the Ruby Crystal Laser, was constructed in 1962 by T.H. Maimann. Since then, a wide variety of lasers have been developed. In 1964, C.K.N. Patel invented the CO_2 laser at Bell Laboratories, and J.E. Guesic, H.M. Marcos, and L.G. Van Uitert invented the Nd:YAG laser in the same year. The power output of the Nd:YAG laser reached 100 W, while the CO_2 laser achieved outputs on the order of 1,000 W. In the late 1980s, the fiber laser technology emerged, utilizing erbium-doped single-mode fibers. By the early 1980s, lasers were no longer confined to laboratory research and had evolved into a multibillion-dollar industry. The manufacturing sector greatly benefited from laser technology, employing it for various applications such as cutting, drilling, marking, engraving, welding, surface hardening, surface alloying, cladding, dressing, and cleaning. These processes catered to a diverse range of materials, including plastics, metals, ceramics, diamonds, and composites.

Laser Beam Machining is a thermal process that relies on a high-energy-density laser beam to induce localized heating on a workpiece. This heating causes the material to melt, vaporize, or ablate, allowing for the removal of material at the micron level. Laser Beam Machining is a non-contact, energy beam-based thermal machining process, offering several unique advantages, such as the potential for micro-machining, elimination of finishing operations, adaptability to automation, improved material utilization, and faster machining speeds. Lasers have a long history of application in machining processes, including cutting, slicing, drilling, grooving, turning, marking, cleaning, and dressing. They are also employed in non-machining applications, such as welding, cladding, and surface modification, across various manufacturing fields and with different types of materials.

The lasers commonly used for material machining include CO_2 lasers, Nd:YAG lasers, Fiber lasers, Diode lasers, Metal Vapor Deposition lasers, and Excimer lasers. Among these, Nd:YAG lasers exhibit superior micro-machining characteristics, providing high accuracy and surface finish. CO_2 lasers operate at a wavelength of 10,640nm, while Fiber lasers and Nd:YAG lasers operate in the infrared (IR) range at a wavelength of 1,064 nm. Excimer lasers, which are gas lasers, emit in the ultraviolet (UV) range, with wavelengths such as ArF (193nm), KrF (248nm), XeCl (308nm), and XeF (351nm). Laser technologies find applications in various fields, including manufacturing, instrumentation, imaging, medicine, communications, and more. While each field holds significant importance, this discussion specifically focuses on different types of laser beam machining technologies and their applications (Dubey and Yadava, 2008; Meijer et al., 2002).

4.10.1 Basics of Laser Generation

A laser beam is an energy beam consisting of coherent, monochromatic photons that exhibit high directionality. To generate a laser beam, a lasing medium is excited using an external energy source. However, the lasing medium cannot remain in the excited state for an extended period and eventually returns to its normal state by emitting energy in the form of a photon beam, known as a laser beam.

The mechanism of light emission can be understood through the behavior of atoms. An atom consists of a nucleus with a positive charge, surrounded by electrons orbiting around it. When these electrons absorb energy, they undergo a transition to a higher energy orbit. However, after a short period, they return to their ground state, releasing the absorbed energy in the form of light or photons. This phenomenon follows Planck's law ($E_2 - E_1 = h\gamma$), where γ represents the frequency of the photon and h is Planck's constant.

Light radiation falls within the spectrum of electromagnetic radiation, with wavelengths ranging from 0.1μm to 100μm. The emission of photons can occur through two processes: spontaneous emission and stimulated emission.

Spontaneous emission occurs when an atom absorbs energy, raising its electrons to higher energy levels. The atom remains in this excited state for a brief period, typically in the range of nanoseconds to milliseconds. Within this timeframe, the atom spontaneously emits a photon in a random direction and returns to its ground state. This is known as spontaneous emission.

On the other hand, stimulated emission involves the stimulation of atoms to release photons by interacting with other passing photons. For stimulated emission to occur, the passing photon must possess the same energy as the difference between the upper and lower energy levels of the atom. These photons are often emitted from lamps or semiconductor diodes. The wavelength (λ) of a photon is determined by its energy ($E = hc/\lambda$, where c is the velocity of light). The passing and emitted photons will have the same wavelength and will be in the same phase. This property of equal wavelength, direction, and phase characterizes laser light as coherent light.

Population inversion refers to a state where there is a higher number of atoms in a higher excited state compared to the lower energy level. At normal room temperature, atoms are primarily in the ground state, with only a few in the higher excited state. However, the population distribution can be altered by externally supplying energy to selectively increase the number of atoms in the higher excited state. This non-equilibrium state is known as population inversion. The external energy source responsible for creating this situation is called an energy pump. Since population inversion is not a stable state, the atoms in the higher excited state quickly transition to lower energy levels, following the population distribution dictated by thermal equilibrium. Therefore, continuous population inversion and subsequent stimulated emission require excitation through an external energy source, which is a fundamental requirement for laser generation (Mishra and Yadava, 2015; Parandoush and Hossain, 2014).

An atom consists of a positively charged nucleus surrounded by orbiting electrons. These electrons revolve in specific orbits called energy levels. When energy is supplied to an electron in a particular orbit, it transitions to a different orbit or energy level and subsequently emits the energy when returning to the ground state.

Laser devices convert electrical energy into a high-energy-density beam of light through the processes of stimulation and amplification. The energy needed to raise an electron from one energy state to another is provided through an excitation process or pumping. This can be achieved by the lasing medium absorbing energy from

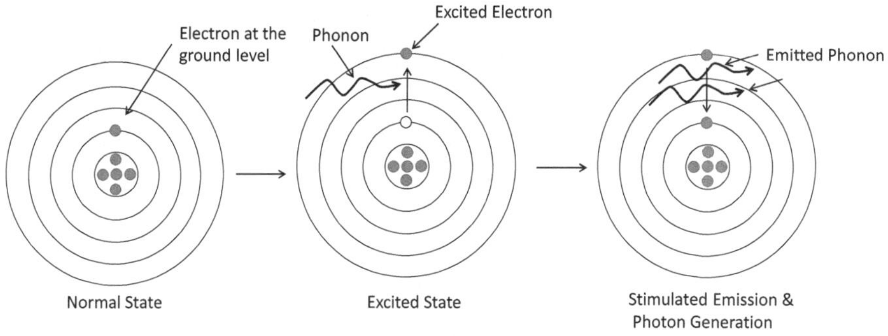

FIGURE 4.20 Schematic representation of stimulated emission and generation of Photon. (Reprinted with permission from Bhattacharyya and Doloi, 2020b. Copyright © 2020, Elsevier.)

mechanical, chemical, electrical, or light sources. The lasing medium typically contains ions, atoms, or molecules with discrete electron energy levels. The wavelength of the laser beam produced is a characteristic of the specific lasing medium. Initially, the emission of photons in a laser is random. However, as each photon stimulates other excited electrons to emit photons, the new photons will have similar characteristics in terms of wavelength, direction, and phase as the initial photon. Eventually, a stream of photons with identical wavelength, direction, and phase is generated (Magdum and Kittur, 2021; Mishra and Yadava, 2015).

The amplification of light in a laser is accomplished using an optical resonator, which consists of a cavity with the lasing medium positioned between two precisely aligned mirrors. One mirror is fully reflective, while the other is partially transmissive to allow for the output of the laser beam. These mirrors redirect light back into the lasing medium. As the photons pass back and forth through the lasing medium, they stimulate further emissions. The mirrors in the resonator selectively amplify only those photons with the proper orientation, while not redirecting photons that are not aligned with the resonator. This process ensures that the cavity amplifies the desired photons, leading to a beam with the desired characteristics. Figure 4.20 illustrates the schematic diagram depicting stimulated emission and the principle of photon generation.

4.10.2 Materials Removal in Laser Beam Machining Process

After passing through a lens, the energized laser beam is focused onto a specific point, resulting in a high concentration of energy at that location. The laser radiation is initially absorbed by the surface of the material, where the optical energy is converted into heat. The amount of laser radiation absorbed by a workpiece depends on factors such as the wavelength of the laser, temperature, reflectivity of the workpiece, angle of incidence of the laser beam, and intensity of the focused laser beam. The intense heat generated by the laser beam melts and vaporizes the material. In some cases, the material vaporizes rapidly, creating a significant mechanical impulse that removes a substantial amount of molten material from the cutting zone. Laser

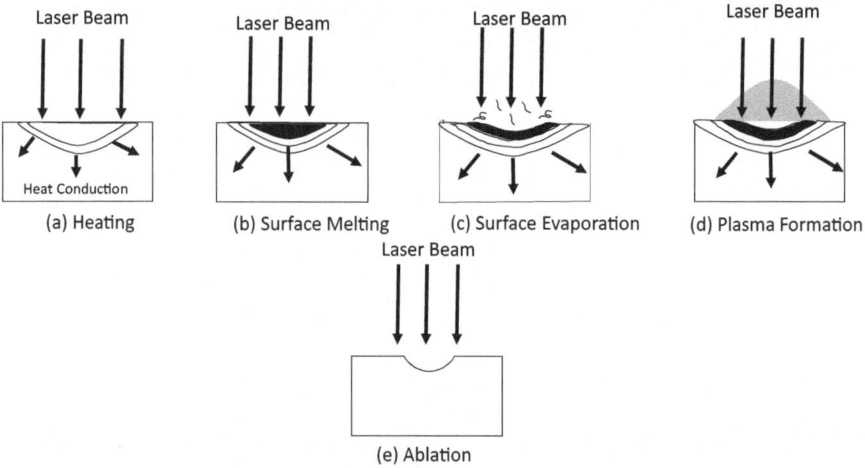

FIGURE 4.21 Representation of various stages in laser-workpiece interaction (a) Heating of workpiece, (b) Melting of workpiece, (c) Vaporization of workpiece, (d) Generation of Plasma, (e) Ablation of workpiece. (Reprinted with permission from "Laser Materials Interactions," 2008. Copyright © 2008, Springer.)

ablation occurs with short-pulse lasers, while thermal spalling takes place for ceramic materials. Additionally, a high-velocity assist gas jet is often used to exert momentum on the material, aiding in the ejection of molten material (Chryssolouris et al., 2013).

During the material removal process in laser beam machining, various effects of laser-material interactions occur, including heating, surface melting, vaporization, plasma formation, and ablation, as depicted in Figure 4.21. When using a long-pulse laser beam for machining, thermal effects such as the formation of a melting zone, heataffected zone, recast layer, and thermal damage to adjacent layers can occur, as shown in Figure 4.22. However, when employing a short-pulse or ultra-short pulse laser beam for machining, the material is removed without melting, thereby avoiding thermal damage during the material removal process.

When the laser beam is incident on the material's surface, various changes in states occur, including heating from room temperature to the melting point, melting of the material, heating of the melt to the vaporization temperature, and vaporization of the molten material. The energy (E) required for these state changes is determined by the following equation:

Equation 4.16 represents the energy required (E) for the laser beam to remove a mass (m) of material. It takes into account the specific heat (Cp) of the material, the room temperature (Tr), the melting point (Tm), the vaporization point (Tv), as well as the latent heat of melting (Lm) and vaporization (Lv). The machining time is denoted by t, and the percentage of incident energy absorbed is represented by A.

Using the energy balance equation, we can express the mass of material removed per unit time (MRR) as:

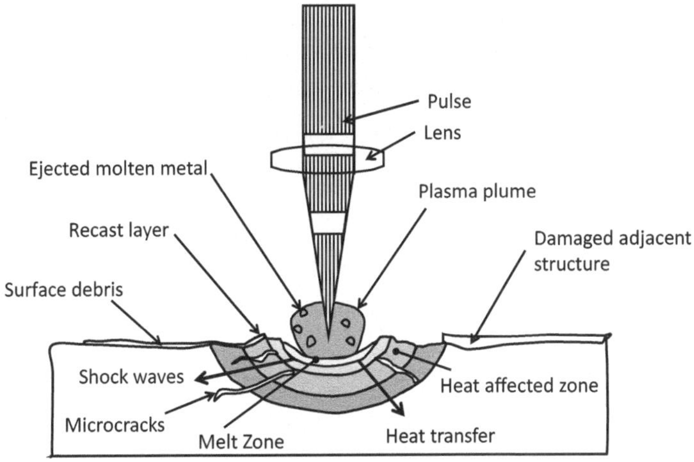

FIGURE 4.22 Representation of laser-work piece interactions. (Reprinted with permission from Pham et al., 2007, Copyright © 2007, Elsevier.)

$$MRR = \frac{m}{t} = EA\left[Cp(Tm - Tr) + Lm + Cp(Tm - Tv) + Lv\right] \tag{4.16}$$

The machining speed (V) during laser cutting of a slot with width (w), depth (d), and material density (p) can be calculated as:

$$\begin{aligned} V &= m/(w.d.p.t) \\ &= EA/\left[\left(Cp(Tm - Tr) + Lm + Cp(Tm - Tv) + Lv\right]w.d.p\right] \end{aligned} \tag{4.17}$$

4.10.3 Components of Laser Beam Machining Process

The laser beam machining system comprises several sub-systems, including the laser generation and beam delivery unit, cooling unit, assist air supply unit, and CNC controller for X-Y-Z axis movement. In Figure 4.23, you can see the schematic view of a typical Nd:YAG laser beam machining system, illustrating the various sub-systems involved.

4.10.3.1 Laser Beam Generation Delivery Systems

The laser generation and beam delivery unit consist of various sub-systems, including:

(a) Laser head
(b) Optical excitation unit
(c) Cavity mirrors
(d) Q-switch and RF drive unit
(e) Laser beam delivery and focusing unit

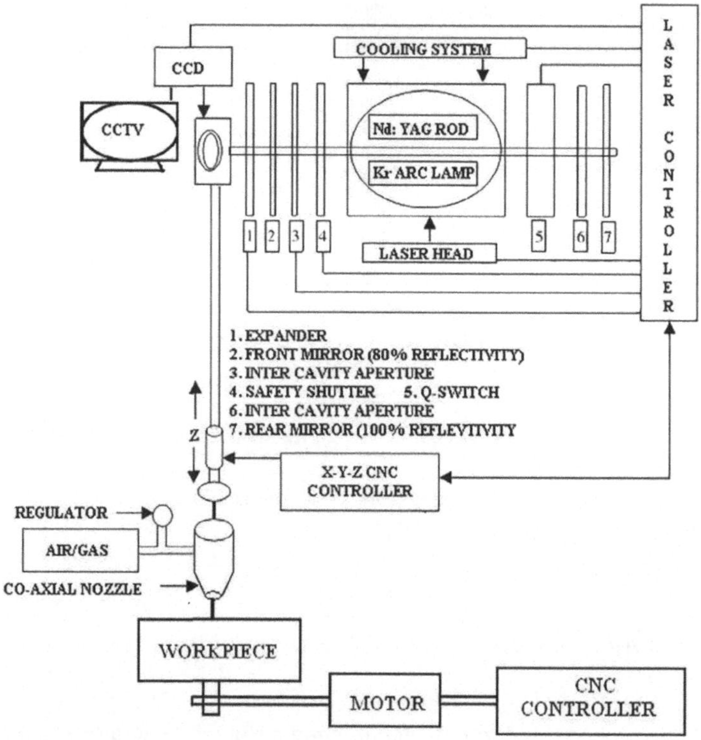

FIGURE 4.23 Schematic representation of Nd:YAG laser beam machining system. (Reprinted with permission from Bhattacharyya and Doloi, 2020b, Copyright © 2020, Elsevier.)

(a) Laser Head

The laser head for a lamp pumped Nd:YAG Laser primarily consists of a Nd:YAG rod, a krypton arc lamp, and a gold-plated elliptical cavity. The krypton arc lamp is positioned at one focus of the ellipsoid, while the Nd:YAG rod is located at the other focus. The inner surface of the cavity is coated with a reflective material, usually gold, to ensure that all the radiation emitted from the lamp is directed towards the active medium. The active medium in the Nd:YAG laser is neodymium ions (Nd^{3+}) doped in a solid host material called yttrium aluminum garnet (YAG). YAG is a synthetic crystal with a garnet-like structure (chemical formula: $Y_3Al_5O_{12}$). Neodymium ions replace yttrium ions in the garnet lattice, as they are similar in size. YAG offers good thermal stability compared to other host materials. Laser light is generated through transitions between energy levels of the neodymium ions. Most solid-state lasers are derived from optically pumped transitions among electronic levels of ions in crystals or other solid-state hosts.

The laser head for a diode pumped fiber laser system comprises a fiber doped with Yb^{3+}, a fiber Bragg grating, and a number of diodes. The number of laser diodes depends on the desired output power of the system. Typically, eight diodes are used for optical pumping, generating an average fiber laser power of 50 W. Fiber Bragg gratings are incorporated at the two ends of the fiber optics to enable laser generation. The length of the fiber can vary up to 3 mm, depending on the output power of the fiber laser. Once the laser is generated within the optical fiber, it is transferred to a collimator. The collimator contains two prisms at its ends, serving as a beam expander to allow the laser to propagate over the required distance. The diameter of the laser beam at the end of the collimator is typically around 9 mm.

(b) Optical Excitation Unit

The optical excitation unit in the Nd:YAG Laser machining system comprises a krypton arc lamp and its power supply unit, responsible for controlling the laser output. Its primary function is to ignite and regulate the intensity of light emitted by the krypton arc lamp, which is used to pump the Nd atoms in the Nd:YAG rod. To initiate the lamp, which has high impedance, a high voltage pulse is applied as a trigger. Once the lamp discharge is established, the intensity of the emitted light can be controlled by adjusting the current flowing through the lamp.

In many applications, lasers are not operated continuously. To address this, the power supply unit includes a standby mode feature, which maintains the lamp discharge. This feature enhances the operational life of both the lamp and the power supply by eliminating the need to generate trigger pulses for lamp ignition. The power supply unit is interlocked with an internal chiller unit to ensure that if the cooling system fails, the lamp is shut down, protecting the expensive lamp and the Nd:YAG rod. The lamp current, adjusted using a knob in the power supply system, is a crucial controllable process parameter for laser machining.

(c) Cavity Mirror

The cavity mirror setup consists of a fully reflecting mirror, also known as the back mirror or rear mirror, located at one end of the rod with 100% reflectivity, and a partially reflecting output mirror, known as the front mirror, with approximately 80% reflectivity. These mirrors provide the necessary feedback for amplification of the light. This arrangement of cavity mirrors, enabling light circulation, is common to most laser generation units.

(d) Q-Switch and RF Drive Unit

The output of a gain-switched, pulsed solid-state laser typically consists of a train of irregular pulses in terms of peak power, pulse width, and repetition frequency. The technique of Q-switching can remove these irregularities and significantly increase the peak power. Q-switched lasers emit giant pulses during an operational cycle, with pulse lengths ranging from less than a microsecond to several nanoseconds, and peak powers increasing from 10^6 W to 10^9 W.

Q-switching is a laser operation mode in which energy is stored in the lasing medium during pumping as excited atoms and then released suddenly in a single, short burst. This is achieved by modifying the optical properties of the laser cavity. The quality factor Q, defined as the ratio of energy stored in the cavity to energy loss per cycle, plays a crucial role. An acousto-optic Q-switch is often used, which includes a shutter-like arrangement placed between the active medium and the highly reflective (HR) mirror. When the shutter is closed, the HR mirror is blocked, preventing oscillation. When the amplifier gain reaches a predetermined value, the shutter is opened, increasing the cavity quality and enabling the emission of a giant pulse. Acousto-optic Q-switches use a transparent element excited with intense, standing acoustic waves generated by piezoelectric crystals. This element diffracts part of the intra-cavity laser beam out of alignment, resulting in low feedback. Upon removing the acoustic wave, the diffraction effect vanishes, realigning the cavity, and producing a giant pulse.

To achieve pulsed laser output with high peak power, the laser operates in Q-switched mode, utilizing an acousto-optic Q-switch. This Q-switch requires a specific power supply for operation, typically utilizing an RF source at 27.2 MHz modulated with frequencies ranging from 0.1 kHz to 19.9 kHz. An RF generator is employed for this purpose. Due to the high-frequency switching, the RF source is cooled by chilled water for repetitive operations. The unit is equipped with an over-temperature interlock for protection.

(e) Laser Beam Delivery and Focusing Unit

The laser beam delivery system includes a beam bender, an intra-cavity shutter, an intra-cavity aperture, and a focusing lens. The beam bender facilitates the redirection of the laser beam by 90 degrees, allowing it to be focused by the focusing lens onto the work surface.

An essential component within the laser cavity is the electrically activated safety shutter, which blocks the path of the laser beam and prevents emission of laser radiation when in the closed position. Unlike the Q-switch, the safety shutter is a mechanical device designed for prolonged blocking of the lasing action. Gravity ensures that the shutter moves into the closed or non-lasing position in the event of an electrical failure. The safety shutter is typically located inside the laser head assembly and actuated by a toggle switch.

The intra-cavity aperture restricts amplification along the off-axis of the resonator and is positioned within the cavity. This arrangement ensures that feedback occurs as the laser beam passes through the two mirrors.

The focusing unit consists of a CCD camera and CCTV. The CCD camera, operating at 12 V and positioned atop the laser head, captures images during laser beam machining. Precise focusing of the laser beam on the work surface is crucial for effective machining. The CCTV, connected to the CCD camera, assists in achieving accurate beam focusing. If the focal point of the beam is well above the job surface, no material removal occurs. Conversely, if the beam is focused well below the surface, thermal shock may cause material cracking. Therefore, it is necessary to adjust the focal point of the laser beam precisely. The experimental setup often incorporates a CNC-controlled arrangement for adjusting the focal point by controlling the movement of the lens position along the Z-axis.

4.10.3.2 Cooling Unit

The cooling system maintains the desired temperature of the laser head and Q-switch. It consists of several subunits, including a three-phase chiller unit, a heat exchanger, and a pump unit. The three-phase chiller unit supplies chilled water to the heat exchanger, ensuring proper temperature control during machining operations. The heat exchanger and pump unit work together to circulate deionized water through a resin filter, water-to-water heat exchanger, laser head, and Q-switch.

When laser generation occurs inside the laser head cavity, a substantial amount of heat is generated due to the low electro-optic efficiency. If this heat is not removed, it can damage both the lamp and the rod. To address this, a proper heat removal arrangement is necessary. Chilled deionized water is supplied through the chiller system to the heat exchanger, and a pump system circulates the deionized water via the water-to-water heat exchanger, resin filter, laser head, and Q-switch. Both the lasing medium and the lamp are immersed in flowing cold deionized water within the laser cavity. Deionized water is preferred due to its high optical transparency and low electrical conductivity. Heat from the deionized water is then extracted using water-to-water heat exchanges connected to an external water source. A solenoid regulates the deionized water temperature by controlling the flow of outside water, contributing to the stability of the laser output. While high-power laser systems typically require water cooling, low-power laser systems often incorporate air cooling arrangements.

4.10.3.3 Assist Gas Supply Unit

The compressor supplies air, which is then compressed and chosen as the assist gas. The primary function of the assist gas is to aid in the removal of molten material from the ablated surface. The assist gas is supplied to the laser marking zone through a co-axial air flashing nozzle. By adjusting the supply pressure of the assist gas, the quality of the machining surface can be improved. The compressed air supply line passes through a moisture separator and is connected to a pressure regulatory valve. It supplies dry pressurized air to the laser machining zone through the co-axial nozzle attached to the beam delivery system. For experimental purposes, a range of air pressure from 0.3 to 2.3 kgf/cm^2 is used. However, at high laser power levels where instantaneous vaporization occurs, gas pressure has little significance.

4.10.3.4 CNC Controller for X-Y-Z Axes Movement

The movement of the workpiece along the X-Y axis and the focusing lens movement along the Z-axis are controlled by a computer numerical control (CNC) system. The CNC system allows for precise machining by properly focusing the laser beam through the focusing lens. The feed rate is controlled by the movement of the X-Y table to ensure appropriate machining. A CCD camera and CCTV monitor are used to view the location of the workpiece and check the proper focusing condition.

4.10.3.5 Diode Pumped Fiber Laser Micro-Machining System Components

A diode pumped fiber laser micro-machining system, specifically a 50 W ytterbium-doped fiber laser system operating at a wavelength of 1,064 nm in TEM00 mode, manufactured by M/S Sahajanand Laser Technology Limited (Model: AKSHAR Fiber Pro-309), consists of the following units:

- Power supply unit
- Laser head
- Collimator

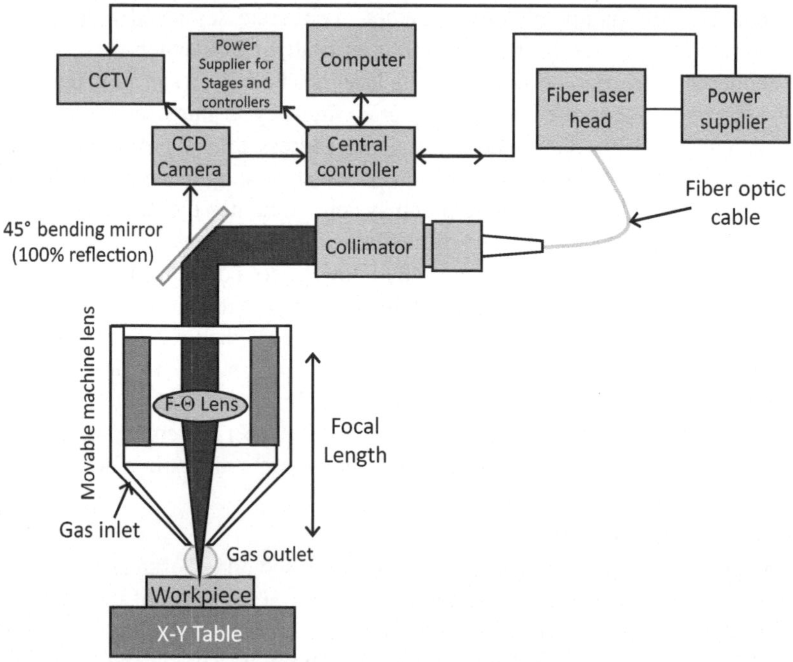

FIGURE 4.24 Schematic diagram of the fiber laser machining system. (Reprinted with permission from Sen et al., 2017, Copyright © 2017, Springer.)

- Beam bender
- Beam delivery unit and focusing lens
- Assist gas supply unit
- CNC controller for X-Y-Z movement

The schematic diagram of the fiber laser machining system is shown in Figure 4.24. The system has a working range of pulse frequency from 50 to 120 kHz and a pulse width of 120 ns. The laser beam, after passing through a 71 mm diameter F-θ lens (focus lens) protected by a nozzle, has a spot diameter of 21 μm. Precise alignment and focusing of the laser beam on the surface are crucial for optimal micro-machining efficiency. A CCD camera placed on top of the laser head captures images during the laser micro-machining operation. The CCTV unit is used to adjust the position of the focused beam on the workpiece and is connected to the CCD camera to obtain precise laser beam focusing conditions. The focal point position of the laser beam needs to be accurately adjusted to ensure efficient material melting and evaporation. The CNC-controlled arrangement allows for effective adjustment of the focal point by controlling the movement of the lens along the Z-axis.

4.10.3.6 Parameters Governing the Performance Characteristics of LBM Process

The critical process parameters that require control during laser beam machining include pulse frequency, Pulse Duration, Peak Power, Scanning Speed, Focused Spot

Size, and Air Pressure, among others. The following section will delve into the discussion of these process parameters for laser beam machining.

1. Pulse Duration, Peak Power, and Pulse Frequency
   The pulse repetition rate governs the time interval between successive bursts of the laser striking a surface. The capability of Q-switching allows for generating short pulses with exceptionally high peak powers, which facilitates the machining of a wide range of materials. Laser parameters, including pulse duration, peak power, and average power, can be adjusted according to the material and type of machining process needed. The relationship among average power, peak power, pulse frequency, and pulse duration is as follows:

$$\text{Peak Power} = \text{Average Power} / (\text{Pulse frequency} \times \text{Pulse duration}) \quad (4.18)$$

2. Scanning Speed
   In the laser beam machining process, the scanning speed of the laser beam concerning the workpiece is a crucial variable that significantly impacts machining performance across different operations. When aiming for deeper machining (typically > 0.03 mm), each point on the material necessitates exposure to several pulses to attain the desired depth. To achieve this depth, the scanning speed must be reduced accordingly.

   The combination of pulse frequency and beam speed results in the phenomenon known as laser spot overlap. As a general guideline, for a seamless appearance in laser beam machining, pulses should overlap by at least 50–60%. However, to ensure effective machining, the overlap should ideally be between 75% to 90%. This ensures a thorough and precise material removal process.
3. Focal Spot Size
   The focal spot size plays a critical role in material processing, as it determines the maximum energy density achievable when the laser beam power is set. Consequently, it significantly impacts the machining results. The diameter of the focused laser beam on the work surface dictates both the width and depth of the machining process.

   The focal spot diameter (d) is influenced by two key factors: the focal length (f) of the focusing lens and the laser beam's divergence angle (α). The divergence angle (α) relies on the wavelength of the laser and the beam diameter (D) at the entry point to the lens. The relationship between these parameters can be expressed as follows:

$$d = f \times \alpha = f\,.2\lambda = D \quad (4.19)$$

4. Air Pressure
   Air pressure plays a crucial role in laser beam machining by serving two primary functions. Firstly, it aids in material removal and helps to minimize the heataffected zone during the machining process. Additionally, it acts as

a protective barrier for the focusing optics, preventing damage caused by vapor or spatter generated due to the high heat produced during machining. However, it's essential to strike a balance with the air pressure to avoid excessive vibrations in machine parts.

The performance criteria for laser beam machining in various operations are outlined as follows:

1. Laser Cutting Operation:
    a. Cut or kerf width
    b. Cut depth
    c. Taper formation and surface roughness
    d. Heat-Affected Zone (HAZ)
    e. Dross formation
2. Laser Drilling Operation:
    a. Hole diameter on both the entry and exit sides
    b. Drilling depth
    c. Surface characteristics
    d. Heat-Affected Zone (HAZ)
    e. Aspect ratio and taper
3. Laser Marking Operation:
    a. Mark width
    b. Mark depth
    c. Mark continuity
    d. Mark intensity or contrast
4. Laser Turning Operation:
    a. Machining depth
    b. Turning length
    c. Surface finish
    d. Heat-Affected Zone (HAZ)
5. Laser Grooving Operation:
    a. Groove dimensions
    b. Groove geometry
    c. Heat-Affected Zone (HAZ)
    d. Surface quality

By considering these performance criteria, laser beam machining can be optimized to achieve desired results for each specific operation.

4.10.3.7 Advantages and Disadvantages of LBM Processes

Laser beam machining (LBM) is an exceptional material processing technique, particularly suited for a wide range of advanced materials, including challenging-to-machine advanced ceramics and composites. LBM offers several advantages over traditional machining processes, making it an attractive choice for various industrial applications:

1. LBM yields superior finished machining surfaces, often eliminating the need for post-finishing operations.
2. It allows the laser beam to be directed remotely, making it suitable for machining hazardous materials like nuclear fuel rods.
3. LBM can selectively remove material from specific regions of the workpiece by precisely focusing the laser beam.
4. It is a clean machining process as no solvents or chemicals are involved.
5. LBM is adaptable to automation, enabling high productivity and speed.
6. Being a non-contact process, there is no physical tool involved, eliminating machining forces, tool wear, material damage, and machine component vibration.
7. LBM can easily cut complex profiles with high accuracy.
8. The process depends on thermal and optical properties of the material rather than its mechanical properties, enabling machining of materials with varying brittleness and hardness. Low thermal diffusivity and conductivity materials are well-suited for laser machining.
9. LBM is not constrained by the electrically conductive nature of the workpiece and can process a wide range of materials, from plastics to diamonds.
    a. The processing speed with lasers is very fast, resulting in minimal heat-affected zones (HAZ).
10. LBM can produce large aspect ratio micro-features with acceptable accuracy and dimensions.
11. Less material wastage occurs in LBM, as it selectively removes material from chosen portions.
12. LBM is highly flexible and can be combined with multi-axis work positioning systems or robots, making it amenable to automation.
13. Laser beam machining offers various techniques, such as drilling, cutting, grooving, scribing, marking, cleaning, and turning.

However, as with any machining process, laser beam micro-machining has some limitations, including:

1. High initial investment cost.
2. High maintenance costs and a need for skilled operators.
3. Relatively lower efficiency compared to the input electrical energy to the LBM machine. Research and development efforts may enhance energy conversion efficiency.
4. Limitations in processing highly reflective and transparent materials under certain conditions.
5. Less suitable for thicker materials but highly applicable to thin cross-sectional materials.
6. Relatively low metal removal rate.

4.10.3.8 Applications of LBM

Laser beam machining finds diverse applications in various industries and processes. Some of its key uses include:

1. Drilling micro and macro holes in hard metals like tungsten and ceramics.
2. Trimming sheet metal, plastic parts, carbon resistors, and wafer labeling in microelectronics.
3. Machining intricate geometrical profiles on thin and hard metals.
4. Engraving patterns on thin films.
5. Fabricating integrated chips (IC) and computer peripherals.
6. Utilization in communication and space applications.
7. Aeronautics and guided missiles in defense.
8. Medical device fabrication, creating micro holes in nozzles, and artificial ceramic hip joints.
9. Non-machining applications like heat treatment, surface hardening, alloying, cladding, welding, etc.
10. A wide range of machining processes, such as cutting, grooving, drilling, trepanning, turning, milling, cleaning, dressing, and marking.
11. Laser Assisted Machining, where laser beams enhance traditional machining processes.
12. Laser Assisted Oxygen Flame Cutting (LASOX), which combines laser and oxygen flame for cutting applications.
13. Laser Assisted Hybrid Machining, a combination of laser machining with other machining methods to improve performance and efficiency.
14. These applications demonstrate the versatility and significance of laser beam machining in modern manufacturing and material processing technologies.

4.11 ELECTRON BEAM MACHINING (EBM) PROCESS

Currently, one of the most advanced material processing techniques is electron beam machining (EBM), which involves the focused beam of high-velocity electrons. EBM is a thermal material removal process where the workpiece material is melted and vaporized to achieve the desired shape and size. The pioneering work of Dr. K.H. Steigerwald and his team in 1952 marked the first application of electron beams for machining and welding . Subsequently, the development of the prototype of EBM machining setup in Germany further advanced the technology (Gentili et al., 2005).

Over the last few decades, EBM technology has witnessed rapid growth, owing to its high processing speed and easy control mechanism. The electron beam finds application not only in machining, but also in welding, heat treatment, deposition, purification, and material inspection. Since 1960, EBM machines have been commercially adopted in the industry for manufacturing, primarily for welding applications in nuclear and aerospace sectors (Okada, 2019).

Furthermore, EBM is employed for high-speed drilling of small holes, cutting, and milling of various materials, particularly in the semiconductor domain of microelectronics. Key features of EBM include high power density, precise controllability of spot diameter, and adjustability of parameters. These characteristics make it a promising technology for micro-machining applications due to its very small focused spot, flexibility, and high level of accuracy. The energy density of the electron beam can be precisely regulated, ranging from low to extremely high power densities, depending on the specific requirements of diverse applications. These applications encompass a wide array of processes, including hardening, surface modification, welding, and various machining operations such as perforation, drilling, and milling of both metals and plastics.

4.11.1 Generation of Electron Beam

The generation of free electrons is achieved by heating a tungsten or tantalum filament to a high temperature, using an electron beam gun. This gun is a crucial component of any electron beam machining facility and comprises two main parts as shown in Figure 4.25. The first part is responsible for producing electrons, while the second part generates the focused electron beam for various purposes.

The electron gun's detailed configuration is shown in Figure 4.26. It consists of a cathode that generates free electrons, a negatively biased grid cup with respect to the cathode, and a grounded anode. The source of electrons is a highly negatively biased cathode cartridge containing a heated tungsten hairpin filament. Thermo-ionic emission of electrons occurs due to the filament's heat, with a magnitude of 25–100mA and current densities ranging from 5 to 15 A/cm^2, depending on the cathode material and cross-sectional area. Low vacuum in the electron beam gun chamber enhances electron emission, and the generated electrons are repelled by the negatively biased grid.

The electron beam remains focused and approaches as a beam towards the highly negatively biased annular grid, situated just below the cathode. This annular grid is also used for controlling the beam current. The electron beam is then attracted by an

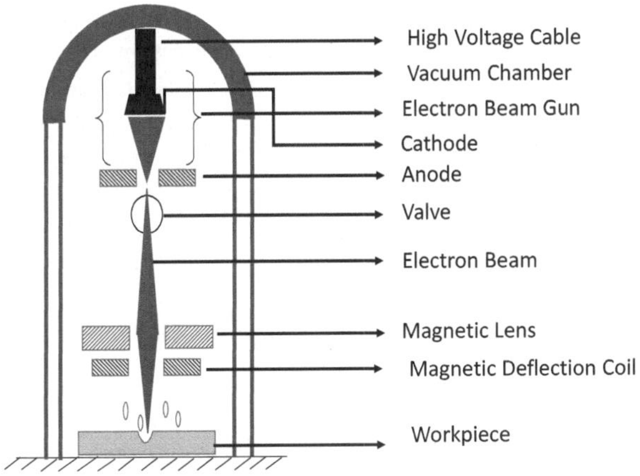

FIGURE 4.25 Schematic diagram of the electron beam machining process. (Bhattacharyya and Doloi, 2020b, Copyright © 2020, Elsevier.)

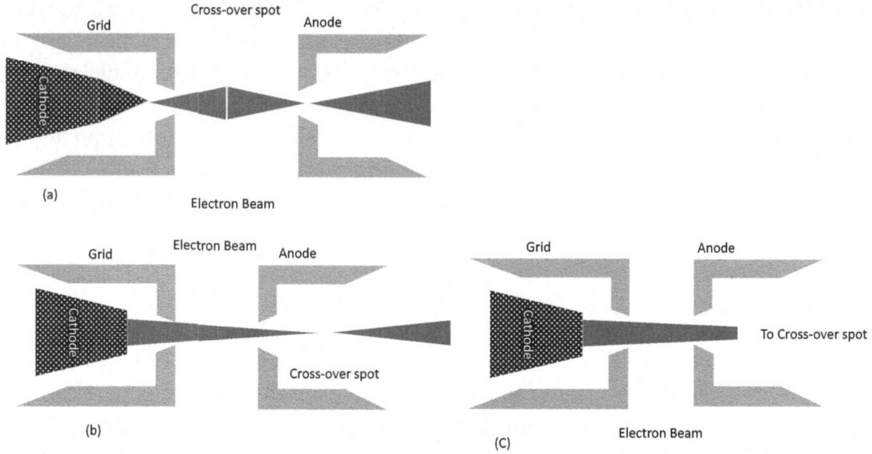

FIGURE 4.26 Schematic diagram of (a) Normal triode setup (b) Pierce gun setup, (c) Steigerwald gun setup. (Bhattacharyya and Doloi, 2020b, Copyright © 2020, Elsevier.)

annular anode, maintaining an accelerating voltage in the range of 100 to 150 kV, which gives the electrons high velocity.

To separate this part of the electron beam generation unit from the rest of the beam focusing unit, a tungsten diaphragm with a central hole is employed. Proper vacuum maintenance (in the range of 10^{-4} to 10^{-6} Torr) is crucial for retaining electron energy and prolonging the life of the cathode cartridge.

Various configurations of electron guns are available for generating electron beams. Three types of electron gun systems are commonly used for different purposes. The normal triode system with cylindrical annular grid and annular anode, particularly the closed anode configuration, is widely utilized due to its smaller column length and critical alignment dependency. Other systems, such as the Pierce type and Steigerwald type, require longer machine column lengths and perfect alignment, making the normal triode system preferable for electron gun applications (Pierce, 2004).

After passing through magnetic lenses and apertures, the electron beam's divergence is reduced, and its shape is controlled. Apertures allow only convergent electrons to pass through, making the beam narrower and improving its quality. The focused beam achieves its spot size with the help of a final set of electromagnetic lenses, and a deflection coil allows precise manipulation of the beam within a small range.

To protect the electron beam gun from damage by metal spatter and vapor during various operations, a series of slotted disks are mounted at the gun's exit. The typical triode electron beam gun is widely used for various machining applications, generating high-accuracy features.

4.11.2 Materials Removal Mechanism in EBM Process

The electron beam, generated from an electron gun, possesses high velocity and strikes the workpiece. Electrons, being the smallest stable elementary particles with a negative charge of 1.602 x 10^{-19}C and a mass of 9.109 x 10^{-31}kg, have negligible mass, making their kinetic energy primarily dependent on their velocity. Therefore, the electron gun is used to maintain a very high velocity for the focused electron beam. The beam's high velocity is achieved through an acceleration voltage applied to the annular anode, and it is further focused by a set of electromagnetic lenses in the electron gun. This focused beam, with high power density reaching approximately 10^8 W/cm^2, impacts the workpiece in a well-defined area of 2 to 5 μm in diameter.

Upon striking the workpiece, the high kinetic energy of the electron beam rapidly converts into heat, resulting in a localized increase in temperature. This temperature rise is sufficient to cause rapid melting and vaporization of any material, regardless of its melting point and thermal conductivity. Consequently, metal removal occurs from the workpiece due to this rapid melting and vaporization process. Electron Beam Machining (EBM) can be used on a wide range of engineering materials due to its high power density.

Initially, the high-velocity electron beam penetrates the surface of the work material without significantly disturbing the molecules, forming a transparent layer. However, beyond this transparent layer, the electrons begin to collide with the molecules, losing their kinetic energy, and generating heat through thermal agitation. Some incident electrons are also backscattered, producing secondary electrons, some of which escape the target material (reflection), and others remain within the material's lattice (intrusion).

The depth of beam penetration into the material depends on the energy of the incident electrons and the atomic number of the workpiece material. The collisions between the incident electrons and the molecules in the workpiece material lead to high temperatures at the colliding spot, resulting in melting and vaporization. As a consequence, the beam and evaporating molecules exert pressure on the liquid interface, creating a phenomenon known as recoiling pressure.

The dynamic complexities at the liquid interface during machining are not yet fully understood. Recoiling pressure, along with surface tension at the liquid interface, plays a significant role in the mechanisms of machining. Due to the rapid ejection and vaporization of the liquid molten material, material removal occurs at the striking spot on the work piece.

EBM finds extensive application in generating multiple high aspect ratio holes of different cross-sections. In drilling, an auxiliary backing plate made of organic or synthetic material supports the work piece. During the process, only a small amount of material is vaporized, forming a capillary channel inside the molten material. The electron beam and the capillary channel penetrate the work piece and a certain depth of the backing plate. The part of the backing material in direct contact with the beam is almost vaporized, causing a violent discharge of backing material vapor. This high-pressure vapor forces the molten material wall of the generated capillary to be expelled. As a result, the molten material leaves the channel as a shower of sparks, forming a hole in the work piece and creating a small cavity in the backing material (Martins et al., 2018; Saptaji et al., 2018; Taniguchi, 1983).

These mechanisms of EBM, including the formation of high aspect ratio holes and the perforation process, are depicted in Figure 4.27.

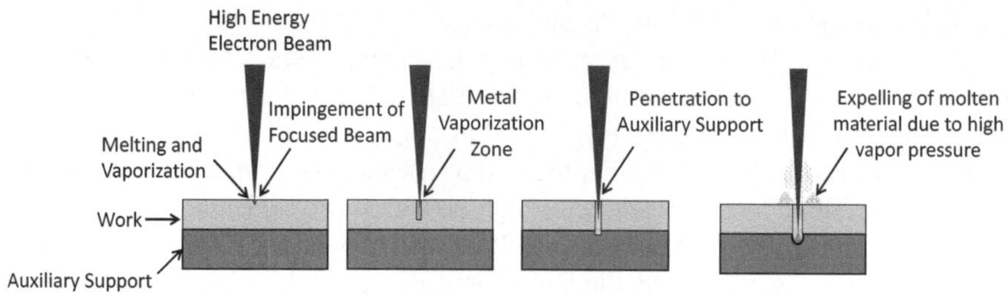

FIGURE 4.27 Schematic diagram of the perforation process.

4.11.3 Advantages and Disadvantages of Electron Beam Machining Processes

Electron Beam Machining (EBM) offers numerous advantages, making it an attractive choice for drilling small holes with a high aspect ratio and machining various materials, regardless of their mechanical properties. However, EBM also has some limitations. Here are the advantages of EBM:

- *High Drilling Rates:* EBM enables exceptionally high drilling rates, ranging from 1,000 to 10,000, depending on the hole diameter and work piece thickness.
- *Cost-Effective Work Holding:* EBM operations require less work holding and fixturing cost since there is zero mechanical cutting force involved.
- *Machining Brittle Materials:* EBM can effectively machine fragile and brittle materials, even in thin workpieces.
- *Minimal Heat-Affected Zone:* Shorter pulses in EBM result in a smaller heat-affected zone, reducing the risk of thermal damage to the workpiece.
- *Versatility in Hole Shapes:* EBM can generate holes of any shape with high accuracy by combining beam deflection using electromagnetic coils and CNC table control.
- *High Accuracy and Repeatability:* The process achieves high accuracy and repeatability, with a precision of 0.1 mm for hole position and 5% for hole diameter.
- *Moderate Surface Finish:* Compared to other nonconventional machining processes, EBM produces moderate-level surface finish with high accuracy.
- *No Mechanical or Thermal Distortion:* During EBM, there is no mechanical or thermal distortion produced, ensuring dimensional stability of the machined parts.
- *Easy Parameter Control:* Machining parameters are easily controlled due to computer interfacing with the EBM system, allowing precise adjustments and optimization.

Despite its advantages, EBM also comes with certain limitations and considerations that need to be taken into account during its application. Nonetheless, the benefits make it a valuable option for various manufacturing processes.

EBM, despite its advantages, has some limitations that should be considered. The primary limitations include the high capital cost of the equipment and the need for regular maintenance, which is typical for any system utilizing vacuum technology. Additionally, EBM involves a significant non-productive pump-down period required to achieve the desired vacuum, but this can be somewhat mitigated by employing vacuum load locks. Although EBM results in a relatively small heat-affected zone, it still leads to the formation of recast layers, which cannot be entirely avoided (Shahsavari et al., 2022).

Other disadvantages of EBM are as follows:

- *High Operator Skill Requirement:* EBM demands a high level of operator expertise to ensure accurate and effective machining.
- *Thickness Limitation:* EBM is not suitable for materials with a thickness greater than 10 mm; it is better suited for thinner workpieces.
- *Extended Production Time:* The time needed to generate and maintain a vacuum can result in long production cycles.
- *Auxiliary Backing Material:* During perforation, EBM necessitates the use of auxiliary backing material, adding to the complexity of the process.
- Despite these limitations, EBM remains a valuable machining method, but careful consideration should be given to the specific requirements and challenges associated with its implementation.

4.11.4 Applications of Electron Beam Machining Process

Electron Beam Machining (EBM) finds extensive applications across various industries, serving purposes such as milling, cutting, welding, and heating materials for vacuum deposition and purification. It is also utilized for inspecting, examining, and analyzing materials, where techniques like electron microscopy and X-ray microanalysis come into play.

EBM has gained widespread acceptance in industries such as aerospace, insulation, food processing, chemicals, and clothing. One of its most popular applications is drilling thousands of small holes (< 1 mm) in thin plates used for turbine engine combustor domes. Additionally, EBM is employed in the following areas:

Space Nuclear Reactors: EBM is used to create fine gas orifices in space nuclear reactors.
Wire Drawing Dies: It is applied for perforating holes in wire drawing dies.
Turbine Blades: EBM is utilized to create cooling holes in turbine blades.
Diesel Engines: It helps in generating metering holes in injector nozzles of diesel engines.

Mainly, EBM is applied in drilling, perforation, and pattern generation for fabricating intricate circuits and other components. Its versatility and precision make it suitable for a wide range of industrial applications.

REFERENCES

A, D.D., K, S.K., R, P., 2022. Investigating the Eeffect of Wwire cut EDM of Ttitanium Aalloy 6242 Uusing TOPSIS. Adv. Mater. Process. Technol. 8, 2824–2836.

Abu Qudeiri, J.E., Saleh, A., Ziout, A., Mourad, A.-H.I., Abidi, M.H., Elkaseer, A., 2019. Advanced Electric Discharge Machining of Stainless Steels: Assessment of the State of the Art, Gaps and Future Prospect. Materials 12, 907. https://doi.org/10.3390/ma12060907

A.J. Bard, L.R. Faulkner, 2001. Electrochemical- Methods Fundamentals and Applications, second ed. ed. John Wiley & Sons,.

Alam, M.N., Siddiquee, A.N., Khan, Z.A., Khan, N.Z., 2022. A Comprehensive Review on Wire EDM Performance Evaluation. Proc. Inst. Mech. Eng. Part E J. Process Mech. Eng. 236, 1724–1746. https://doi.org/10.1177/09544089221074843

Allen, D.M., 2004. Photochemical Machining: From 'Manufacturing's Best Kept Secret' to a $6 Billion per Annum, Rapid Manufacturing Process. CIRP Ann. 53, 559–572. https://doi.org/10.1016/S0007-8506(07)60029-8

Atkinson, J., Noble, C.F., 1979. Residual Stresses in Workpieces After Peripheral Electrochemical Grinding, in: Davies, B.J. (Ed.), Proceedings of the Nineteenth International Machine Tool Design and Research Conference: Held in Manchester, 13th–15th September 1978. Macmillan Education UK, London, pp. 525–532. https://doi.org/10.1007/978-1-349-81412-1_63

Bähre, D., Weber, O., Rebschläger, A., 2013. Investigation on Pulse Electrochemical Machining Characteristics of Lamellar Cast Iron using a Response Surface Methodology-Based Approach. Procedia CIRP 6, 362–367. https://doi.org/10.1016/j.procir.2013.03.028

Banu, A., Ali, M.Y., 2016. Electrical Discharge Machining (EDM): A Review. Int. J. Eng. Mater. Manuf. 1, 3–10. https://doi.org/10.26776/ijemm.01.01.2016.02

Bard, A.J., Faulkner, L.R., 2001. Electrochemical-Methods Fundamentals and Applications, second ed. John Wiley & Sons, New York.

Benedict, G.F., 1987. Nontraditional Manufacturing Processes, first ed. CRC Press, Boca Raton.

Bhattacharyya, B., 2015. Electrochemical Micromachining for Nanofabrication, MEMS and Nanotechnology, 1st ed. Elsevier, William Andrew, Oxford.

Bhattacharyya, B., Doloi, B., 2020a. Chapter Five – Machining processes utilizing chemical and electrochemical energy, in: Bhattacharyya, B., Doloi, B. (Eds.), Modern Machining Technology. Academic Press, London, pp. 365–460. https://doi.org/10.1016/B978-0-12-812894-7.00005-0

Bhattacharyya, B., Doloi, B., 2020b. Chapter Four – Machining processes utilizing thermal energy, in: Bhattacharyya, B., Doloi, B. (Eds.), Modern Machining Technology. Academic Press, London, pp. 161–363. https://doi.org/10.1016/B978-0-12-812894-7.00004-9

Black, J.T., Kohser, R.A., 2008. Materials & Processes in Manufacturing, 10th ed. John Wiley & Sons Hoboken, NJ.

Bijoy Bhattacharyya, 2015. Electrochemical Micromachining for Nanofabrication, MEMS and Nanotechnology, 1st Edition. ed. Elsevier.

Chamarthi, S., Reddy, N.S., Elipey, M.K., Reddy, D.V.R., 2013. Investigation Analysis of Plasma Arc Cutting Parameters on the Unevenness Surface of Hardox-400 Material. Procedia Eng. 64, 854–861. https://doi.org/10.1016/j.proeng.2013.09.161

Choi, I.-H., Kim, J.-D., 1998. Electrochemical Deburring System Using Electroplated CBN Wheels. Int. J. Mach. Tools Manuf. 38, 29–40. https://doi.org/10.1016/S0890-6955(97)00027-8

Chryssolouris, G., Stavropoulos, P., Salonitis, K., 2013. Process of Laser Machining, in: Nee, A. (Ed.), Handbook of Manufacturing Engineering and Technology. Springer, London, pp. 1–25. https://doi.org/10.1007/978-1-4471-4976-7_74-2

Chryssolouris, G., Wollowitz, M., Sun, N.P., 1984. Electrochemical Hole Making. CIRP Ann. 33, 99–104. https://doi.org/10.1016/S0007-8506(07)61388-2

Das A D, Kumar K S, Prasanna R. 2022. Investigating the effect of wire cut EDM of titanium alloy 6242 using TOPSIS. Adv. Mater. Process. Technol. 8(3):2824–2836. https://doi.org/10.1080/2374068X.2021.1945261

Dubey, A.K., 2008. Experimental Investigations on Electrochemical Honing. Proc. Inst. Mech. Eng. Part B J. Eng. Manuf. 222, 413–426. https://doi.org/10.1243/09544054JEM905

Dubey, A.K., Yadava, V., 2008. Laser Beam Machining—A Review. Int. J. Mach. Tools Manuf. 48, 609–628. https://doi.org/10.1016/j.ijmachtools.2007.10.017

Dwivedi, A.P., Choudhury, S.K., 2016. Increasing the Performance of EDM Process Using Tool Rotation Methodology for Machining AISI D3 Steel. Procedia CIRP 46, 131–134. https://doi.org/10.1016/j.procir.2016.03.207

Fadaei Tehrani, A., Imanian, E., 2004. A New Etchant for the Chemical Machining of St304. J. Mater. Process. Tech 149, 404–408. https://doi.org/10.1016/j.jmatprotec.2004.02.055

Gentili, E., Tabaglio, L., Aggogeri, F., 2005. Review on Micromachining Techniques, in: Kuljanic, E. (Ed.), AMST'05 Advanced Manufacturing Systems and Technology. Springer, Vienna, pp. 387–396.

Ghosh, A., 2012. Electrochemical Machining (ECM), in: Bhushan, B. (Ed.), Encyclopedia of Nanotechnology. Springer, Netherlands Dordrecht, pp. 684–688. https://doi.org/10.1007/978-90-481-9751-4_363

He, T., Huang, S., Fang, S., Deng, H., Ding, Q., Zhang, M., 2022. Study on the Electrochemical Deburring for the External Surface of the Microhole Caused by Mechanical Drilling Process. Machines 10, 726. https://doi.org/10.3390/machines10090726

Ilhan, R.E., Sathyanarayanan, G., Storer, R.H., Phillips, R.E., 1992. A Study of Wheel Wear in Electrochemical Surface Grinding. J. Eng. Ind. 114, 82–93. https://doi.org/10.1115/1.2899762

Ilii S-M, Cotea M, Munteanu A., n.d. Experimental results concerning the variation of surface roughness parameter (Ra) at plasma arc cutting of a stainless steel workpiece. Int J Modern Manuf Technol. 2, 31–36.

Ilii S-M, Cotea M, Munteanu A., n.d. Experimental results concerning the variation of surface roughness parameter (Ra) at plasma arc cutting of a stainless steel workpiece. Int J Modern Manuf Technol. 2, 31–36.

Ilii, S.-M., Cotea, M., Munteanu, A., n.d. ExperimentalEXPERIMENTAL ResultsRESULTS ConcerningCONCERNING theTHE VariationVARIATION ofOF SurfaceSURFACE RoughnessROUGHNESS ParameterPARAMETER (Ra) atAT PlasmaPLASMA ARC CuttingCUTTING ofOF aA StainlessSTAINLESS SteelSTEEL Workpiece WORKPIECE.

J T. Black, Ronald A. Kohser, 2008. Materials & processes in Manufacturing, 10th Edition. ed. John Wiley & Sons, Inc.

Kadam, S.P., Mitra, S., 2021. Electrochemical Deburring – A Comprehensive Review. Mater. Today Proc. 46, 141–148. https://doi.org/10.1016/j.matpr.2020.07.059

Keerthivasan, T., Kaushik, P., Mirunalini, P., Surendhiran, J., 2021. Design and Analysis of Erosion in Electrochemical Machining Tool. Mater. Today Proc. 37, 182–186. https://doi.org/10.1016/j.matpr.2020.05.012

Koryta, J., Dvořák, J., Kavan, L., 1993. Principles of Electrochemistry, 2nd ed. Wiley, Chichester; New York.

Kozak, J., Rajurkar, K.P., Makkar, Y., 2004. Study of Pulse Electrochemical Micromachining. J. Manuf. Process. 6, 7–14. https://doi.org/10.1016/S1526-6125(04)70055-9

Laser Materials Interactions, 2008. in: Laser Fabrication and Machining of Materials. Springer US, Boston, MA, pp. 34–65. https://doi.org/10.1007/978-0-387-72344-0_2

Leese, R.J., Ivanov, A., 2016. Electrochemical Micromachining: An Introduction. Adv. Mech. Eng. 8, 1687814015626860. https://doi.org/10.1177/1687814015626860

Li, H., Gao, C., Wang, G., Qu, N., Zhu, D., 2016. A Study of Electrochemical Machining of Ti-6Al-4V in $NaNO_3$ solution. Sci. Rep. 6, 35013. https://doi.org/10.1038/srep35013

Liu, G.X., Zhang, Y.J., Jiang, S.Z., Liu, J.W., Gyimah, G.K., Luo, H.P., 2016. Investigation of Pulse Electrochemical Sawing Machining of Micro-inner Annular Groove on Metallic Tube. Int. J. Mach. Tools Manuf. 102, 22–34. https://doi.org/10.1016/j.ijmachtools.2015.12.001

Machinability Data Center, 1980. Machining Data Handbook. Metcut Research Associates, Cincinnati.

Magdum, V.B., Kittur, J.K., 2021. A Review on Laser Beam Machining Process Modeling and Optimization. IOP Conf. Ser. Mater. Sci. Eng. 1104, 012015. https://doi.org/10.1088/1757-899X/1104/1/012015

Manjunathan, K., Palanivel, A., Vasudevan, B., Megaraj, M., Dillibabu, S.P., Durvasulu, R., 2023. A Review: Study of Plasma-Assisted Machining to Inconel-718. AIP Conf. Proc. 2766, 020030. https://doi.org/10.1063/5.0139436

Martins, M., Godina, R., Pimentel, C., Silva, F.J.G., Matias, J.C.O., 2018. A Practical Study of the Application of SMED to Electron-beam Machining in Automotive Industry. Procedia Manuf. 17, 647–654. https://doi.org/10.1016/j.promfg.2018.10.113

Meijer, J., Du, K., Gillner, A., Hoffmann, D., Kovalenko, V.S., Masuzawa, T., Ostendorf, A., Poprawe, R., Schulz, W., 2002. Laser Machining by Short and Ultrashort Pulses, State of the Art and New Opportunities in the Age of the Photons. CIRP Ann. 51, 531–550. https://doi.org/10.1016/S0007-8506(07)61699-0

Mishra, S., Yadava, V., 2015. Laser Beam MicroMachining (LBMM) – A Review. Opt. Lasers Eng. 73, 89–122. https://doi.org/10.1016/j.optlaseng.2015.03.017

Nemchinsky, V.A., Severance, W.S., 2009. Plasma Arc Cutting: Speed and Cut Quality. J. Phys. Appl. Phys. 42, 195204. https://doi.org/10.1088/0022-3727/42/19/195204

Okada, A., 2019. Electron Beam Machining, in: Chatti, S., Laperrière, L., Reinhart, G., Tolio, T. (Eds.), CIRP Encyclopedia of Production Engineering. Springer, Berlin; Heidelberg, pp. 582–587. https://doi.org/10.1007/978-3-662-53120-4_6480

Pandilov, Z., 2018. Application of Electro Chemical Machining for Materials Used in Extreme Conditions. IOP Conf. Ser. Mater. Sci. Eng. 329, 012014. https://doi.org/10.1088/1757-899X/329/1/012014

Parandoush, P., Hossain, A., 2014. A Review of Modeling and Simulation of Laser Beam Machining. Int. J. Mach. Tools Manuf. 85, 135–145. https://doi.org/10.1016/j.ijmachtools.2014.05.008

Patel, J.D., Maniya, K.D., 2018. A Review on: Wire Cut Electrical Discharge Machining Process for Metal Matrix Composite. Procedia Manuf. 20, 253–258. https://doi.org/10.1016/j.promfg.2018.02.037

Peças, P., Henriques, E., 2009. Intrinsic Innovations of Die Sinking Electrical Discharge Machining Technology: Estimation of Its Impact. Int. J. Adv. Manuf. Technol. 44, 880–889. https://doi.org/10.1007/s00170-008-1902-x

Pham, D.T., Dimov, S.S., Petkov, P.V., 2007. Laser Milling of Ceramic Components. Int. J. Mach. Tools Manuf. 47, 618–626. https://doi.org/10.1016/j.ijmachtools.2006.05.002

Pierce, J.R., 2004. Rectilinear Electron Flow in Beams. J. Appl. Phys. 11, 548–554. https://doi.org/10.1063/1.1712815

Rajurkar, K.P., Sundaram, M.M., Malshe, A.P., 2013a. Review of Electrochemical and Electrodischarge Machining. Procedia CIRP 6, 13–26. https://doi.org/10.1016/j.procir.2013.03.002

Rajurkar, K.P., Sundaram, M.M., Malshe, A.P., 2013b. Review of Electrochemical and Electrodischarge Machining. Procedia CIRP 6, 13–26. https://doi.org/10.1016/j.procir.2013.03.002

Rajurkar, K.P., Zhu, D., McGeough, J.A., Kozak, J., Silva, A.D., 1999. New Developments in Electro-Chemical Machining. CIRP Ann. 48, 567–579. https://doi.org/10.1016/S0007-8506(07)63235-1

Rao, P.S., Jain, P.K., Dwivedi, D.K., 2017. Optimization of Key Process Parameters on Electro Chemical Honing (ECH) of External Cylindrical Surfaces of Titanium alloy Ti_6Al_4V. Mater. Today Proc. 4, 2279–2289. https://doi.org/10.1016/j.matpr.2017.02.076

Saptaji, K., Gebremariam, M.A., Azhari, M.A.B.M., 2018. Machining of Biocompatible Materials: A Review. Int. J. Adv. Manuf. Technol. 97, 2255–2292. https://doi.org/10.1007/s00170-018-1973-2

Saravanan, K.G., Prabu, R., Venkataramanan, A.R., Beyessa, E.T., 2021. Impact of Different Electrolytes on the Machining Rate in ECM Process. Adv. Mater. Sci. Eng. 2021, 1432300. https://doi.org/10.1155/2021/1432300

Sarkar, S., Mitra, S., Bhattacharyya, B., 2004. Mathematical Modeling for Controlled Electrochemical Deburring (ECD). J. Mater. Process. Technol. 147, 241–246. https://doi.org/10.1016/j.jmatprotec.2004.01.002

Selvarajan, L., Rajavel, J., Prabakaran, V., Sivakumar, B., Jeeva, G., 2018. A Review Paper on EDM Parameter of Composite material and Industrial Demand Material Machining. Mater. Today Proc. 5, 5506–5513. https://doi.org/10.1016/j.matpr.2017.12.140

Sen, A., Doloi, B., Bhattacharyya, B., 2017. Fiber Laser Micro-machining of Engineering Materials, in: Kibria, G., Bhattacharyya, B., Davim, J.P. (Eds.), Non-Traditional Micromachining Processes: Fundamentals and Applications. Springer International Publishing, Cham, pp. 227–252. https://doi.org/10.1007/978-3-319-52009-4_6

Senbel, H.A., 2000. Factors Affecting Electrochemical Honing, in: Hayhurst, D.R., Hinduja, S., Atkinson, J., Burdekin, M., Hannam, R.G., Li, L., Labib, A.W. (Eds.), Proceedings of the 33rd International MATADOR Conference. Springer, London, pp. 327–332.

Shahsavari, M., Imani, A., Setavoraphan, A., Schaller, R.F., Asselin, E., 2022. Electron Beam Surface Remelting Enhanced Corrosion Resistance of Additively Manufactured Ti-6Al-4V as a Potential in-situ Re-finishing Technique. Sci. Rep. 12, 11589. https://doi.org/10.1038/s41598-022-14907-2

Singh, A., Ghosh, A., 1999. A Thermo-electric Model of Material Removal during Electric Discharge Machining. Int. J. Mach. Tools Manuf. 39, 669–682. https://doi.org/10.1016/S0890-6955(98)00047-9

Taniguchi, N., 1983. Current Status in, and Future Trends of, Ultraprecision Machining and Ultrafine Materials Processing. CIRP Ann. 32, 573–582. https://doi.org/10.1016/S0007-8506(07)60185-1

Tarng, Y.S., Tseng, C.M., Chung, L.K., 1997. A Fuzzy Pulse Discriminating System for Electrical Discharge Machining. Int. J. Mach. Tools Manuf. 37, 511–522. https://doi.org/10.1016/S0890-6955(96)00033-8

Wilson, J.F., 1971. Practice and Theory of Electrochemical Machining. Wiley-Interscience, New York.

Xu, Z., Wang, Y., 2021. Electrochemical Machining of Complex Components of Aero-engines: Developments, Trends, and Technological Advances. Chin. J. Aeronaut. 34, 28–53. https://doi.org/10.1016/j.cja.2019.09.016

Yuan, C.J., 2021. Electrochemical Machining (ECM) and Its Recent Development. mytribos 28, 20–31.

5 Renewable Energy Applications

5.1 BATTERIES

Batteries are critical to various mobile electronic gadgets as an energy storage medium. It is an essential component of any renewable energy source to store energy. The battery is an electrochemical device continuously striving to improve energy density and charging/discharging rate. Lithium-ion batteries (LIBs) are the most studied batteries due to their huge potential. LIBs are popular power sources for large-scale production, such as electronic devices, power tools, backup power units, and automobiles, where high power and high energy density are the primary focus. In LIBs, the charge/discharge process commenced with the movement of lithium ions from the cathode to the anode and the subsequent intercalation of ions in the electrodes and *vice versa*. The anode and cathode are separated by an ion-conductive membrane called a separator, and an ion-conductive electrolyte is used for the smooth progression of ions during the charge-discharge process, as shown in Figure 5.1. Nanotechnology can significantly enhance the performance of Li-ion batteries due to the increased contact area of the electrode and the electrolyte, leading to high energy density and minimal self-discharge.

The application of nanomaterials in Li-ion batteries can be classified into three categories: electrodes, electrolytes, and separators. Due to the high surface area of the nanostructured electrodes, the electrode-electrolyte interaction is improved; thereby, faster ion and electron transport occurs, ultimately, the charge density and the charge/discharge rates. The nanostructured electrodes are composed of nanostructures with dimensions of 1–100 nm. Carbon allotropes, layered V_2O_5 and MnO_2 spinels, nanoparticles of $LiMn_2O_4$ or $LiCo_2O_4$, and graphene hybrids are common nanostructured electrode materials. The hybrid electrodes containing the conductive carbon nanomaterials increase the electrodes' overall conductivity, reducing the electrodes' internal resistance, thereby creating more efficient charge and discharge processes.

The safety of the batteries is yet another issue. Batteries operated using liquid electrolytes, especially Li-ion batteries, can lead to short circuits, the failure of the cell, and sometimes explosion when overheated. Nanotechnology allows the use of nanomaterials, gels, etc., with high surface area and ionic conductivity to replace liquid electrolytes. The nanostructured ceramic oxides such as Al_2O_3, SiO_2, or ZrO_2

DOI: 10.1201/9781003317272-5

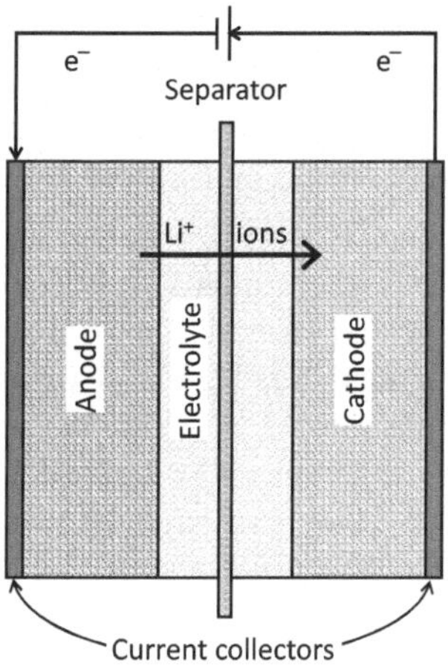

FIGURE 5.1 The various components of a lithium-ion battery.

are incorporated into polymeric gel to form composite electrolytes to enhance the ionic conductivity of the electrolyte. Nanostructured solid ceramic electrolytes offer high-temperature resistance, and the solid-state batteries are helpful for large vehicles and power stations. Nanostructured polymer electrolytes hold good attention due to their high electrochemical stability and affinity to the electrolytes. Porous Polymer gel electrolytes are established in lightweight, high-performance batteries. Electrospun nanofibers can have porosities above 90%, making them the ideal material for battery electrolytic applications. All these improvements are due to the electrolyte's high surface area and the electrolyte solution's wettability.

Separators are commonly used in batteries with liquid electrolytes to prevent physical contact between the anode and cathode. In the meantime, separators act as the electrolyte reservoir to enable free ionic transport. The structure of the separator can be either a membrane or a fibrous mat. The separator must have good porosity and wettability to the electrolyte. The separator does not participate in electrochemical reactions; still, its properties affect energy density, power density, cycle life, and battery safety.

Using nanostructured materials in batteries increases the charge/discharge speed due to the improved kinetics. Rapid charge/discharge cycle is sought in electric vehicles and portable electronic devices. Nanomaterials replaced the liquid electrolytes with solid electrolytes to create solid-state batteries. Nanostructured coatings on the electrodes enhance the cycle life of batteries by reducing electrode degradation and preventing unwanted side reactions. The introduction of nanotechnology to battery technology resulted in the creation of flexible and stretchable batteries suitable for flexible and wearable electronic devices. In such batteries, the electrodes and electrolytes are flexible and can withstand any mechanical deformation. Introducing conductive nanomaterials in the battery electrodes helps dissipate heat more effectively, thus preventing overheating and avoiding the potential risk of explosion.

5.2 METAL-AIR BATTERIES

Metal-air batteries use pure metals such as lithium, aluminum, magnesium, and zinc as anodes, ambient air as cathodes, and suitable electrolytes. The ambient air is reduced at the porous cathode, the metal is oxidized at the anode, and electricity is generated during this reaction. The construction of a metal-air battery is shown in Figure 5.2. Metal-air batteries can potentially replace LIBs in electric vehicles due to their high specific capacity and energy density. The performance of the metal-air batteries is based on the effective reduction of oxygen in the air at the cathode and the electrolyte as the anode is fixed. Selecting the appropriate reduction catalyst for air at the cathode can improve metal-air batteries' capacity, energy efficiency, rechargeability, and cycle life. Many nanostructured oxide, hydroxide, and layered materials are identified as cathode materials for metal-air batteries to achieve superior performance.

Based on the anode material, metal-air batteries are known as Li-air, Zn-air, Mg-air, Na-air, and Al-air batteries. The abundant elements such as Zn, Al, Mg, and Na are considered suitable anode materials. The performance of a metal-air battery majorly depends on the transport and electrochemical properties of cathode materials. The suitability of several nanostructured oxygen reduction catalysts has been studied as cathodes for metal-air batteries. They are noble metals and their alloys, transition metal oxides and composites, and pure and metal-doped carbon allotropes such as carbon nanotubes, graphene, and metal-organic frameworks (MOFs). The nanomaterials exhibit large surface areas, offering numerous highly active catalytic centers compared to their bulk counterparts. The cathode for a metal-air battery should be a good conductor, have high structural stability allowing the deposition of discharge products, excellent permeability to oxygen and electrolyte, and excellent ORR and OER catalytic activity. The hydrophobic air-breathing membrane separates

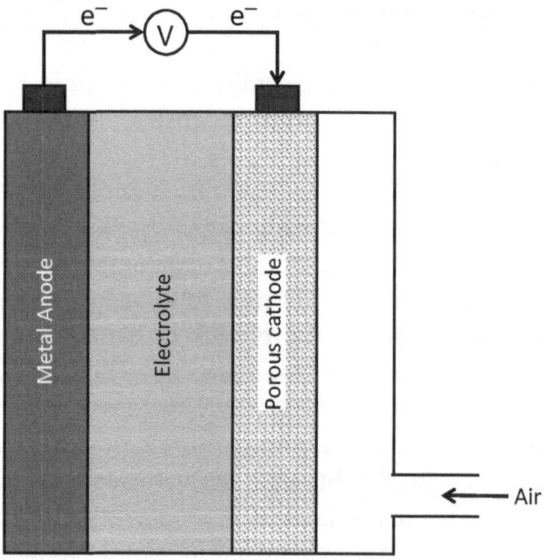

FIGURE 5.2 Construction of metal-air batteries.

oxygen from air, filtering CO_2, H_2O, and other gaseous components in air entering the cell. A rechargeable metal-air battery requires an effective OER catalyst to lower the overpotential and increase the energy efficiency.

Nanostructured complex oxides with the transition metal ions of Co, Ni, Mn, Fe, etc. exhibit excellent ORR catalytic performance in both acid and alkaline media. These compounds with transition metal cations can adopt different valence states, which makes them suitable as cathodes in metal-air batteries. Perovskites are one such complex with catalytic activity close to noble metal catalysts for oxygen reduction. The substitution of A-site elements in perovskites enhances the ability to absorb oxygen, and B-site substitution influences the activity of adsorbed oxygen. In general, the transition metal containing oxides exhibits a well-defined oxygen transport, and the ORR activity is based on the σ^*-antibonding orbital filling of surface transition metal cations and the metal-oxygen covalent bonding.

5.3 PHOTOVOLTAIC CELLS

Photovoltaic cells are solid-state devices that convert light to electrical energy, also known as solar cells. Conventional solar cells are p-n junction devices made of silicon-based materials. These solar cells are bulk, rigid, and expensive, and their efficiency is remarkably low. Dye-sensitized solar cells (DSSCs) are emerging alternatives for p-n junction photovoltaic cells. The significant components of DSSC are the working electrode, sensitizer (dye), redox-mediator (electrolyte), and counter electrode. These components are deposited on transparent conducting oxide (TCO) layer such as fluorine-doped tin oxide (FTO) or indium-doped tin oxide (ITO) glass. The working electrode is composed of nanostructured TiO_2, Nb_2O_5, ZnO, SnO_2 (n-type), or NiO (p-type), and it is deposited on the transparent conducting glass electrode made of FTO or ITO.

The heart of a dye-sensitized solar (DSSC) cell is a nanocrystalline metal oxide mesoporous layer on which the dye molecules are attached. The dye is essential for DSSC, and the maximum absorption of the incident light depends on the dye used. Dye is a luminescent material with absorption ranges in the ultraviolet-visible (UV-vis) and near-infrared (NIR) regions. The commonly used metal oxide is porous TiO_2 nanostructures. As the light falls on the dye, photoexcitation will occur, leading to the transfer of electrons from the ground to an excited state. The construction of a dye-sensitized solar cell is shown in Figure 5.3. The electrons from the excited state of the dye will be transferred to the conduction band of the nanostructured metal oxide, which is, in turn, connected to the electrodes of the cell. The electrons from the metal oxide are transferred to the electrode, then pass through the external circuit and reach the counter electrode. The restoration of dye to its original state is taking place by the electron transfer from the redox mediator in the electrolyte, during which I^- oxidizes to I_3^-. These I_3^- ions diffuse towards the counter electrode and reduce to I^- ions. The electrolyte comprises a redox couple, solvent, additives, ionic liquids, and cations. The redox couples such as I^- / I_3^-, Br^- / Br_2^-, SCN^- / SCN_2^-, etc. are a few examples of redox couples (Sharma et al., 2018). The counter electrode is made of platinum or carbon.

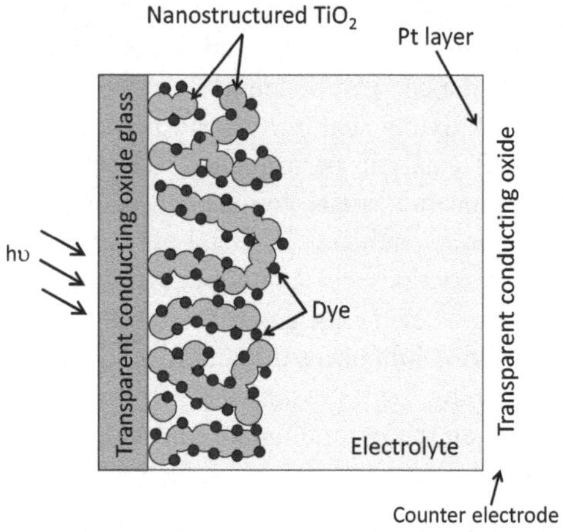

FIGURE 5.3 Construction of a dye-sensitized solar cell.

Nanostructures with different architectures, such as nanorods or wires as working electrodes, increase the efficiency of DSSC due to their unique geometrical structure and interfacial effects in electron transport. The volatility of the organic solvents carrying the electrolyte in DSSC is high. Therefore, the gel polymer electrolytes composed of electrospun polymer fibers are fabricated. Several companies present flexible solar cells attached to clothes to power low-energy devices in the market. Silicon and conjugated polymers are used in such devices. Electrospinning is used to fabricate non-woven solar cloth, which is an innovative approach in this area.

Thin-film solar cells are another important type of solar cell under research as an alternative to silicon-based solar cells. These solar cells are flexible and lightweight compared to conventional silicon-based solar cells. Thin-film solar cells have a reported conversion efficiency of 25%. The thin-film solar cells are known as perovskite solar cells (PSCs). In PSCs, a hybrid organic-inorganic perovskite is used for harvesting light. The mineral $CaTiO_3$ is called perovskite, and the compounds adopting the formula ABX_3 are generally called perovskites. The perovskites used in PSCs are lead or tin-based organic or inorganic halides (e.g., $CsPbI_3$, $CsSbI_3$, $CsPbBr_3$, $CsSnBr_3$, $CH_3NH_3PbI_3$, $CH_3NH_3PbBr_3$, etc.).

A schematic diagram showing the construction of PSC is shown in Figure 5.4. There are multiple thin films of different materials and thicknesses. The topmost layer is called the substrate. The substrate is made of a transparent material, which can be either glass or plastic, provided it must transmit a large spectrum of light and protect the underlying thin-film layers. Behind the glass substrate, a conductive layer of TCO is placed as the substrate is non-conducting, and it serves as an anode in PSCs. Below the TCO layer, an electron transport layer (ETL) is deposited, which enables the transport of only electrons (blocking holes) from the perovskite layer to the TCO anode. The materials used for ETL are nanostructured TiO_2, SnO_2, or ZnO. This inert layer prevents the electrode from degrading the perovskite layer. Below ETL, the perovskite layer is deposited, and it is the active layer that absorbs sunlight and generates electron-hole pairs (excitons). It is the uniqueness of the perovskite materials to

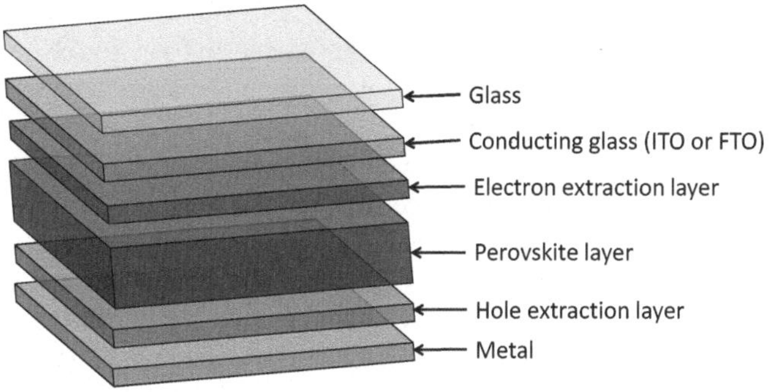

FIGURE 5.4 Different layers of PSCs.

generate electron-hole pairs when excited by visible light. A Hole Transport Layer (HTL) is provided below the perovskite layer to transport only holes to the counter electrode, blocking electrons. Materials for HTL poly(3,4-ethylenedioxythiophene) polystyrene sulfonate (PEDOT: PSS), NiOx, etc. This layer is critical for the improved efficiency and stability of the solar cell. The next layer is a top electrode, anode, or counter electrode. This layer is made of a highly conducting metal such as gold, silver, or aluminum. In transparent PSCs, ITO is used as the counter electrode. An inert layer finally encapsulates the cell to prevent the interaction of moisture and oxygen from the atmosphere. The perovskite layer is highly unstable and undergoes degradation after a short period of operation, which is a major challenge in PSCs.

As the sunlight hits the perovskite layer, the energy from the photons excites the electrons in the perovskite layer to elevate them from the valence band to the conduction band, creating electron-hole pairs (excitons). The ETL and HTL separate the electrons and holes from the exciton and transport them to the cathode and anode, respectively. The movement of charge carriers generates an electric current if an external load is connected. The electrons from the cathode travel through the external load, reach the anode and recombine with the holes. In PSCs, most of the components are designed based on nanotechnology principles.

In general, nanotechnology has made advancements possible in photovoltaic devices. Nanotechnology offers methods to improve efficiency, reduce costs, and improve overall performance. The potential manipulation of materials at the nanoscale improves light absorption and electron transport, which is critical for solar cells. Similarly, quantum dots can significantly enhance the absorption spectrum of the solar cell, thereby an increase in overall efficiency. Plasmonic nanomaterials, such as gold or silver nanoparticles, can improve light trapping, provided they must be placed strategically in the solar cells. Nanoscale thin films can be applied to Si-based solar cells as antireflective coatings to minimize reflection and maximize light absorption.

5.4 SUPERCAPACITORS

Supercapacitors are energy storage systems with high power and energy density, high cyclic stability, rapid charging and discharging abilities, and long cycle life.

Supercapacitors can be used for energy storage systems' high volumetric and gravimetric energy density requirements. The power and energy characteristics of supercapacitors fall between electrolytic capacitors and batteries with a higher power density than batteries but lower than capacitors, at the same time, lower energy densities than batteries but higher than conventional capacitors. The high surface area of the electrode and the smaller distance between the electrodes in supercapacitors allow more charge storage than traditional electrolytic capacitors. In supercapacitors, energy is electrostatically stored, resulting in fast charge and discharge rates and a longer cycle life than batteries. Supercapacitors can undergo thousands to millions of charge-discharge cycles without significant degradation in the charge storage capacities, making them ideal for applications where frequent cycling is required. In batteries, the energy is stored through chemical reactions, so the charge-discharge rates are slower than supercapacitors. The supercapacitors are intended to be used in applications that require power surges rather than a high energy storage capacity. Moreover, a significant reduction in the size of energy storage systems is possible by combining supercapacitors with batteries while retaining the power and energy requirements with an increased lifetime.

Supercapacitors can be divided into two categories based on the charge storage mechanism: electrochemical double-layer capacitors (EDLC) and pseudocapacitors. EDLCs store charges through surface adsorption called electrochemical double layers. Most of the nanostructured allotropes of carbon and carbonaceous materials like graphene, reduced graphene oxide (rGO), carbon nanotube (CNT), and activated carbon are potential EDLC electrodes. Pseudocapacitive materials store charges by faradic redox reaction, and inorganic electrodes and conjugated polymers exhibit pseudocapacitance. Irrespective of the type of capacitance, the capacitance of supercapacitors can be improved by increasing either the energy density or the operating voltages. Increasing the capacitor electrodes' surface area can improve supercapacitors' capacitance based on EDLC and pseudocapacitance. Moreover, electrode materials' size, morphology, and composition can significantly enhance the supercapacitance. At the nanoscale, the charge storage and transport through the electrode improve tremendously, and as a result, the supercapacitors' overall electrical conductivity and reaction kinetics are enhanced.

The construction of an EDLC supercapacitor is shown in Figure 5.5. A supercapacitor consists of two porous electrodes immersed in an electrolyte solution. The electrodes are typically composed of nanomaterials with a large surface area for the accumulation of charges. An ion-permeable membrane separates the electrodes, and the electrolyte solution fills the remaining space. The electrolyte is an ionic solution that can carry charge between the electrodes. As an external voltage is applied across the electrodes, the dissociated ions in the electrolyte solution accumulate on the surface of the electrodes. This process forms two layers of opposite charges, as shown in Figure 5.5, one at each electrode-electrolyte interface called the electric double layer, and the capacitor is known as EDLC. Carbon nanomaterials are identified as excellent materials for the EDLC supercapacitor electrodes.

The charge storage mechanism in pseudocapacitors the charge storage takes place through a fast surface-dependent process that occurs at the electrode surface. Nanostructured electrodes based on transition metal oxides, perovskites, spinels,

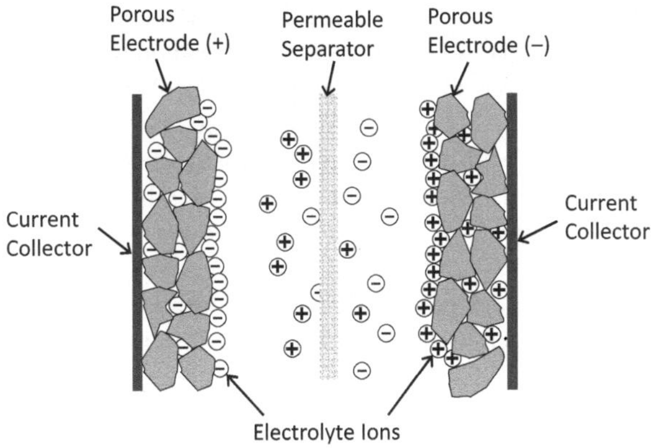

FIGURE 5.5 Schematic of a supercapacitor.

hydroxides, and conducting polymers exhibit pseudocapacitance. Additionally, materials that exhibit EDLC and pseudocapacitive behavior can be hybridized to improve the overall electrochemical performance of supercapacitors. Combining EDLC and pseudocapacitive electrode materials is done by forming a composite or growing the pseudocapacitive materials on the carbon materials. Moreover, combining materials with different dimensionalities enhances the surface area and the active electrochemical sites. Additionally, tuning the bandgap through doping can improve the tunneling probability and mobility of the charge carrier. As a result, one can obtain quick charge-discharge rates and stability of the electrode even at a high current density.

Complex oxides with more than one metal cation, such as perovskites and spinels, exhibit a charge storage mechanism different from the transition metal oxides. By controlling the ratio of metal cations, we can improve the performance of the supercapacitors. Additionally, these materials provide intrinsic 3D diffusion channels formed by large interstitial cavity chains for ionic transport. The optimization of electrical conductivity, structural flexibility, band gap, and charge carrier mobility are the key points to solve in the electrochemical charge storage mechanism of supercapacitors. Semiconducting heterostructured nanomaterials are the best choice for storing energy by near-surface ion adsorption, along with an additional contribution from fast reversible faradic reactions. Creating active sites and defects in the grain boundary of the heterostructure materials results in multiple redox activity, superior ionic conductivity, and a short diffusion path. Therefore, sufficient research enrooted to the doped and nano heterostructure electrode materials needs to be performed to exploit the high power and energy storage applications.

5.5 FUEL CELLS

Fuel cells are widely recognized as clean energy sources that convert chemical energy into electrical energy, offering great potential for transportation and portable power applications while mitigating the negative impacts of fossil fuels on the environment, such as pollution and climate change. The three main components of a fuel cell are the cathode, anode, and electrolyte. Typically, the cathode and anode are made of porous carbon with platinum catalysts. Different types of fuel cells, named after the electrolytes they use, include Phosphoric Acid fuel cells, Molten Carbonate fuel cells, Solid Oxide fuel cells, Direct Methanol fuel cells, Regenerative fuel cells, Microbial fuel cells, Alkaline fuel cells, and Proton Exchange Membrane fuel cells (PEMFCs). Among these, PEMFCs stand out due to their high power density and wide operating temperature range.

The construction of a fuel cell involves three primary components: the anode, cathode, and electrolyte membrane. Fuel is supplied to the anode side, while air is introduced to the cathode side. At the anode, fuel undergoes oxidation, producing ions and electrons. For example, H_2 gas is oxidized to H^+ ions and electrons in a polymer electrolyte membrane fuel cell. The H^+ ions migrate through the electrolyte to the cathode while electrons flow through an external circuit, generating an electric current. At the cathode, oxygen from the air combines with H^+ ions and electrons to form water. The working of a fuel cell is schematically shown in Figure 5.6. Multiple cells are stacked together to meet power production requirements, with interconnects preventing chemical short circuits and facilitating current transfer.

Nanomaterials are crucial in fuel cell fabrication, particularly in electrodes and electrolytes. Their high surface area and tunable porosity make them suitable for electrode applications. Catalysts based on noble metals like Pt or Au are commonly used for oxygen reduction and evolution reactions. However, replacing these noble metals with transition metals can reduce costs significantly. Converting noble metals nanostructures grown on catalytic support materials such as carbon nanotubes, graphene, etc., can reduce the price of the fuel cells. Many nanomaterials exhibit excellent catalytic activity comparable to noble metals. For instance, metal-doped carbon allotropes and inorganic nanomaterials show promising oxygen reaction catalytic behavior. Integrating nanomaterials in polymer matrices can enhance ionic conductivity, making them suitable for use as electrolytes in PEMFCs. For instance, the polymer nanocomposite membranes of sulfonated poly (ether ether ketone) (SPEEK)

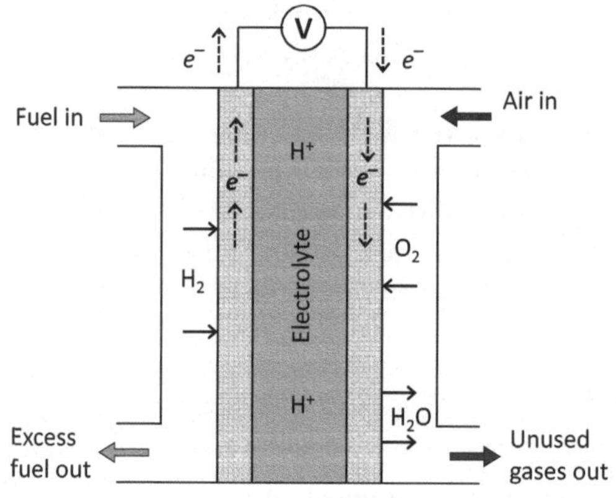

FIGURE 5.6 Working of a fuel cell.

reinforced with nanoparticles of silica, titanium dioxide, and zirconium oxide are excellent electrolytes with good mechanical properties, water retention, and prevent fuel crossover.

In solid oxide fuel cells (SOFCs), nanostructured materials are utilized for cathodes, anodes, electrolytes, and interconnectors. If one can lower the operating temperature and replace the noble metals in SOFCs, then the overall cost of SOFCs can be significantly reduced. Nanostructured cathode materials like $La_{1-x}Sr_xMnO_{3-\delta}$ and $La_{1-x}Sr_xCoO_{3-\delta}$ offer high electronic conductivity despite low ionic conductivity. Yttria-stabilized zirconia (YSZ) is a promising electrolyte material due to its good electronic and ionic conductivity. Perovskite materials with various compositions are also being explored for SOFC electrolytes, with options for doping to enhance their properties. Proton-conducting oxides like $BaZrO_3$ and $BaCeO_3$ are also promising for electrolyte applications, offering flexibility for doping with precious metals to improve oxygen mobility. The interconnects of SOFCs must be stable for oxidizing and reducing atmospheres, highly electronic conductive, low ionic conductive, gas-tight, and compatible with thermal and chemical behavior with other components of SOFCs. Nanostructured lanthanum and yttrium chromates, both *p*-type semiconductors, are identified as suitable candidates for SOFCs.

In direct methanol fuel cells (DMFC), methanol is used as a fuel, oxidized electrochemically at the anode and generates electricity, and oxygen is reduced (ORR) at the cathode. Nanostructured platinum supported on carbon nanomaterials is a promising candidate for methanol oxidation. Complex oxides such as perovskites with electroactive metals at the B-sites can also be a promising non-noble metal cathode in DMFCs. Lanthanum-based perovskites with the general formula $LaMO_3$ (M = Co, Mn, Ni, Fe) are promising candidates for ORR in an alkaline medium with the characteristics of noble metal catalysts. The partial substitution of A-site cations with Ca and Sr enhances the current densities at a lower overpotential; however, the transition metals at B-site significantly improve the ORR catalysis. Pt-containing DMFC anodes exhibit poor reaction kinetics and can be contaminated by carbon species such as CO. Metal oxides are an anticipated alternative as anode due to their chemical stability. The presence of the noble metal Ru in the nanocatalyst can enhance the methanol oxidation reaction (MOR). The activity towards MOR is further improved by controlling the particle size and doping noble metals, such as Pd, or depositing Pt on the surface of the catalysts.

Many polymers are suitable for proton-conducting membranes and electrolytes, commonly called polymer electrolyte fuel cells. These fuel cells are operated at a low temperature. The effectiveness of these polymeric membranes can be improved by converting them to nanofiber architectures. Electrospinning, phase separation, etc., produce such porous structures.

5.6 CATALYSTS FOR WATER SPLITTING

Water is a renewable source of energy. Hydrogen can be produced by splitting water into hydrogen and oxygen. For splitting water, efficient catalysts, such as electrocatalysts and photocatalysts, are required. Noble metals are the most efficient electrocatalysts for water splitting, but the cost of such catalysts is very high. There have been numerous attempts to reduce the price of catalysts for water splitting. Using the nanostructured materials, researchers achieved efficiency close to that of noble metal catalysts. The nanomaterials' high surface area, active catalytic centers, and tunable electronic properties make them effective electrocatalysts and photocatalysts for water splitting. Nanomaterials can enhance efficiency and reduce the cost of electrocatalytic splitting of water. Nanostructured materials containing transition metals such as Co, Ni, Fe, and a combination of these elements are the forerunners in electrocatalysis for water splitting. Modified carbon nanomaterials, MOFs, hydroxides of different transition elements, complex oxides such as perovskites and spines, etc., are identified as potential candidates for water splitting.

Nanostructured materials are active catalysts for photocatalytic water-splitting reactions. Any nanocatalysts for water splitting are based on photocatalysis, where semiconductor materials are used to absorb light energy and catalyze the water-splitting reaction. The common semiconductor nanomaterials used as photocatalysts are TiO_2 and ZnO. There are different strategies adopted for the control of catalytic activity of nanomaterials; they are size and shape, changing the composition by doping, adding co-catalysts such as Pt, Pd, CoO_x, NiO_x, etc., and the use of highly conducting support materials. Incorporating the above aspects in the same electrode is not trivial. Identifying the suitable composition from the abundance of materials is challenging, and there is still room for improvement in the stability of the nanomaterials in the long run.

5.7 HYDROGEN STORAGE

Hydrogen is considered to be the most dominant alternative fuel, and it has the potential to replace fossil fuels. The advantages of hydrogen as a fuel are it provides a green source of energy, is renewable, has high energy density, is pollution-free, and is abundant in nature. Hydrogen is also used as a catalyst, reagent, etc. In most cases, the application is not fully achieved because of the lack of an efficient storage media. Hydrogen storage using nanomaterials is a promising area of research aimed at overcoming the limitations of conventional hydrogen storage methods, such as compressed gas or liquid hydrogen storage. Nanomaterials offer unique properties such as high surface area, tunable pore sizes, and enhanced reactivity, which can facilitate the adsorption and desorption of hydrogen molecules.

Many nanomaterials can store hydrogen, like metals, intermetallic compounds, ceramics, etc., but stored hydrogen's storage capacity and reversibility at the required environmental conditions are limited. The highly porous ceramics and carbon-based materials such as CNTs, fullerenes, and activated carbon are efficient hydrogen storage media. Still, the cost-effective production of such materials has not been designed to date. Carbon nanomaterials have an extremely high surface area, which helps them to adsorb hydrogen on their surface. Metal-organic frameworks (MOFs) are potential storage mediums for hydrogen because of their high surface area. MOFs can adsorb hydrogen molecules onto their surface through hydrogen bonding. The size and shapes of the pores of nanomaterials can be tuned to achieve the required storage capacity. Moreover, the surface functionality can be adjusted for efficient adsorption and desorption of hydrogen molecules.

Though hydrogen storage is promising using nanomaterials, many challenges must be solved before it can be commercially used. The nanomaterials must offer high storage capacities at ambient conditions, maintaining stability and durability over multiple hydrogen adsorption-desorption cycles.

5.8 PIEZOELECTRIC GENERATORS

Piezoelectric materials harvest mechanical energy and convert it into valuable electrical energy or *vice versa*. It works on the principle of the piezoelectric effect. The piezoelectric effect occurs due to the broken symmetry of the crystal structure of the materials under external forces. The device is a piezoelectric nanogenerator (PENG) when nanostructured materials are used for piezoelectric current. PENGs exhibit piezoelectric properties better than their bulk counterparts. Piezoelectric materials are commonly used in the form of films. Piezoelectric materials find application in sensors, actuators, MEMS systems, muscle and neural scaffolds, and electronics applications. PENGs have become an integral part of wearable 'smart clothes' to produce electrical energy from body movements, enough to run low-power electronic devices efficiently. When a mechanical force is applied, electric charges are generated on the opposite ends of the surface. Piezoelectric materials may be classified into piezoelectric crystals, piezoelectric ceramics, piezoelectric polymers, and piezoelectric composites. Piezoelectric ceramics and polymers are easily produced compared to crystals, and piezoelectric ceramics are the most common materials. $BaTiO_3$, $KNbO_3$, $LiNbO_3$, $NaNbO_3$, GaN, InN, CdS, ZnO, $PbTiO_3$, and $PbZrTiO_3$, and lead zirconate titanate (PZT) are ceramic materials that exhibit piezoelectric properties. Polymers such as poly(vinylidene fluoride) (PVDF), polyvinylidene fluoride-trifluoroethylene (PVDF–TrFE)) poly(vinylidene fluoride-co-hexafluoropropylene) (PVDF-HFP)); PA-11, Nylon11, polypropylene (PP), polyacrylonitrile (PAN), etc., exhibit piezoelectric properties. The piezoelectric composites are the latest addition to the list, which is formed by adding piezoelectric polymers and piezoelectric ceramics together to attain a synergistic effect (e.g., PVDF/$BaTiO_3$ and PVDF/$PbTiO_3$). Among these polymers, PVDF exhibits the best piezoelectric properties. PVDF exhibits different crystalline phases such as α-, β, γ, and δ phases. Among these phases, β-phases exhibit the highest piezoelectric energy conversion rate. One can increase the fraction of β-phase in the PVDF matrix by adding ionic compounds or nanomaterials. Incorporating nanostructured piezoelectric ceramics listed above can significantly enhance the piezoelectric properties of PVDF.

The construction of a PENG is shown in Figure 5.7. Generally, PENG consists of piezoelectric materials and electrodes. The electrodes are conductive materials, and the piezoelectric material is sandwiched between two electrodes (Figure 5.7a). The electrode is sandwiched between two piezoelectric material layers in another construction, as shown in Figure 5.7b. Either way, the operation is the same. At equilibrium, opposite charges are generated on the upper and lower layers of the piezoelectric materials. When an external load is applied to the generator, the volume decreases, and the equilibrium will be lost as the center of charge for the cations and anions changes, and electric dipoles are formed. As a result, piezoelectric potential

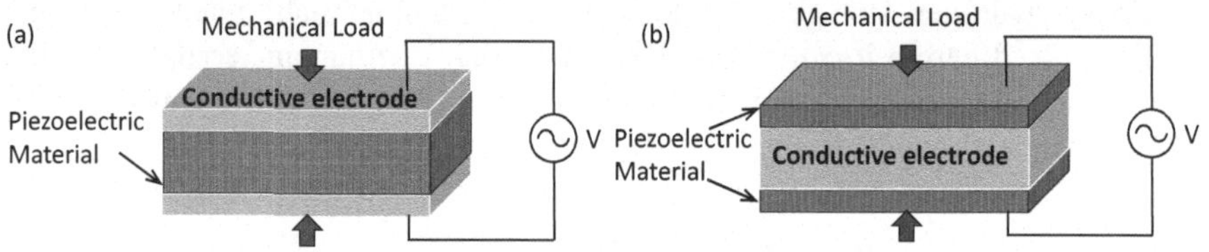

FIGURE 5.7 Construction of PENG operating in two different arrangements of piezoelectric materials.

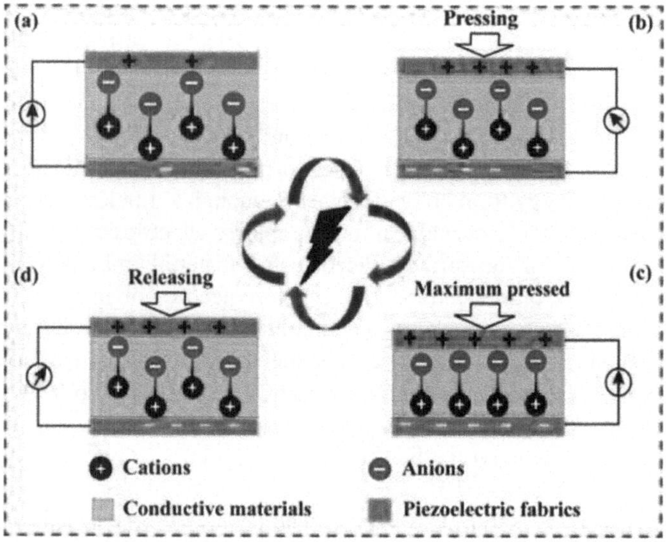

FIGURE 5.8 Working of a PENG. (Reprinted with permission from Zhang et al. (2021). Copyright © 2021, American Chemical Society.)

is generated between the two electrodes. Upon releasing the load, the electrons flow back, the system returns to the initial state, and a reversed piezoelectric electric current can be observed (Zhang et al., 2021). The operation of a PENG is shown in Figure 5.8.

Piezoelectric PVDF electrospun fibers are used as nanogenerators with high energy conversion efficiency. The piezoelectric coefficient of a single PVDF nanofiber is twice as significant as that of a thin film. The applications of piezoelectric materials are not limited to electricity generation alone. Piezoelectric fibers can help in nerve regeneration by acting as simulators for neurite extension. A single piezoelectric polymer PVDF produced by near-field electrospinning can be used as a generator with a consistent production of 7.2 pW with the external strain. Needleless electrospun nanofiber webs are useful in mechanical-to-electrical energy harvest devices. The β crystal phase in PVDF nanofibers leads to an enhanced mechanical-to-electrical energy conversion. Adding silver nanoparticles and CNTs to PVDF nanofibers can enhance the piezoelectric properties of the nanofibers. Poly(γ-benzyl, L-glutamate) and PVDF–TrFE are two other promising polymers that exhibit piezoelectric behaviors in the form of electrospun nanofibers.

5.9 THERMOELECTRIC GENERATORS

Thermoelectric materials produce electricity using the temperature difference between a heat source and a heat sink. Thermoelectric generators (TEGs) provide another environment-friendly solution for power production. The principle behind the working of TEGs is the Seebeck effect, during which electricity is generated when a temperature gradient is applied across dissimilar materials, as shown in Figure 5.9a. In such a condition, the charge carriers migrate from the high-temperature side to the low-temperature due to the temperature difference, called the thermoelectric effect. This is identical to the working of a thermocouple for measuring temperature. TEGs are used in devices with low energy consumption, such as wristwatches, hearing aids, etc., using the body temperature source for power generation. Additionally, TEGs are suitable for waste heat recovery in many applications. TEGs are not only used for electricity generation but also for heating and cooling. Thermoelectric materials produce electricity when exposed to thermal gradient and cooling/heating when voltage is applied, as shown in Figure 5.9b.

Thermoelectric generators require *p*-type and *n*-type junctions. The P-type junction has positive charge carriers, and the n-type junction has negative ones. The *p*-type components are doped so that holes are the primary charge carriers, and *n*-type components have electrons as the charged carriers. As the junctions are connected electrically, the positively charged carrier moves towards the n-junction, and the negatively charged carrier moves to the p-junction.

The thermoelectric properties of the materials can be improved substantially by forming them into nanostructures. The ZnO nanomaterials with different morphologies exhibit piezoelectric and thermoelectric properties. Tellurium (Te) also exhibits

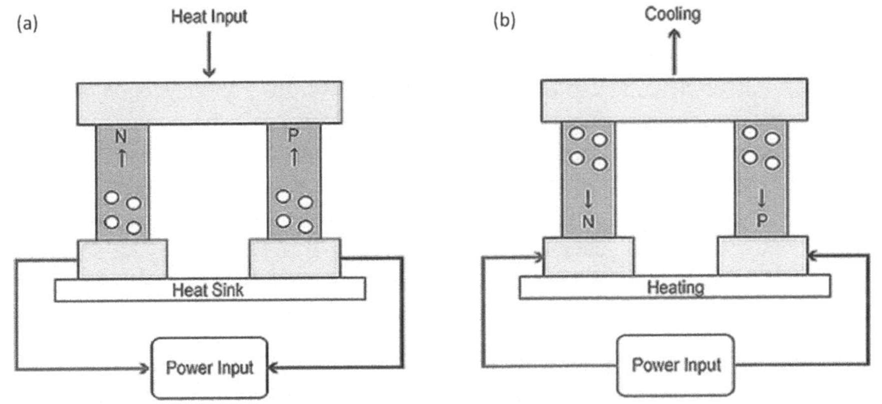

FIGURE 5.9 Application of TEGs for (a) electricity generation and (b) for heating and cooling. (Aksamija et al., 2019)

excellent TE properties. Many compounds based on Te, sulfur (S), and selenium (Se) are an essential class of TE materials. N-type Bi_2Te_3 nanomaterial exhibits remarkable thermoelectric properties. Multi-walled carbon nanotube (MWCNT)/Bi_2Te_3 nanocomposite is better than pure Bi_2Te_3. Te-based nanostructured compounds such as Ag_2Te, SnTe, and Sb_2Te_3 and sulfur compounds such as TiS_2, Ag_2S, and WS_2, and selenium compounds $NbSe_2$ and SnSe are also exhibiting thermoelectric properties suitable for TENGs.

5.10 PYROELECTRIC NANOGENERATORS

Pyroelectric Nanogenerators (PyENGs) can convert heat energy into electricity due to a change of spontaneous polarization within pyroelectric materials, which results from the oscillation of electric dipoles induced by a time-dependent temperature fluctuation. Many organic and inorganic materials exhibit pyroelectric properties. Nanomaterials with a specific high surface area can perform better than commercial pyroelectric materials. Lead zirconate titanate (PZT) performs excellently with a huge pyroelectric coefficient. Many nano/micro-PZT structures with different morphologies were used to fabricate PyENGs for energy harvest and conversion. $KNbO_3$ is an example of a non-toxic pyroelectric material suitable for PyENGs. Embedding these pyroelectric nanomaterials in PDMS can improve the flexibility of PyENGs. Like $KNbO_3$ nanowires, $NaNbO_3$ nanomaterials, nanorods, nanosheets, and nanocubes show superior pyroelectric properties.

$BaTiO_3$/$BiFeO_3$ composite film shows photovoltaic–pyroelectric effect under 405 and 450 nm light illumination. ZnO nanowire array-based PyENG on indium tin oxide (ITO) substrate. Researchers have fabricated a high-performance PyENG based on SbSeI nanowires for low-temperature waste heat recovery. A high-performance $Ba_{0.7}Sr_{0.3}TiO_3$ is another example of pyroelectric nanomaterial.

5.11 TRIBOELECTRIC NANOGENERATORS

Energy harvesting from mechanical movements is possible using triboelectric nanogenerators (TENGs). The electrical energy produced by TENGs is known as triboelectricity. Triboelectricity is the electricity generated while combing hair using a plastic comb or cleaning a grass surface using a fur brush. The conversion of mechanical energy to electrical energy in TENG occurs through triboelectrification and electrostatic induction. TENGs are operated without any fuel, but the relative contact motion of the surfaces of two different dielectric materials in contact, and these surfaces can be called electrodes. As these surfaces are rubbed against each other, electrostatic charges are generated on both surfaces. If these surfaces are separated mechanically, an electric potential will be generated between these two surfaces. There are two types of TENGs: conductor-dielectric type (or single electrode) and dielectric-dielectric type (double electrode). Figure 5.10 shows the working of a dielectric-to-dielectric contact-mode TENG. Here, the dielectric surfaces of thickness d_1 and d_2 are placed face to face. Conducting metal layers are deposited behind the dielectric surfaces to form the electrode. In the original position, a mechanical force

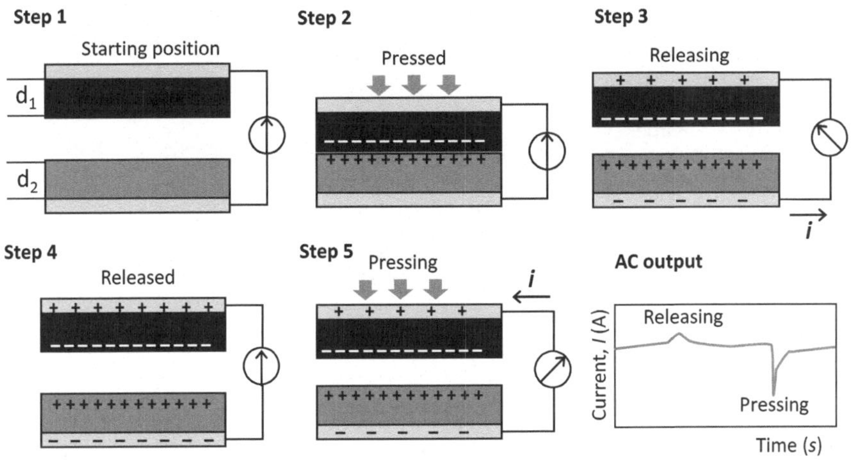

FIGURE 5.10 Working of a triboelectric nanogenerator.

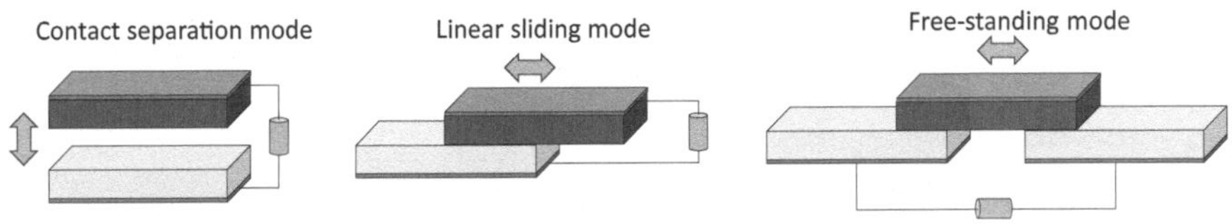

FIGURE 5.11 Schematic representation of different operating modes of TENG.

keeps these two layers separated. When an external mechanical force is applied, the surfaces come into frictional contact. Thus, static charges are generated at the inner surface of two triboelectric layers. As the force is removed, the layers separate, and a potential difference emerges that will drive the electrons to flow under short circuit (SC) conditions. As the surfaces are pressed again, the flow of electrons occurs in the opposite direction, resulting in an alternating current (AC) as in the output shown in Figure 5.10.

The TENG system generally comprises an energy harvesting module, a power management module, and an energy storage module. The TENGs can be operated in three modes, and the appropriate mode is selected depending on the type of application (Vivekananthan et al., 2020). The different modes are schematically shown in Figure 5.11.

Contact Separation Mode: in this mode, electricity is produced by the contact and separation process of two triboelectric materials or layers.

Linear Sliding Mode: In this mode, the electricity is generated by the forward and backward sliding of the dielectric layers.

Free-Standing Mode: In this mode, one electrode moves freely between the two electrodes or triboelectric layers.

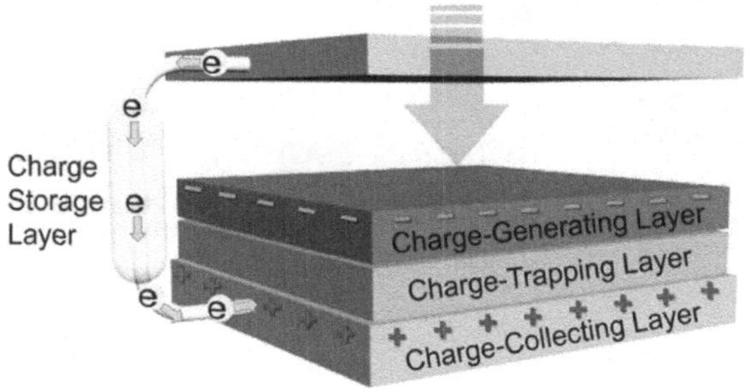

FIGURE 5.12 Representative components of a triboelectric nanogenerator (TENG). (Kim et al., 2020)

Both conductor-dielectric type (or single electrode) and dielectric-dielectric type (double electrode) TENGs can be operated in the above modes.

TENGs can harvest energy from vibration, human walking, body movements, and ocean waves. The characteristics of the electrode play a significant role in the electricity generated. A typical triboelectric module has three layers: the charge-generating layer, the charge-trapping layer, and the charge-collecting layer (Figure 5.12).

5.11.1 Charge-Generating Layer

The charge generation is the main parameter that controls the TENG's performance. Increasing the contact areas of the electrodes will improve the charge generation. Nanostructured surface features and surface functionalization improve the surface charge density straight away. The nanostructured array can be fabricated using forced assembly of colloidal systems, lithography, anodic aluminum oxide (AAO), etc.

The surface potentials of the contact surfaces can increase triboelectricity. Thus, surface functionalization that changes the surface potential of the material surface can increase the performance of TENGs. Surface modification methods are ion doping, radical injection, and plasma treatment. Different materials will develop different charges.

Several materials are potential dielectric materials for TENGs, and these materials are arranged on the triboelectric series. The series arranges materials that generate strong positive charge to those that generate strong negative charge. A triboelectric series is given in Table 5.1. Triboelectric series means if these materials on the series are used as electrodes, the higher will generate a positive charge and the lower will generate a negative charge. These materials can be made into nanostructures to enhance their performance.

5.11.2 Charge-Trapping Layer

The charge-trapping layer is required to prevent the recombination of opposite charges generated during the process. Aromatic polymers, such as polystyrene (PS) and polyimide (PI), have many trapping sites due to the non-uniform energy levels along

TABLE 5.1
Triboelectric Series

Positive	Glass
.	Mica
.	Polyamide (Nylon 6,6)
.	Rock salt (NaCl)
.	Wool
.	Fur
.	Silica
.	Silk
.	Aluminum
.	Poly(vinyl alcohol) (PVA)
.	Poly(vinyl acetate) (PVAc)
.	Paper
.	Cotton
.	Steel
.	Wood
.	Amber
.	Poly(methyl methacrylate) (PMMA)
.	Copper
.	Silver
.	Gold
.	Poly(ethylene terephthalate)(Mylar)
.	Epoxy Resin
.	Natural Rubber Polyacrylonitrile (PAN)
.	Poly(vinylidene chloride) (Saran)
.	Polystyrene (PS)
.	Polyethylene (PE)
.	Polypropylene (PP)
.	Poly(vinyl chloride)(PVC)
Negative	Polytetrafluorethylene (Teflon, PTFE)

their main chains, and they can produce more triboelectric output. Nanostructured materials with large surface-to-volume ratios can provide more charge-trapping sites.

5.11.3 Charge-Collecting Layer

As the triboelectric potential is generated, the generated charges electrostatically flow from one electrode to another. Solid and flexible electrodes, including metals, graphene, and ITO, are potential charge-collecting layers (Kim et al., 2020).

5.12 WIND ENERGY

Wind is a power source of energy that can be converted to electrical energy with the help of wind turbines or mechanical energy using windmills. Nanotechnology helps develop various coatings, lubrication, and fabrication of lightweight and highly durable materials for turbine blades and different sensors. Adding nanomaterials, such as carbon nanotubes or graphene, to the epoxy resins can improve their strength, stiffness, and fatigue resistance, allowing lighter, more durable blades to capture wind energy more efficiently. Using self-healing nanocomposites for turbines can increase the lifespan of wind turbine blades by repairing the cracks as they are formed and prevent the propagation of the same to irrecoverable damage. The nanostructured hydrophobic surface coatings can prevent the buildup of ice and dirt on the turbine blades, which otherwise leads to reduced efficiency and increased maintenance costs.

The turbine blade materials can degrade by exposure to UV light nano coatings, and structured coatings are available to prevent UV damage. Additional coating is required for offshore wind turbines exposed to harsh, corrosive marine environments. Additionally, nanocomposite insulators in electrical wires can reduce transmission losses. Since wind energy is intermittent, efficient storage units are essential to use the produced energy effectively. Moreover, piezoelectric nanomaterials and devices are employed to harvest wind energy at a smaller scale. These nanogenerators can power electronic devices operating in remote locations, such as temperature sensors, humidity sensors, etc., used in weather monitoring.

5.13 GEOTHERMAL ENERGY

The energy obtained from the earth's core in the form of heat is called geothermal energy. The heat is extracted from the hot rock or water to run geothermal power plants. Geothermal energy is used directly to heat buildings, agriculture, industrial uses, etc. Geothermal power plants use steam turbines for electricity generation. In a typical geothermal powerplant, the working fluid (generally water) absorbs heat from the geothermal sources, such as hot rocks or water streams, to form superheated steam. Which then passes through a steam turbine to produce electrical energy. The working fluid at the exit of the steam turbine is then cooled and sent back to the geothermal sources to capture heat and repeat the cyclic process. The arrangement of various components of a geothermal power plant is shown in Figure 5.13. Around 10,000 MW of electrical power is generated using geothermal sources worldwide. Thus, geothermal sources provide a significant contribution to meeting the electricity needs of today's world.

The primary application of nanomaterials in steam-based power plants as a coating is to prevent any corrosion, especially in components in direct contact with the hot working fluids. Another application is nanofluids as working fluids. Coatings composing the nanostructured materials prevent corrosion without significantly compromising the heat transfer rate of the components. Many polymer/nanoparticle composites exhibit characteristics ideal for corrosion prevention. One example is nanoscale montmorillonite (MMT) clay dispersed in polyphenylene sulfide (PPS) polymer. The MMT clay enhances the polymer's thermal properties; here, PPS's melting point is increased from 40 °C to 290 °C (Sugama, 2006). The composite coating of polytetrafluoroethylene (PTFE) and carbon nanotube exhibits endurance at

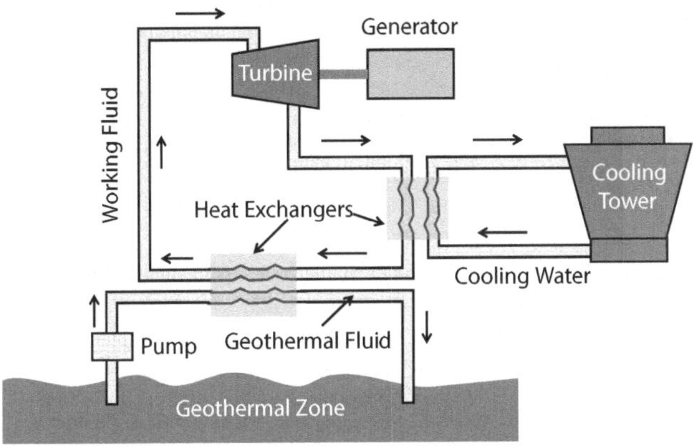

FIGURE 5.13 Working of a geothermal power plant.

high temperatures, prevents corrosion, and exceptional flame retardance. Nano-ZrO_2–TiO_2 is a ceramic non-corrosive coating that effectively reduces corrosion (Cai et al., 2016). Nano coatings of SiO_2 and TiO_2 also exhibit antifouling and anticorrosive properties (Song et al., 2018; Zhao et al., 2023).

The nanofluids are used as the working fluids in geothermal power plants. The nanofluids enhance the heat extracted from the geothermal sources. The CuO-water nanofluids exhibit a higher heat transfer coefficient than water. The heat transfer coefficient obtained for water-CuO nanofluid is about 800 W/m^2 K at a volume fraction of 6 wt.% of CuO nanoparticles (Daneshipour and Rafee, 2017). The CuO-water nanofluid is also used as the solar collector of novel solar-geothermal systems. They are replacing nanofluid with water in such systems to improve the thermal and exergy efficiency of the system. The base fluids for the nanofluids can be water, ethylene glycol, or propylene glycol, and pyrogenic nanoparticles or carbon nanotubes. If the dispersion of the nanoparticles is uniform, the efficiency is enhanced remarkably (Yu and Xie, 2011).

5.14 TIDAL/WATER ENERGY

Waves offer a tremendous source of power. Tidal barrages, tidal current turbines, and wave turbines are used to generate electricity from waves. However, other potential technologies can be used to generate electricity. For instance, salinity gradient energy (works based on the motion of water due to differences in salt concentration) or electrodialysis (salt movement from higher to lower concentrated solution). Additionally, triboelectric nanogenerators (TENGs). The working of TENGs and the different operation modes are discussed in this chapter. TENGs can generate electricity from mechanical movements, including ocean waves. TENGs can be deployed in a large area in the ocean for large-scale energy generation (LSEG). An interconnected network of LSEGs can produce a remarkable amount of electricity from the ocean waves.

REFERENCES

Aksamija, A., Aksamija, Z., Counihan, C., Brown, D., Upadhyaya, M., 2019. Experimental Study of Operating Conditions and Integration of Thermoelectric Materials in Facade Systems. Front. Energy Res. 7. https://doi.org/10.3389/fenrg.2019.00006

Cai, Y., Quan, X., Li, G., Gao, N., 2016. Anticorrosion and Scale Behaviors of Nanostructured ZrO_2–TiO_2 Coatings in Simulated Geothermal Water. Ind. Eng. Chem. Res. 55, 11480–11494. https://doi.org/10.1021/acs.iecr.6b02920

Daneshipour, M., Rafee, R., 2017. Nanofluids as the Circuit Fluids of the Geothermal Borehole Heat Exchangers. Int. Commun. Heat Mass Transf. 81, 34–41. https://doi.org/10.1016/j.icheatmasstransfer.2016.12.002

Kim, D.W., Lee, J.H., Kim, J.K., Jeong, U., 2020. Material Aspects of Triboelectric Energy Generation and Sensors. NPG Asia Mater. 12, 1–17. https://doi.org/10.1038/s41427-019-0176-0

Sharma, K., Sharma, V., Sharma, S.S., 2018. Dye-Sensitized Solar Cells: Fundamentals and Current Status. Nanoscale Res. Lett. 13, 381. https://doi.org/10.1186/s11671-018-2760-6

Song, J., Liu, M., Sun, X., Wang, J., Zhu, J., 2018. Antifouling and anticorrosion behaviors of modified heat transfer surfaces with coatings in simulated hot-dry-rock geothermal water. Appl. Therm. Eng. 132, 740–759. https://doi.org/10.1016/j.applthermaleng.2017.12.071

Sugama, T., 2006. Polyphenylenesulfied/Montomorillonite Clay Nanocomposite Coatings: Their Efficacy in Protecting Steel Against Corrosion. Mater. Lett. 60, 2700–2706. https://doi.org/10.1016/j.matlet.2006.01.111

Vivekananthan, V., Chandrasekhar, A., Alluri, N.R., Purusothaman, Y., Khandelwal, G., Kim, S.-J., Vivekananthan, V., Chandrasekhar, A., Alluri, N.R., Purusothaman, Y., Khandelwal, G., Kim, S.-J., 2020. Triboelectric Nanogenerators: Design, Fabrication, Energy Harvesting, and Portable-Wearable Applications, in: Nanogenerators. IntechOpen. https://doi.org/10.5772/intechopen.90951

Yu, W., Xie, H., 2011. A Review on Nanofluids: Preparation, Stability Mechanisms, and Applications. J. Nanomater. 2012, e435873. https://doi.org/10.1155/2012/435873

Zhang, C., Fan, W., Wang, S., Wang, Q., Zhang, Y., Dong, K., 2021. Recent Progress of Wearable Piezoelectric Nanogenerators. ACS Appl. Electron. Mater. 3, 2449–2467. https://doi.org/10.1021/acsaelm.1c00165

Zhao, H., Huang, Y., Deng, S., Wang, L., Peng, H., Shen, X., Ling, D., Liu, L., Liu, Y., 2023. Research Progress on Scaling Mechanism and Anti-scaling Technology of Geothermal Well System. J. Dispers. Sci. Technol. 44, 1657–1670. https://doi.org/10.1080/01932691.2022.2033625

6 MEMS and Robotics

6.1 INTRODUCTION

At the forefront of modern engineering lie two dynamic disciplines: Micro-Electro-Mechanical Systems (MEMS) and Robotics. These fields converge at the nexus of miniature technology and intelligent automation, promising transformative advancements across industries. MEMS, with its integration of mechanical elements, sensors, actuators, and electronics on a microscopic scale, has reshaped sectors from healthcare to telecommunications. Meanwhile, Robotics encompasses the design, construction, and operation of robots for tasks ranging from industrial assembly to hazardous environment exploration. This chapter embarks on an exploration of the symbiotic relationship between MEMS and robotics, unraveling their synergies, fundamental principles, and profound implications. From their historical roots to contemporary applications and future trajectories, we delve into how this convergence is revolutionizing industries and propelling engineering into uncharted territories (Judy, 2001).

6.2 EVOLUTION OF MEMS

In the annals of engineering history, the birth of Micro-Electro-Mechanical Systems (MEMS) stands as a testament to humanity's relentless pursuit of innovation. Emerging from the crucible of mid-20th-century technological fervor, MEMS traces its lineage to a pivotal moment in the quest for miniaturization in electronics and mechanics.

The groundwork for MEMS was laid amidst the burgeoning landscape of integrated circuits (ICs)—a revolutionary advancement that heralded a new era of compact electronic devices. Spearheaded by visionaries such as Jack Kilby and Robert Noyce, the invention of the integrated circuit in the late 1950s marked a watershed moment in engineering history. By ingeniously integrating multiple electronic components onto a single silicon wafer, ICs paved the way for the miniaturization of electronic systems, setting the stage for the emergence of MEMS.

The integration of mechanical components into this miniature electronic ecosystem marked a paradigm shift in engineering thinking. Engineers envisioned a future where sensors, actuators, and electronics could coexist seamlessly on a

DOI: 10.1201/9781003317272-6

microscopic scale, enabling a myriad of innovative applications. This vision was not without its challenges, as it necessitated breakthroughs in materials science, fabrication techniques, and system integration.

One of the earliest manifestations of MEMS can be traced back to the pioneering work of Harvey C. Nathanson and his colleagues at Bell Laboratories in the early 1960s. Nathanson's groundbreaking research on electrostatically actuated micromechanical resonators laid the foundation for MEMS technology, demonstrating the feasibility of integrating mechanical structures with electronic circuits on a silicon substrate.

As the field of MEMS began to take shape, researchers and engineers across academia and industry embarked on a journey of exploration and experimentation. The nascent field witnessed a flurry of activity, with researchers pushing the boundaries of microfabrication techniques to create increasingly intricate and sophisticated MEMS devices.

The 1980s marked a watershed moment for MEMS, with the advent of surface micro-machining techniques that enabled the mass production of MEMS devices. Innovations such as the bulk micro-machining and surface micro-machining processes revolutionized the fabrication of MEMS devices, making them more accessible and cost-effective.

Throughout its formative years, MEMS found applications in diverse fields, from automotive airbag sensors to inkjet printer heads. However, it was the telecommunications industry that provided a fertile ground for MEMS to flourish. The emergence of MEMS-based optical switches and tunable capacitors revolutionized telecommunications infrastructure, paving the way for the era of optical communication networks.

6.3 THE RISE OF ROBOTICS

The evolution of robotics stands as a testament to humanity's quest for innovation and technological advancement. From the rudimentary industrial manipulators of yesteryears to the sophisticated autonomous systems of today, robotics has traversed a remarkable trajectory of growth and transformation.

In the early days of robotics, machines were predominantly employed as industrial manipulators, tasked with repetitive and labor-intensive tasks on assembly lines. One of the most significant milestones in the history of robotics occurred in the 1960s with the introduction of the Unimate robot. Developed by George Devol and Joseph Engelberger, the Unimate became the world's first industrial robot to be deployed on a General Motors assembly line. Its ability to perform repetitive tasks with precision and consistency revolutionized manufacturing processes, laying the groundwork for the automation revolution that would follow.

As the field of robotics continued to mature, advancements in computing, sensor technology, and artificial intelligence (AI) paved the way for the development of increasingly sophisticated robotic systems. These systems transcended their role as mere industrial tools and began to find applications across a diverse array of industries.

In the automotive sector, robots emerged as indispensable assets in manufacturing plants, undertaking tasks ranging from welding and painting to assembly and quality control. The automotive industry witnessed a paradigm shift as robotic automation enabled faster production cycles, improved quality control, and enhanced worker safety. Beyond manufacturing, robotics found applications in fields as varied as healthcare, logistics, agriculture, and space exploration. In healthcare, surgical robots such as the da Vinci Surgical System have revolutionized surgical procedures, offering surgeons enhanced precision, dexterity, and control during minimally invasive surgeries. In logistics, robots are deployed in warehouses and fulfillment centers to streamline order fulfillment processes, increasing efficiency and reducing operational costs.

In agriculture, robots equipped with advanced sensing and imaging technologies are employed for tasks such as crop monitoring, harvesting, and precision agriculture. These robots help farmers optimize crop yields, reduce resource consumption, and mitigate environmental impact. Moreover, space exploration has benefited immensely from robotics, with robotic rovers and probes venturing into the depths of space to explore distant planets, asteroids, and celestial bodies. Robotic systems such as the Mars rovers have expanded our understanding of the cosmos, uncovering new insights into the origins and evolution of the universe. The evolution of robotics has been characterized by a relentless pursuit of innovation and a commitment to pushing the boundaries of what is possible. From humble beginnings on factory floors to the far reaches of outer space, robots have transcended their mechanical origins to become indispensable companions in humanity's quest for progress and exploration. As we look to the future, the possibilities for robotics are boundless, promising new frontiers of discovery and innovation that will shape the course of human history for generations to come.

6.4 THE UNION OF MEMS AND ROBOTICS

In the ever-evolving landscape of engineering, the convergence of Micro-Electro-Mechanical Systems (MEMS) and Robotics represents a monumental shift, heralding a new era of technological innovation and advancement. This chapter delves into the intricate interplay between MEMS and robotics, unraveling the synergies that emerge when these two dynamic fields intersect.

At the heart of this intersection lie the enabling technologies that drive the fusion of MEMS and robotics. MEMS sensors, with their ability to detect and measure various physical phenomena with unprecedented precision, serve as the sensory backbone of robotic systems. These miniature sensors, ranging from accelerometers and gyroscopes to pressure sensors and temperature detectors, provide robots with a sophisticated awareness of their surroundings, enabling them to perceive and interact with the environment in real-time. Complementing MEMS sensors are MEMS actuators, which bestow robots with the ability to manipulate their surroundings with finesse and agility. These microscale actuators, such as microvalves, micropumps, and microactuators, translate electrical signals into mechanical motion, enabling robots to perform a myriad of tasks with remarkable precision and control. Whether it's delicately grasping objects in surgical procedures or navigating complex terrain in search and rescue missions, MEMS actuators empower robots to adapt and respond to dynamic environments with unparalleled dexterity.

Furthermore, advanced control algorithms serve as the neural network of MEMS-enabled robotic systems, orchestrating the seamless integration of MEMS sensors and actuators to achieve optimal performance and functionality. Leveraging techniques from machine learning, artificial intelligence, and adaptive control, these algorithms enable robots to interpret sensory data, make informed decisions, and execute complex tasks autonomously or in collaboration with human operators. The fusion of MEMS and robotics not only enhances the capabilities of individual robots but also catalyzes the emergence of new classes of robotic systems with unprecedented versatility and intelligence. Microscopic giants, as these MEMS-enabled robots are often referred to, transcend the limitations of traditional robotics by virtue of their miniature scale and multifunctional capabilities. From swarms of microrobots navigating intricate biological environments for medical diagnostics and drug delivery to autonomous drones equipped with MEMS-based sensors for environmental monitoring and disaster response, the applications of MEMS-enabled robotics are as diverse as they are transformative.

As we journey deeper into the realm of MEMS-enabled robotics, we uncover a world brimming with possibilities, where the marriage of miniature technology and intelligent automation unlocks new frontiers of exploration and discovery. The synergy between MEMS and robotics holds the promise of revolutionizing industries, improving quality of life, and shaping the future of humanity in ways we have yet to imagine. By harnessing the power of MEMS sensors, actuators, and control algorithms, robots are poised to transcend their mechanical origins and become indispensable companions in our quest for progress and innovation.

6.5 FABRICATION TECHNIQUES FOR MEMS MODULES

The development of micro-electromechanical systems (MEMS) has transformed various industries by assisting the creation of miniature devices with extraordinary functionality. MEMS modules, including intricate components and systems at microscale dimensions, are at the forefront of this technological innovation. Fabrication techniques tailored for MEMS modules play a pivotal role in comprehending these miniaturized marvels, offering precise control over their design, structure, and performance.

In this exploration of fabrication techniques for MEMS modules, we explore into the diverse types of methods employed to develop these miniature yet powerful devices. From lithography and thin-film deposition to etching and bonding processes, each technique contributes uniquely to the intricate assembly of MEMS modules. These fabrication methods not only enable the miniaturization of devices but also facilitate the integration of mechanical, electrical, and optical functionalities onto a single platform.

As the demand for MEMS modules continues to surge across industries such as healthcare, telecommunications, automotive, and consumer electronics, the importance of efficient and scalable fabrication techniques cannot be overstated. Therefore, understanding the principles and intricacies of these fabrication methods is essential for advancing MEMS technology and unlocking its full potential in addressing modern engineering challenges.

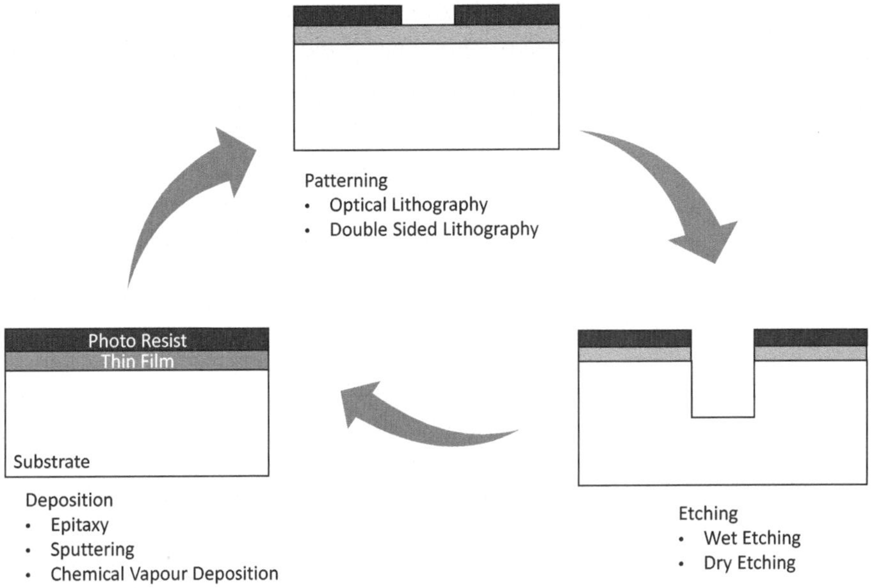

FIGURE 6.1 Schematic representation of different steps in micro-machining process.

In this context, this chapter aims to provide an insightful overview of the various fabrication techniques tailored for MEMS modules, highlighting their principles, applications, and recent advancements. By elucidating the underlying processes and innovations driving the fabrication of MEMS modules, we aim to contribute to the broader discourse surrounding microscale engineering and its transformative impact on society. Figure 6.1 shows the steps involved in micro-machining process.

6.5.1 The Basic Manufacturing Processes

Micro-electromechanical systems (MEMS) devices encompass a diverse array of miniature structures and components designed to integrate mechanical, electrical, and/or optical functionalities on a microscale. These devices find applications across various industries, from healthcare and automotive to consumer electronics and aerospace. Central to the fabrication of MEMS devices are several basic fabrication methods tailored to manipulate materials at the microscale. Photolithography stands as a fundamental technique in MEMS fabrication, enabling precise patterning of thin films on substrates. This process involves applying a photoresist layer onto a substrate, selectively exposing it to ultraviolet light through a photomask, and then etching away the unexposed regions. Photolithography is pivotal in defining intricate patterns and features crucial for MEMS structures. Thin-film deposition techniques, such as physical vapor deposition (PVD) and chemical vapor deposition (CVD), are employed to deposit thin layers of materials onto substrates. These layers serve as structural elements, conductive electrodes, or insulating coatings within MEMS devices, offering versatility in device design and functionality. Etching processes play a crucial role in MEMS fabrication, allowing for the precise removal of material to define MEMS structures. Wet etching, using chemical solutions, and dry etching methods like reactive ion etching (RIE) or deep reactive ion etching (DRIE) offer control over feature shapes and dimensions, enabling the creation of complex MEMS components.

Lithography techniques, including optical lithography, electron beam lithography (EBL), and nanoimprint lithography (NIL), are utilized to pattern substrates with high precision at the nanometer scale. These methods facilitate the creation of intricate features and structures critical for MEMS devices' functionality. Bonding processes are employed to assemble multiple layers or components of MEMS devices. Techniques such as anodic bonding, fusion bonding, and adhesive bonding enable the integration of diverse materials and functionalities within MEMS structures, enhancing device performance and reliability. Doping techniques, including ion implantation and diffusion, are utilized to introduce impurities into semiconductor materials, modifying their electrical properties. This enables the creation of semiconductor-based MEMS devices with tailored electronic functionalities. These basic fabrication methods form the foundation of MEMS device manufacturing, providing the necessary tools and techniques to create intricate, high-performance microscale structures and components. As MEMS technology continues to advance, further innovations in fabrication methods are poised to drive the development of increasingly sophisticated MEMS devices with enhanced capabilities and applications.

6.5.1.1 Physical Vapor Deposition Process

Physical Vapor Deposition (PVD) includes a variety of methods employed to apply thin films of materials onto substrates using physical mechanisms like evaporation or sputtering. These methods are chosen based on factors such as the desired properties of the film, the type of substrate, and the conditions of deposition. There are different types of PVD techniques such as Thermal Evaporation, Electron Beam Evaporation, Sputtering, Ion Plating and Pulsed Laser Deposition method.

6.5.1.2 Thermal Evaporation

The thermal evaporation process is a widely used technique in physical vapor deposition (PVD) for depositing thin films of materials onto substrates. Its working principle involves heating the material to be deposited within a vacuum chamber until it reaches its vaporization temperature, as shown in Figure 6.2. Once vaporized, the material condenses onto a substrate, forming a thin film. The working principle of the thermal evaporation process is as follows:

Vacuum Chamber: The process takes place within a vacuum chamber to eliminate air and other contaminants, ensuring a clean environment for deposition.

Material Source: The material to be deposited, often in the form of a solid rod or pellets, is placed in a crucible or boat within the vacuum chamber. This material is typically a metal or semiconductor with a relatively low melting point.

Heating Element: A heating element, such as a resistive coil or electron beam gun, is used to heat the material to its vaporization temperature. The heating

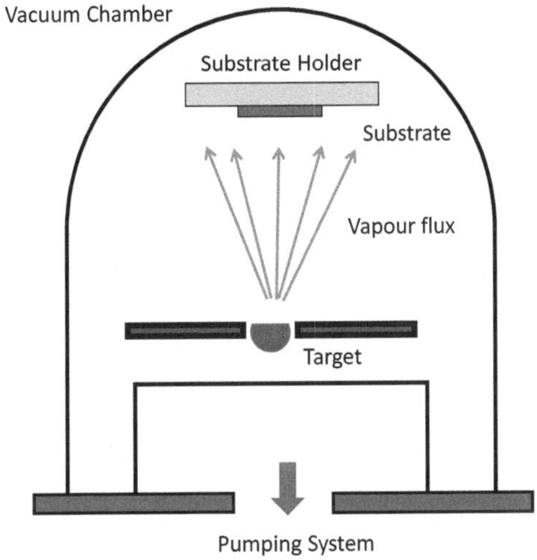

FIGURE 6.2 Schematic representation of thermal evaporation process.

mechanism raises the temperature of the material to the point where it transitions directly from a solid to a vapor phase, bypassing the liquid phase (sublimation).

Vaporization: As the material is heated, atoms or molecules are released from the surface of the solid and enter the gas phase, forming a vapor cloud within the vacuum chamber. This vaporization process is essential for creating a flux of material that can be deposited onto the substrate.

Substrate Preparation: Meanwhile, the substrate onto which the thin film will be deposited is prepared. The substrate is typically cleaned and positioned in the vacuum chamber to receive the evaporated material.

Deposition: The vaporized material travels across the vacuum chamber and condenses onto the substrate surface. The substrate is positioned in a way that allows uniform deposition of the material across its surface, resulting in the formation of a thin film with the desired properties.

Film Growth: As deposition continues, the thin film grows in thickness until the desired film thickness is achieved. Control over deposition parameters such as temperature, pressure, and deposition rate ensures the uniformity and quality of the deposited film.

6.5.1.3 Electron Beam Evaporation

In the electron evaporation process, a material is transformed from its solid state directly into vapor through the application of electron beam energy. This procedure is integral to various applications, including thin-film deposition in microfabrication and the creation of coatings with unique properties. The principle behind electron evaporation is rooted in the interaction between the electron beam and the target material, which is meticulously outlined in the following sections.

Generation of Electron Beam

The generation of an electron beam, a pivotal step in processes like electron evaporation, involves converting electrical energy into a focused stream of high-energy electrons. This process is executed within an electron gun, a specialized device designed to produce and accelerate electrons. The initial stage involves the emission of electrons from a cathode, typically made of materials with a low work function to facilitate easy electron release. When heated or subjected to a strong electric field (a process known as thermionic emission or field emission, respectively), electrons gain sufficient energy to escape the surface of the cathode. Once emitted, the electrons are accelerated towards an anode through a vacuum by a high-voltage electric field established between the cathode and anode. The strength of this electric field determines the kinetic energy and, consequently, the velocity of the electrons. After acceleration, the electron beam is focused using electromagnetic lenses. These lenses, functioning similarly to optical lenses but for electrons, narrow the beam to achieve a fine, concentrated stream. The focusing can be adjusted to control the beam's diameter and intensity, which is crucial for applications requiring precision. In some applications, additional electromagnetic fields may be employed to steer or shape the beam to specific geometries or to ensure it precisely targets a particular area or material. This fine control over the beam's path and focus is essential for achieving desired outcomes in various industrial and scientific applications.

Interaction With Target Material

Upon directing the electron beam towards the target material, also known as the evaporant, the high-energy electrons collide with the atoms of the target. This collision transfers energy from the electrons to the atoms of the material, increasing their kinetic energy. The energy imparted to the atoms causes them to vibrate more vigorously, which manifests as an increase in the material's temperature. This heating effect is localized to the area of impact, allowing for precise control over which part of the target material is altered. As the temperature of the target material rises due to the continuous influx of energy from the electron beam, it eventually reaches a critical point where the atoms or molecules at the surface gain enough kinetic energy to overcome their mutual attraction. This leads to their release from the solid phase directly into the gas phase, a process known as sublimation in the case of solids that do not melt under normal conditions, or evaporation if the material first transitions into a liquid phase. he atoms or molecules that escape the target material then either deposit on a nearby substrate, forming a thin film in the case of deposition processes, or undergo further reactions if the purpose is material synthesis or modification. This phase is essential for tailoring the properties of materials or for the fabrication of thin films with specific characteristics. This interaction between the electron beam and the target material demonstrates the delicate balance of energy transfer and material transformation. It highlights the electron beam's role as a precise tool for engineering material properties at the atomic level, facilitating advancements in fields ranging from microelectronics to materials science.

6.5.1.4 Heating and Phase Transition

The accumulation of kinetic energy within the atoms of the target material leads to an increase in temperature. As the temperature rises to a point exceeding the material's evaporation temperature, atoms or molecules on the surface of the material gain enough energy to overcome their bonding forces. Consequently, they escape the solid phase and enter the vapor phase. As the target material interacts with the high-energy electron beam, electrons from the beam transfer their kinetic energy to the atoms of the material upon collision. This transfer of energy manifests as an increase in the vibrational energy of the atoms, effectively raising the temperature of the material. The continued absorption of energy causes the temperature of the target material to increase steadily. Different materials have specific temperatures at which they undergo phase transitions, such as melting (transition from solid to liquid) or sublimation (transition from solid to gas without passing through a liquid phase). For the electron evaporation process, the focus is often on reaching the evaporation temperature. Upon reaching the evaporation temperature, atoms or molecules at the surface of the material gain sufficient kinetic energy to overcome intermolecular forces binding them to the surface. This results in the liberation of these atoms or molecules into the surrounding vacuum as a gas. The phase transition from solid directly to gas, without passing through a liquid phase, is termed sublimation in materials that exhibit this behavior under normal atmospheric conditions. In vacuum conditions, however, even materials that typically melt can directly vaporize from the solid phase, given sufficient energy input. The rate of this phase transition and the subsequent rate of vaporization can be precisely controlled by adjusting the intensity and focus of the electron beam. This allows for meticulous control over the thickness and composition of the material deposited in vapor deposition processes, which is crucial for applications that require thin films of precise thickness and uniformity. The vaporized atoms or molecules then travel away from the target material, moving towards a substrate where they condense, cooling down and returning to a solid state. This results in the formation of a thin film of the target material on the substrate. The electron evaporation process allows for precise control over the rate of evaporation and the thickness of the deposited film. By adjusting parameters such as the intensity of the electron beam and the distance between the target material and the substrate, it is possible to achieve high degrees of accuracy in the coating process. Figure 6.3 represents the schematic of electron beam evaporation.

6.5.2 Sputtering

The sputtering process is a widely used method for depositing thin films of materials onto substrates, an essential step in the fabrication of electronic, optical, and various other high-tech devices. The mechanism involves ejecting atoms from a solid target material and depositing them onto a substrate within a vacuum chamber. Sputtering is carried out in a vacuum chamber to reduce the presence of unwanted gases that could interfere with the deposition process. The quality of the vacuum is critical, with pressures typically ranging from about 0.1 to 100 millitorr (mTorr). This low pressure is essential for allowing the sputtered atoms to travel from the target to the substrate without colliding with gas molecules. Within this vacuum chamber, a gas (often argon, due to its inert nature) is introduced and ionized to create a plasma. Electric fields are used to accelerate these positively charged ions towards the negatively charged target material. Upon colliding with the target, the kinetic energy of the ions is transferred

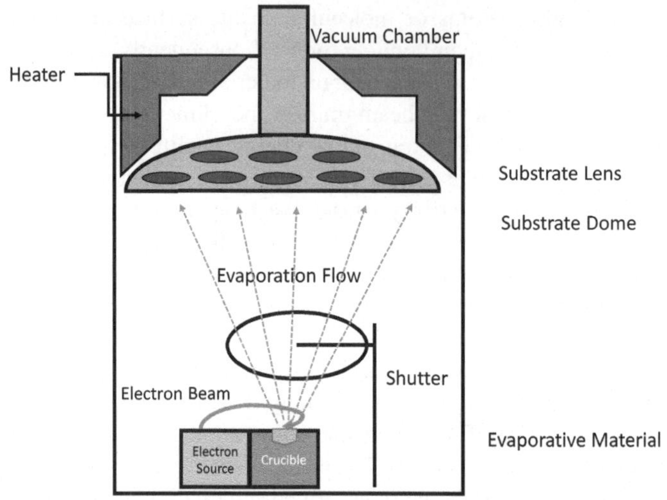

FIGURE 6.3 Schematic representation of electron beam evaporation process.

to atoms in the target material. This energy transfer causes some atoms to be ejected from the target's surface in a process known as sputtering. The pressure within the chamber affects both the mean free path of the ions (the average distance an ion travels before colliding with another particle) and the efficiency of sputtering. Optimizing pressure is crucial for achieving a stable plasma and maximizing the deposition rate without compromising film quality. The distance between the target and the substrate influences the deposition rate and film uniformity. A shorter distance increases the deposition rate but may result in less uniform films, necessitating careful optimization based on specific application requirements. The temperature of the substrate can significantly impact the mobility of the sputtered atoms on the substrate surface, thereby affecting the film's microstructure and physical properties. Controlled heating of the substrate is often employed to enhance film quality. The power applied to generate the plasma and accelerate ions towards the target affects the sputtering rate and the energy of the sputtered atoms. Higher power levels increase the deposition rate but can also lead to increased substrate heating and changes in film properties. DC (Direct Current) and RF (Radio Frequency) sputtering are two distinct methods employed within the broader technique of sputtering, a process used to deposit thin films of material onto substrates. These methods differ primarily in the type of electrical current they use to ionize the gas in the sputtering chamber, which in turn affects their applicability to various materials. DC sputtering utilizes a direct current power supply to maintain an electrical field between the target material (cathode) and the substrate (anode). This field ionizes the gas (usually argon) in the chamber, creating a plasma. The positively charged ions in the plasma are attracted to the negatively charged target. Upon impact, these ions knock atoms off the target, which then deposit onto the substrate. DC sputtering is particularly effective for conductive materials due to its reliance on a direct current. This method is straightforward and can be highly efficient, making it suitable for large-scale industrial applications. Its main drawback is that it cannot be used for insulating materials because they cannot sustain a steady DC plasma without charging up and disrupting the sputtering process. RF sputtering employs an alternating current at radio frequencies to generate the plasma. Unlike DC sputtering, RF sputtering can be used with both conductive and insulating materials. This versatility comes from the fact that the alternating current prevents charge buildup on the target surface, allowing for the sputtering of non-conductive materials. RF sputtering is suitable for a wide range of materials, including insulators, because the RF power supply oscillates too quickly for charge accumulation. While offering greater material flexibility, RF sputtering systems are more complex and typically more expensive than their DC counterparts due to the need for specialized RF power supplies and matching networks. Generally, RF sputtering has a lower deposition rate compared to DC sputtering, which is a consideration for applications requiring rapid material deposition. The choice between DC and RF sputtering is governed by the type of material to be sputtered and the specific requirements of the application. DC sputtering is preferred for conductive targets due to its simplicity and efficiency, while RF sputtering is the go-to choice for insulating materials due to its ability to accommodate a broader range of materials. Both methods, however, play crucial roles in the fabrication of electronic devices, coatings, and various nanotechnology applications, showcasing the versatility and adaptability of sputtering as a thin-film deposition technique. Figure 6.4 shows the schematic representation of sputtering process.

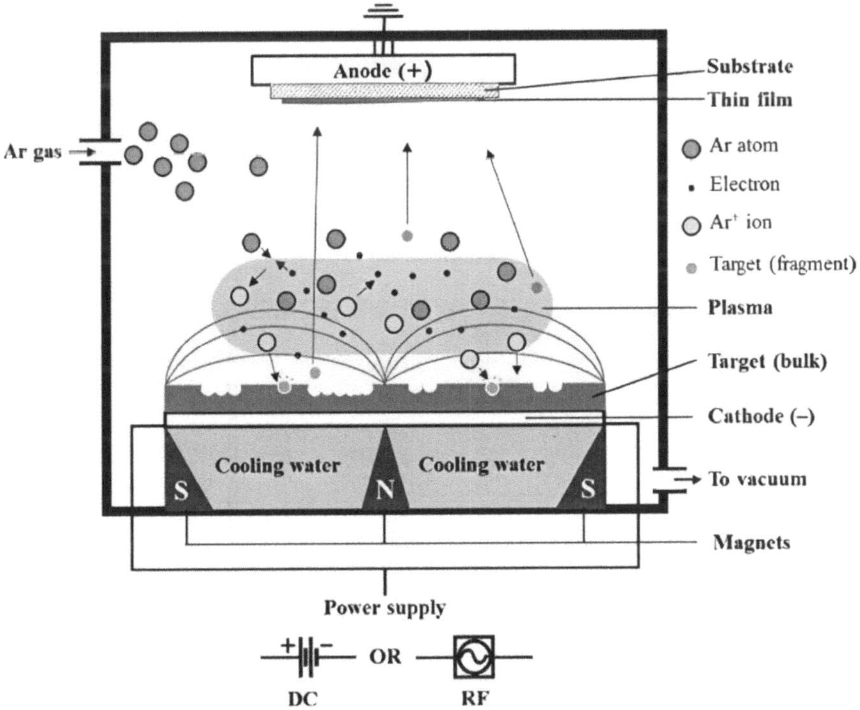

FIGURE 6.4 Schematic representation of sputtering deposition process. (Hishimone et al., 2020)

6.5.3 Ion Plating

Ion plating is a sophisticated physical vapor deposition (PVD) technique used to deposit thin films of various materials onto substrates. It is widely employed in industries such as electronics, aerospace, automotive, and medical devices for applications requiring high-performance surface properties. The ion plating process involves several key steps and mechanisms, as shown in Figure 6.5:

- Preparation of the Substrate:
  Before the ion plating process begins, the substrate to be coated undergoes thorough cleaning and preparation. This step is crucial to ensure good adhesion between the substrate and the deposited film. Common cleaning methods include ultrasonic cleaning, solvent degreasing, and surface etching.
- Creation of a Vacuum Environment:
  Ion plating is conducted in a vacuum chamber to eliminate interference from air molecules and ensure precise control over the deposition process. The chamber is evacuated to a high vacuum level using vacuum pumps.
- Introduction of the Target Material:
  The material to be deposited, known as the target material, is introduced into the vacuum chamber. The target material is typically in the form of a solid metal or alloy. It can also be a compound or ceramic material depending on the desired properties of the coating.
- Generation of Energetic Ions:
  In ion plating, energetic ions are generated within the vacuum chamber to facilitate the deposition process. These ions are typically produced by a glow discharge or plasma created using radio frequency (RF) or direct current

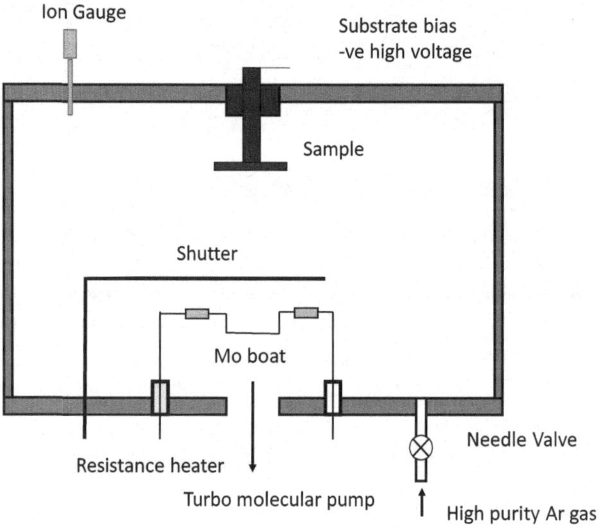

FIGURE 6.5 Schematic representation of ion plating process.

(DC) power sources. The discharge ionizes the gas molecules present in the chamber, creating a plasma of positively charged ions and free electrons.

- Acceleration of Ions Towards the Target:
  Once the plasma is established, an electric field is applied to accelerate the positively charged ions towards the target material. The ions gain kinetic energy as they travel through the plasma, colliding with gas atoms and acquiring positive charges.
- Sputtering of Target Material:
  When the accelerated ions collide with the surface of the target material, they transfer their kinetic energy to the target atoms through a process known as sputtering. This causes atoms from the target material to be ejected from its surface. These sputtered atoms travel in a straight line trajectory and deposit onto the substrate, forming a thin film. In addition to sputtering, the energetic ions in the plasma assist in the deposition process by enhancing adhesion and controlling the microstructure of the deposited film. The ions bombard the substrate surface, promoting diffusion and intermixing of atoms between the film and substrate. The ion plating process parameters, including the gas composition, pressure, temperature, applied voltage, and deposition rate, are carefully controlled to achieve the desired film properties such as thickness, composition, adhesion, and microstructure. As the deposition proceeds, a thin film of the target material gradually forms on the substrate surface. The film thickness and morphology are continuously monitored and controlled throughout the process. After deposition, the film properties are characterized using techniques such as scanning electron microscopy (SEM), X-ray diffraction (XRD), and surface profilometry.

6.5.4 Pulsed Laser Deposition (PLD)

Pulsed Laser Deposition (PLD) is a versatile thin-film deposition technique used in various fields such as materials science, electronics, optics, and nanotechnology. It involves the use of a high-energy pulsed laser to ablate material from a target, which then condenses onto a substrate to form a thin film, as shown in Figure 6.6. Below is a detailed explanation of the mechanism involved in the PLD process:

- Preparation of Target Material:
  PLD begins with the preparation of the target material, which can be a metal, ceramic, semiconductor, or even a complex compound. The target material is typically a high-quality single crystal or polycrystalline material with desired properties for the thin film to be deposited.
- Creation of a Vacuum Environment:
  The PLD process is conducted in a high vacuum chamber to prevent interference from air molecules and ensure the purity of the deposited film. The vacuum chamber is evacuated to a high vacuum level using vacuum pumps to remove any residual gases.

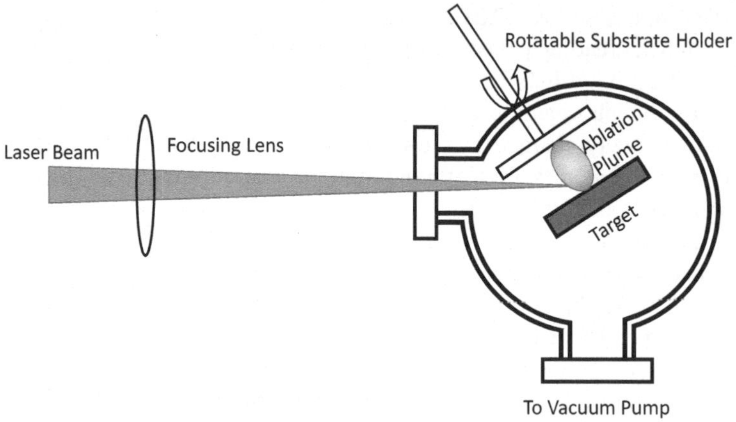

FIGURE 6.6 Schematic representation of pulsed laser deposition process.

- Placement of Substrate:
  A clean substrate, onto which the thin film will be deposited, is carefully mounted inside the vacuum chamber. The substrate material and surface preparation are chosen based on the desired properties of the thin film and its intended application.
- Focusing of Laser Beam:
  A high-energy pulsed laser beam, typically generated by a laser system such as an excimer laser or a Nd:YAG laser, is directed onto the surface of the target material. The laser beam is focused to a small spot size on the target surface to achieve high-energy density and efficient ablation.
- Ablation of Target Material:
  When the pulsed laser beam strikes the target material, it rapidly heats the surface to temperatures exceeding the material's vaporization threshold. This leads to a process called laser ablation, where material is ejected or vaporized from the target surface in the form of a plume or plasma. The ablated material forms a plume or plasma consisting of atoms, ions, clusters, and nanoparticles. The plume expands rapidly in all directions from the target surface due to the high-energy imparted by the laser pulse. The particles in the plume have high kinetic energy and travel towards the substrate.
- Deposition Onto Substrate:
  The particles in the plume impinge upon the substrate surface and condense to form a thin film. The substrate acts as a nucleation site for the deposition process, and the deposited material grows layer by layer to form a uniform film. The properties of the deposited thin film, such as thickness, composition, microstructure, and morphology, are controlled by adjusting various process parameters including laser energy, pulse duration, repetition rate, target-substrate distance, and substrate temperature. After deposition, the thin film may undergo post-deposition treatments such as annealing, surface modification, or film patterning to further enhance its properties or tailor it for specific applications.

6.5.5 Epitaxial Growth

Epitaxial growth is a crucial process in materials science and semiconductor technology, particularly for the production of high-quality crystalline films with well-defined atomic structures, as shown in Figure 6.7. Below is a detailed explanation of the mechanism involved in epitaxial growth:

- Selection of Substrate:
  Epitaxial growth begins with the selection of a suitable substrate material. The substrate serves as a template for the growth of the epitaxial film and provides a crystalline lattice structure on which the film will nucleate and grow. The choice of substrate material depends on factors such as lattice match, thermal expansion coefficient, and chemical compatibility with the epitaxial film material.
- Cleaning and Preparation of Substrate:
  Prior to epitaxial growth, the substrate surface must be thoroughly cleaned to remove any contaminants or oxide layers that may interfere with the epitaxial growth process. Common cleaning methods include solvent cleaning, acid etching, and in-situ treatments such as in-situ annealing or plasma cleaning.
- Introduction of Precursor Species:
  In epitaxial growth, precursor species are introduced into a deposition chamber where they react to form the desired epitaxial film. These precursor species can be in the form of molecular gases, metal-organic precursors, or solid sources such as evaporation boats or effusion cells. Upon reaching the substrate surface, the precursor species adsorb onto the substrate and diffuse across the surface. Surface diffusion allows the precursor species to explore the substrate surface and find suitable sites for nucleation and growth.
- Nucleation:
  Once a critical concentration of precursor species is reached on the substrate surface, nucleation occurs, leading to the formation of small clusters or islands of the epitaxial material. Nucleation can be homogeneous, where islands form randomly across the substrate surface, or heteroepitaxial, where islands nucleate preferentially at lattice-matched sites on the substrate. Following nucleation,

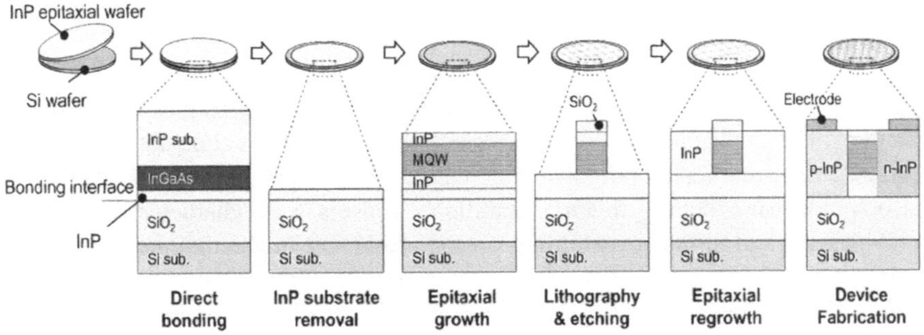

FIGURE 6.7 Steps involved in epitaxial growth process. (Fujii et al., 2021)

the epitaxial film grows layer by layer as additional precursor species are supplied to the substrate surface. The growth process is typically controlled by kinetic factors such as the arrival rate of precursor species, surface diffusion rates, and surface reaction rates. Epitaxial growth results in the alignment of the crystal lattice of the growing film with that of the substrate, leading to epitaxial relationship between the two materials. This alignment ensures that the atoms in the epitaxial film adopt the same crystallographic orientation as those in the substrate, resulting in high-quality crystalline films. The growth parameters, including substrate temperature, deposition rate, precursor flux, and growth atmosphere, are carefully controlled to optimize the epitaxial growth process and achieve the desired film properties such as thickness, composition, crystallinity, and surface morphology. The epitaxial growth is a sophisticated thin-film deposition technique that enables the fabrication of high-quality crystalline films with precise control over their properties. The process involves nucleation and layer-by-layer growth of the epitaxial film on a crystalline substrate, with careful control of growth parameters and in-situ monitoring to achieve desired film properties and performance characteristics. Epitaxial growth plays a critical role in the development of advanced semiconductor devices, optoelectronic devices, and materials for emerging technologies.

6.6 ADVANCED MANUFACTURING PROCESS

The advanced manufacturing processes for MEMS device fabrication, such as chemical milling/ chemical etching, electrochemical etching and thermal based etching, have been already discussed in Chapter 5.

6.7 MINIATURIZATION OF SENSORS AND TRANSDUCERS

In the realm of modern technology, miniaturization plays a pivotal role in enhancing the capabilities of electronic devices. One of the most significant areas of advancement is in sensors and transducers, which are fundamental components in various applications ranging from consumer electronics to industrial machinery, healthcare, and environmental monitoring. This section explores the evolution, challenges, and future prospects of miniaturization in sensors and transducers.

6.7.1 Overview

The journey of miniaturization in sensors and transducers is a testament to the relentless pursuit of innovation in the field of electronics. It finds its roots in the mid-20th century, a time marked by burgeoning demands for compact, portable devices across various industries. During this era, traditional sensors were characterized by their bulky designs and limited capabilities, posing challenges in meeting the increasingly sophisticated requirements of emerging applications. The turning point came with the advent of microelectronics and MEMS (Micro-Electro-Mechanical Systems) technology, heralding a new era of miniaturization. Microelectronics revolutionized the way electronic circuits were designed and fabricated, allowing for the integration of complex functionalities into smaller and smaller form factors. MEMS technology, on the other hand, brought forth a paradigm shift by enabling the fabrication of miniature mechanical and electromechanical devices on semiconductor substrates. These technological advancements paved the way for significant strides in shrinking sensor size while simultaneously enhancing performance metrics such as sensitivity, precision, and reliability. By leveraging the principles of MEMS fabrication, sensors and transducers could be miniaturized to dimensions previously thought unattainable, opening doors to a plethora of novel applications and functionalities.

The miniaturization of sensors and transducers brought about a paradigm shift across various industries. In the consumer electronics sector, it enabled the development of compact yet powerful devices such as smartphones, wearables, and IoT gadgets. These devices seamlessly integrate a myriad of miniaturized sensors, empowering users with unprecedented levels of interactivity, convenience, and functionality. Moreover, the industrial and automotive sectors witnessed a transformative impact, with miniaturized sensors playing pivotal roles in monitoring and control systems. From pressure sensors and accelerometers to temperature sensors and gyroscopes, these compact devices enable real-time data acquisition, facilitating predictive maintenance, optimizing operational efficiency, and ensuring safety across diverse applications.

In the realm of healthcare and biomedical engineering, miniaturized sensors have revolutionized diagnostics, monitoring, and treatment modalities. Implantable sensors enable continuous monitoring of physiological parameters, offering invaluable insights into patient health and enabling personalized interventions. Furthermore, miniaturized sensors find applications in medical imaging, prosthetics, and drug delivery systems, driving advancements in precision medicine and patient care. In essence, the journey of miniaturization in sensors and transducers epitomizes the relentless pursuit of technological advancement and innovation. From humble beginnings marked by bulky, cumbersome devices to the present-day landscape characterized by compact, high-performance sensors, this evolutionary trajectory underscores the transformative impact of miniaturization across diverse industries and applications.

6.7.2 Applications in Consumer Electronics

The miniaturization of sensors and transducers has sparked a revolution in the consumer electronics landscape, reshaping the way we interact with technology and enhancing the capabilities of devices across various categories. In particular, the integration of miniaturized sensors has ushered in a new era of innovation, enabling functionalities that were once considered futuristic. This section delves into the transformative impact of miniaturized sensors in consumer electronics, focusing on their applications in smartphones, wearable devices, and IoT gadgets. Smartphones stand as one of the most prominent examples of how miniaturized sensors have redefined user experience and expanded the functionalities of handheld devices. Within the compact confines of these devices lie an array of miniaturized sensors, each serving a distinct purpose and collectively augmenting the device's capabilities. Accelerometers, gyroscopes, magnetometers, and ambient light sensors are among the key components integrated into modern smartphones, enabling a plethora of features that enhance user interaction and facilitate novel applications.

Accelerometers, for instance, play a crucial role in enabling motion sensing capabilities within smartphones. These sensors detect changes in device orientation and movement, allowing for features such as screen rotation, gaming enhancements, and step tracking. By accurately capturing movements in three-dimensional space, accelerometers empower users to interact with their devices in intuitive ways, enhancing usability and enriching the overall user experience. Similarly, gyroscopes complement accelerometers by providing additional orientation data, enabling more precise motion tracking and gesture recognition. Together, these sensors form the backbone of various motion-controlled features found in modern smartphones, including gesture-based navigation, interactive gaming experiences, and augmented reality applications. By translating subtle movements into actionable commands, gyroscopes empower users to interact with their devices in immersive and intuitive ways, transcending traditional input methods.

Moreover, magnetometers play a vital role in enhancing the spatial awareness of smartphones, enabling features such as compass applications and location-based services. By detecting changes in magnetic fields, these sensors provide valuable orientation information, guiding users in navigation tasks and enriching location-aware applications. Whether exploring unfamiliar surroundings or navigating crowded urban environments, magnetometers empower users to navigate with confidence, enhancing the utility and versatility of smartphones as indispensable companions in daily life. Beyond motion sensing and spatial awareness, miniaturized sensors also enable environmental sensing capabilities within smartphones, enhancing user awareness and facilitating context-aware interactions. Ambient light sensors, for instance, adjust screen brightness based on surrounding lighting conditions, optimizing visibility and conserving battery life. By dynamically adapting to ambient lighting, smartphones ensure optimal viewing experiences across diverse environments, from bright outdoor settings to dimly lit interiors. Figure 6.8a shows touch sensor based on polyvinylidene fluoride (PVDF) on PED and paper substrates and Figure 6.8b shows the schematic representation of RTD sensor.

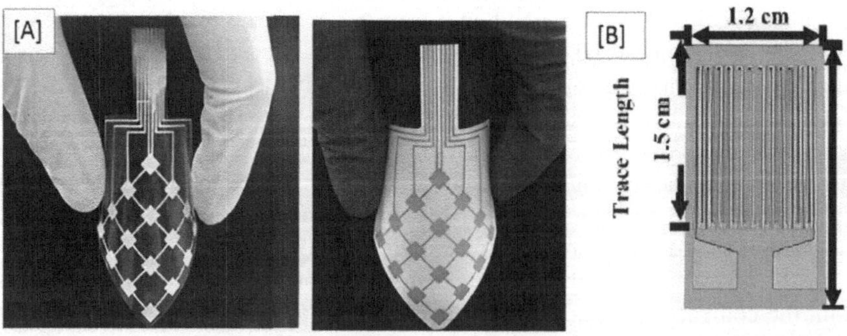

FIGURE 6.8 (a) Touch sensor based on polyvinylidene fluoride (PVDF) on PED and paper substrates, reprinted with permission from Emamian et al. (2017), Copyright © 2071, Elsevier. (b) Schematic representation of RTD sensor, from Turkani et al. (2019).

Furthermore, miniaturized sensors pave the way for innovative applications in health and wellness monitoring, transforming smartphones into personal health assistants capable of tracking various physiological parameters. Heart rate monitors, for example, leverage optical sensors to measure heart rate variability and monitor stress levels, providing insights into overall well-being and enabling proactive health management. Similarly, SpO_2 sensors enable non-invasive blood oxygen saturation measurements, offering valuable health insights and facilitating early detection of respiratory issues. Wearable devices represent another frontier where miniaturized sensors have catalyzed innovation, enabling seamless integration into everyday life and empowering users with personalized insights and actionable data. From smartwatches and fitness trackers to augmented reality glasses and smart clothing, wearables leverage miniaturized sensors to monitor biometric data, track physical activity, and deliver contextually relevant information in real-time. Figure 6.9 shows the sensing bandage made of flexible paper for detecting oxygen in treating chronic

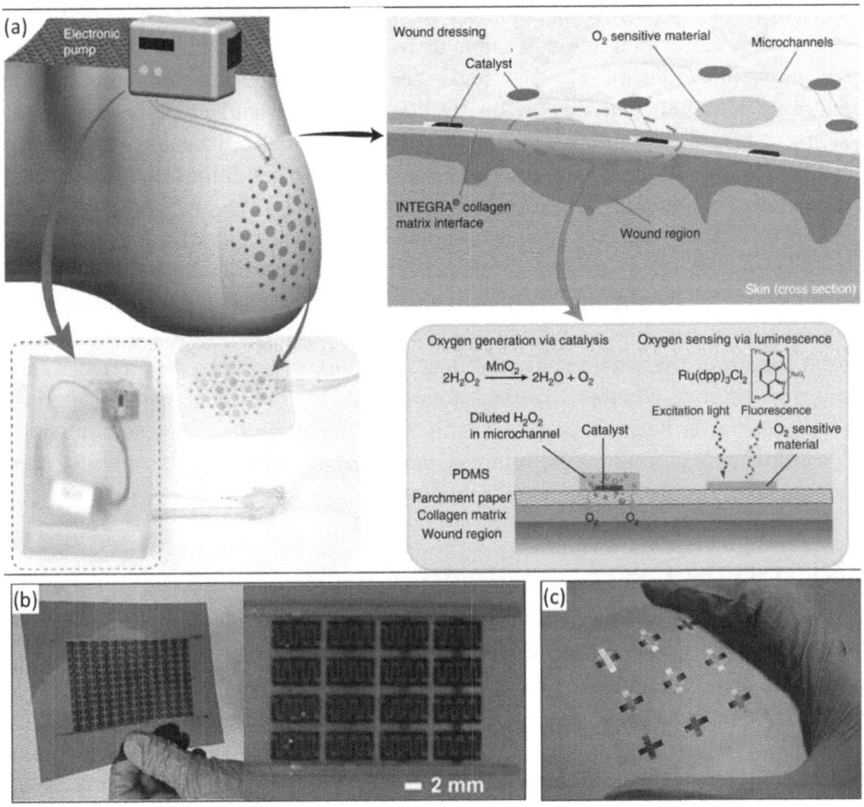

FIGURE 6.9 (a) Represents the bandage made up of flexible paper for sensing and delivery of oxygen in the treatment of chronic wounds from Ochoa et al. (2020), (b) Supercapacitors based on graphene used as electrodes, reprinted with permission from Li et al. (2017), Copyright © American Chemical Society, (c) Metal-insulator–Metal capacitor fabricated using inkjet printing from Mikolajek et al. (2019).

wounds, supercapacitors based on graphene used as electrodes, and metal-insulator—metal capacitor fabricated using inkjet printing.

Accelerometers and gyroscopes serve as foundational sensors in wearable devices, enabling activity tracking, gesture recognition, and motion-based interactions. By accurately capturing movements and gestures, these sensors facilitate gesture-based navigation, fitness tracking, and immersive gaming experiences, enhancing user engagement and promoting active lifestyles. Whether counting steps, monitoring workouts, or navigating virtual environments, wearables empower users to stay connected, informed, and motivated in their health and fitness journeys.

Additionally, physiological sensors such as heart rate monitors and electrodermal activity sensors enable continuous health monitoring, providing insights into vital signs, stress levels, and overall well-being. By leveraging biometric data, wearables offer personalized feedback and actionable insights, empowering users to make informed decisions about their health and lifestyle choices. Whether managing stress, improving sleep quality, or tracking fitness goals, wearables serve as invaluable companions in promoting holistic well-being and fostering healthier habits.

Furthermore, the integration of miniaturized sensors into wearables enables novel applications in healthcare, wellness, and assistive technologies. From remote patient monitoring and fall detection to medication adherence and emergency response systems, wearables leverage sensors to deliver personalized healthcare solutions and enhance quality of life for individuals with chronic conditions or special needs. By providing continuous monitoring and timely interventions, wearables empower users to proactively manage their health and maintain independence in their daily lives. Hence, the integration of miniaturized sensors has revolutionized the consumer electronics industry, driving innovation and unlocking new possibilities across diverse applications. From smartphones and wearable devices to IoT gadgets, these sensors empower users with enhanced functionalities, personalized experiences, and actionable insights. Whether enabling motion tracking, gesture recognition, environmental sensing, or health monitoring, miniaturized sensors have become indispensable components in shaping the future of consumer electronics, enriching lives, and redefining the way we interact with technology.

6.7.3 Industrial and Automotive Applications

In both industrial and automotive settings, the integration of miniaturized sensors and transducers has become indispensable for ensuring efficient operation, enhancing safety, and enabling advanced functionalities. These compact yet powerful devices are deployed across a wide array of applications, ranging from manufacturing plants and production lines to vehicles and transportation systems. Pressure sensors, temperature sensors, and flow sensors are among the key components that have undergone miniaturization to meet the stringent space constraints inherent in industrial machinery and automotive systems. These sensors are meticulously engineered to fit into compact spaces within complex assemblies, such as engines, turbines, hydraulic systems, and production equipment. Despite their diminutive size, these sensors deliver exceptional performance and reliability, enabling real-time monitoring of critical operational parameters.

Pressure sensors are integral to industrial and automotive applications, providing vital information about fluid pressure levels within hydraulic systems, pneumatic systems, and engines. By accurately measuring pressure variations, these sensors enable precise control of processes, ensuring optimal performance and preventing equipment failures. In industrial settings, pressure sensors play a crucial role in monitoring hydraulic presses, pneumatic actuators, and process pipelines, facilitating efficient production and minimizing downtime. In the automotive sector, pressure sensors are deployed in engine management systems, tire pressure monitoring systems (TPMS), and brake systems, enhancing vehicle safety, performance, and fuel efficiency. Figure 6.10 shows the schematic representation of MEMS pressure sensor. Temperature sensors are equally essential in industrial and automotive environments, enabling precise monitoring of temperature variations across critical components and systems. These sensors are deployed in engines, exhaust systems, cooling systems, batteries and manufacturing equipment to ensure optimal operating temperatures and prevent overheating or thermal damage. In industrial applications, temperature sensors enable precise temperature control in furnaces, ovens, and chemical reactors, optimizing energy efficiency and product quality. In the automotive sector, temperature sensors play a vital role in engine management, climate control systems, and thermal management of battery packs in electric vehicles, ensuring reliable performance and longevity. Figure 6.11 shows the flexible microsensor for detection of temperature inside lithium batteries.

Flow sensors are another crucial component in industrial and automotive applications, providing accurate measurements of fluid flow rates within pipelines, cooling systems, and hydraulic circuits. These sensors enable precise control of flow rates, ensuring efficient operation and preventing issues such as flow obstructions or leaks. In industrial settings, flow sensors are deployed in process control systems, water treatment plants, and HVAC (heating, ventilation, and air conditioning) systems, optimizing resource utilization and maintaining optimal operating conditions. In the automotive sector, flow sensors are utilized in fuel injection systems, engine cooling

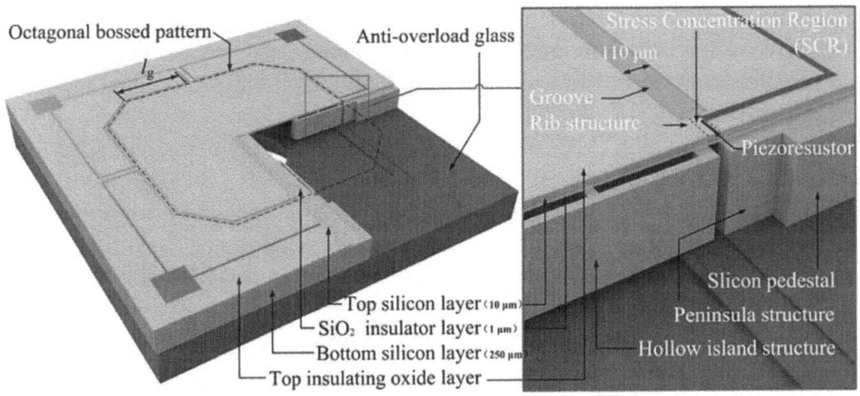

FIGURE 6.10 Schematic representation of MEMS pressure sensor. (Reprinted with permission from Zhao et al. (2016), Copyright © 2016 IOP Publishing Ltd.)

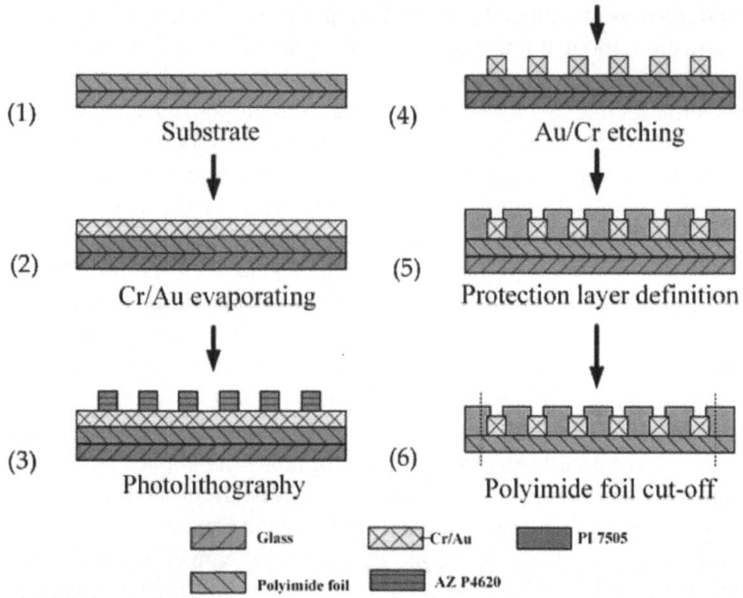

FIGURE 6.11 Three in one flexible microsensor for detection of temperature inside lithium batteries. (Lee et al., 2015)

systems, and air intake systems, enabling precise fuel delivery, thermal management, and combustion control, thereby enhancing vehicle performance, fuel efficiency, and emissions control.

The real-time monitoring capabilities enabled by miniaturized sensors and transducers play a pivotal role in facilitating predictive maintenance strategies in industrial and automotive environments. By continuously monitoring critical operational parameters such as pressure, temperature, and flow rates, these sensors provide early warnings of potential issues or anomalies, allowing maintenance teams to proactively address them before they escalate into costly failures or downtime. Predictive maintenance minimizes unplanned outages, reduces maintenance costs, and prolongs equipment lifespan, thereby improving overall operational efficiency and productivity.

Furthermore, miniaturized sensors contribute to the optimization of performance and safety in industrial and automotive applications. By providing accurate and timely data on operational parameters, these sensors enable real-time adjustments and optimizations to enhance efficiency, productivity, and safety. In industrial settings, sensors facilitate process optimization, quality control, and automation, improving production throughput, and product consistency. In the automotive sector, sensors enable advanced driver assistance systems (ADAS), vehicle stability control, and adaptive cruise control, enhancing driver safety and comfort while reducing the risk of accidents.

In summary, miniaturized sensors and transducers are indispensable components in industrial and automotive settings, enabling real-time monitoring, predictive maintenance, performance optimization, and safety enhancement. Pressure sensors, temperature sensors, and flow sensors, among others, play critical roles in ensuring efficient operation, preventing failures, and enhancing productivity across diverse applications. By leveraging the capabilities of these compact yet powerful devices, industries can achieve greater efficiency, reliability, and safety in their operations, driving innovation and progress in the industrial and automotive sectors. The integration of miniaturized sensors and transducers not only enables precise monitoring and control of critical parameters but also facilitates the implementation of predictive maintenance strategies, reducing downtime and maintenance costs. Moreover, these sensors contribute to the optimization of performance and safety, enhancing operational efficiency and minimizing risks in industrial and automotive environments. As technology continues to advance, miniaturized sensors are expected to play an increasingly pivotal role in shaping the future of industrial automation, smart manufacturing, and autonomous vehicles, ushering in a new era of efficiency, reliability, and safety in industrial and automotive applications.

6.7.4 Healthcare and Biomedical Applications

Healthcare and biomedical applications represent a frontier where the miniaturization of sensors and transducers has led to transformative advancements, revolutionizing patient care, diagnostics, and medical treatments. The integration of miniaturized sensors has ushered in a new era of precision medicine, enabling personalized interventions tailored to individual patient needs and enabling continuous remote health monitoring.

Implantable sensors stand at the forefront of biomedical innovation, offering unprecedented insights into physiological parameters and enabling real-time monitoring of vital signs without the need for invasive procedures. These miniature devices are designed to be implanted within the body, where they continuously monitor key biomarkers such as blood glucose levels, blood pressure, and neural activity. By providing continuous, accurate data, implantable sensors empower healthcare professionals to make informed decisions about patient care, enabling early detection of health issues and proactive interventions. For instance, implantable glucose sensors revolutionize diabetes management by providing continuous glucose monitoring, helping patients to better manage their condition and reducing the risk of complications.

Moreover, miniaturized sensors enable remote health monitoring, allowing patients to stay connected with their healthcare providers and receive timely interventions from the comfort of their homes. By wirelessly transmitting data to cloud-based platforms, these sensors enable continuous monitoring of vital signs and disease progression, facilitating early detection of health issues and timely interventions. Remote monitoring devices equipped with miniaturized sensors enable patients with chronic conditions to manage their health more effectively, reducing the need for frequent hospital visits and improving overall quality of life.

In addition to implantable sensors, miniaturized sensors are integrated into a wide range of diagnostic devices, prosthetics, and medical imaging equipment, enhancing accuracy and reducing invasiveness. Diagnostic devices, such as wearable ECG monitors, pulse oximeters, and portable ultrasound devices, leverage miniaturized sensors to provide rapid, non-invasive assessments of patient health. These devices enable healthcare professionals to diagnose and monitor a variety of medical conditions with greater accuracy and efficiency, leading to improved patient outcomes and reduced healthcare costs.

Furthermore, miniaturized sensors play a crucial role in the development of advanced prosthetics and medical implants, enabling greater functionality, comfort, and integration with the human body. Prosthetic limbs equipped with miniaturized sensors enable more natural movement and better control for amputees, enhancing their quality of life and independence. Similarly, medical implants such as pacemakers, defibrillators, and neurostimulators utilize miniaturized sensors to monitor patient health and deliver targeted therapies, improving treatment outcomes and reducing the need for invasive procedures.

Moreover, miniaturized sensors contribute to the advancement of medical imaging technologies, enabling higher resolution, faster imaging, and improved patient comfort. Miniaturized sensors integrated into MRI, CT, and PET scanners enhance image quality and reduce scan times, enabling more accurate diagnoses and treatment planning. Additionally, miniaturized sensors enable the development of portable, point-of-care imaging devices that can be used in remote or resource-limited settings, expanding access to essential healthcare services and improving patient outcomes.

In summary, miniaturized sensors and transducers have revolutionized healthcare and biomedical fields, enabling continuous remote health monitoring, personalized medicine, and advanced diagnostic and therapeutic interventions. Implantable sensors provide unprecedented insights into physiological parameters, enabling early detection of health issues and proactive interventions. Moreover, miniaturized sensors integrated into diagnostic devices, prosthetics, and medical imaging equipment enhance accuracy and reduce invasiveness, leading to improved patient outcomes and enhanced quality of life. As technology continues to advance, miniaturized sensors are poised to play an increasingly pivotal role in shaping the future of healthcare, driving innovation, and improving patient care.

6.7.5 Environmental Monitoring and Agriculture Applications

Environmental monitoring and agriculture represent two critical domains where the integration of miniaturized sensors and transducers has yielded profound benefits, enabling more efficient resource management, environmental protection, and sustainable agricultural practices.

In the realm of environmental monitoring, miniaturized sensors play a pivotal role in safeguarding air and water quality, detecting pollutants, greenhouse gases, and particulate matter in real-time. These compact sensors are deployed in monitoring stations, drones, and autonomous sensor networks, enabling continuous surveillance of environmental parameters across diverse landscapes. By providing timely data on air and water quality, miniaturized sensors empower policymakers, environmental agencies, and communities to make informed decisions about pollution control measures and resource management strategies. For instance, air quality sensors detect pollutants such as nitrogen dioxide, sulfur dioxide, and particulate matter, enabling authorities to implement targeted interventions to mitigate health risks and improve air quality. Similarly, water quality sensors monitor parameters such as pH, dissolved oxygen, and turbidity, helping to identify sources of contamination and safeguard aquatic ecosystems.

Moreover, miniaturized sensors facilitate the development of compact, portable devices for on-site environmental monitoring in remote or hard-to-reach areas. These devices enable researchers, conservationists, and environmentalists to conduct field studies and monitor environmental parameters in real-time, enhancing our understanding of ecosystem dynamics and facilitating timely interventions to protect biodiversity and natural resources. By leveraging the capabilities of miniaturized sensors, environmental monitoring efforts can be scaled up and extended to previously inaccessible areas, enabling more comprehensive and effective conservation strategies.

In the agricultural sector, miniaturized sensors are instrumental in optimizing resource management and enhancing crop productivity through precision agriculture techniques. Soil moisture sensors, for instance, provide real-time data on soil moisture levels, enabling farmers to precisely tailor irrigation schedules and optimize water usage. By avoiding overwatering and ensuring adequate soil moisture levels, farmers can conserve water resources, minimize water wastage, and mitigate the risk of soil erosion and nutrient leaching. Additionally, soil moisture sensors facilitate site-specific irrigation management, allowing farmers to apply water where it is needed most, thereby maximizing crop yields and reducing input costs.

Furthermore, miniaturized sensors enable precise monitoring of environmental conditions such as temperature, humidity, and light intensity, facilitating better crop management decisions and pest control strategies. For example, temperature sensors help farmers monitor microclimatic conditions within their fields, enabling early detection of frost events and heat stress in crops. Similarly, humidity sensors provide insights into atmospheric moisture levels, helping farmers assess disease risk and optimize fungicide applications. Additionally, light intensity sensors enable farmers to monitor light levels and optimize planting densities and row spacing for maximum sunlight exposure, promoting optimal crop growth and development.

Moreover, miniaturized sensors support the adoption of precision agriculture technologies such as variable rate fertilization and automated pest monitoring, enabling farmers to optimize fertilizer applications and minimize chemical inputs while reducing environmental impact. By integrating data from sensors, GPS technology, and remote sensing platforms, farmers can create detailed maps of soil fertility, crop health, and pest infestations, enabling targeted interventions and reducing reliance on conventional farming practices. This shift towards precision agriculture not only improves resource efficiency and environmental sustainability but also enhances crop quality, resilience, and profitability in the face of changing climate conditions and market pressures.

In summary, miniaturized sensors and transducers play a pivotal role in environmental monitoring and agriculture, enabling real-time monitoring of air and water quality, optimizing resource management, and enhancing crop productivity through precision agriculture techniques. By providing timely data on environmental parameters and enabling targeted interventions, these compact yet powerful devices contribute to sustainable development, environmental protection, and food security. As technology continues to advance, miniaturized sensors are expected to play an increasingly pivotal role in addressing global environmental challenges and ensuring the resilience and sustainability of agricultural systems in the face of growing population pressures and climate change.

REFERENCES

Emamian, S., Narakathu, B.B., Chlaihawi, A.A., Bazuin, B.J., Atashbar, M.Z., 2017. Screen Printing of Flexible Piezoelectric Based Device on Polyethylene Terephthalate (PET) and Paper for Touch and Force Sensing Applications. Sens. Actuators Phys. 263, 639–647. https://doi.org/10.1016/j.sna.2017.07.045

Fujii, T., Hiraki, T., Aihara, T., Nishi, H., Takeda, K., Sato, T., Kakitsuka, T., Tsuchizawa, T., Matsuo, S., 2021. Development of an Epitaxial Growth Technique Using III-V on a Si Platform for Heterogeneous Integration of Membrane Photonic Devices on Si. Appl. Sci. 11. https://doi.org/10.3390/app11041801

Hishimone, P.N., Nagai, H., Sato, M., 2020. Methods of Fabricating Thin Films for Energy Materials and Devices, in: Sato, M., Lu, L., Nagai, H. (Eds.), Lithium-Ion Batteries. IntechOpen, Rijeka. https://doi.org/10.5772/intechopen.85912

Judy, J.W., 2001. Microelectromechanical Systems (MEMS): Fabrication, Design and Applications. Smart Mater. Struct. 10, 1115. https://doi.org/10.1088/0964-1726/10/6/301

Lee, C.-Y., Peng, H.-C., Lee, S.-J., Hung, I.-M., Hsieh, C.-T., Chiou, C.-S., Chang, Y.-M., Huang, Y.-P., 2015. A Flexible Three-in-One Microsensor for Real-Time Monitoring of Internal Temperature, Voltage and Current of Lithium Batteries. Sensors 15, 11485–11498. https://doi.org/10.3390/s150511485

Li, J., Sollami Delekta, S., Zhang, P., Yang, S., Lohe, M.R., Zhuang, X., Feng, X., Östling, M., 2017. Scalable Fabrication and Integration of Graphene Microsupercapacitors through Full Inkjet Printing. ACS Nano 11, 8249–8256. https://doi.org/10.1021/acsnano.7b03354

Mikolajek, M., Reinheimer, T., Bohn, N., Kohler, C., Hoffmann, M.J., Binder, J.R., 2019. Fabrication and Characterization of Fully Inkjet Printed Capacitors Based on Ceramic/Polymer Composite Dielectrics on Flexible Substrates. Sci. Rep. 9, 13324. https://doi.org/10.1038/s41598-019-49639-3

Ochoa, M., Rahimi, R., Zhou, J., Jiang, H., Yoon, C.K., Maddipatla, D., Narakathu, B.B., Jain, V., Oscai, M.M., Morken, T.J., Oliveira, R.H., Campana, G.L., Cummings, O.W., Zieger, M.A., Sood, R., Atashbar, M.Z., Ziaie, B., 2020. Integrated Sensing and Delivery of Oxygen for Next-Generation Smart Wound Dressings. Microsyst. Nanoeng. 6, 46. https://doi.org/10.1038/s41378-020-0141-7

Turkani, V.S., Maddipatla, D., Narakathu, B.B., Altay, B.N., Fleming, P.D., Bazuin, B.J., Atashbar, M.Z., 2019. Nickel Based RTD Fabricated via Additive Screen Printing Process for Flexible Electronics. IEEE Access 7, 37518–37527.

Zhao, L., Xu, T., Hebibul, R., Jiang, Z., Ding, J., Peng, N., Guo, X., Xu, Y., Wang, H., Zhao, Y., 2016. A Bossed Diaphragm Piezoresistive Pressure Sensor with a Peninsula–Island Structure for the Ultra-Low-Pressure Range with High Sensitivity. Meas. Sci. Technol. 27, 124012. https://doi.org/10.1088/0957-0233/27/12/124012

7 Nanotribology

7.1 INTRODUCTION TO NANOTRIBOLOGY

Friction, an omnipresent force in our everyday lives, governs the motion of objects on various surfaces. From the squeak of a door hinge to the smooth glide of a vehicle on the road, friction influences almost every mechanical interaction we encounter. However, the study of friction extends far beyond macroscopic observations. At the nanoscale, where surfaces are consisting of atoms and molecules, a realm known as nanotribology emerges. Nanotribology delves into the intricate mechanisms and behaviors of friction, wear, and lubrication at nanometer scales, offering profound insights into fundamental physical phenomena and promising transformative applications in various fields.

The fascination with understanding friction at the nanoscale stems from the realization that surface interactions drastically differ from those at the macroscale. Traditional theories of friction, such as Amontons' laws and Coulomb's law, break down when applied to nanoscale systems. At this level, quantum effects, surface roughness, adhesion, and molecular dynamics play pivotal roles, challenging our conventional understanding of frictional behavior. Consequently, nanotribology emerges as a multidisciplinary field, drawing expertise from physics, chemistry, materials science, engineering, and nanotechnology to unravel the complexities of interfacial interactions.

One of the fundamental aspects of nanotribology is the exploration of frictional forces between individual atoms and molecules. Unlike macroscopic surfaces, where irregularities are measured in micrometers, nanoscale surfaces possess atomic-scale roughness. As two surfaces come into contact, adhesion forces dominate, leading to intricate patterns of frictional behavior. Theoretical models and advanced experimental techniques, such as atomic force microscopy (AFM) and scanning tunneling microscopy (STM), enable researchers to probe these interactions with unprecedented precision, providing valuable insights into the origins of friction at the atomic level.

Moreover, nanotribology encompasses the study of wear mechanisms at nanometer scales. Wear, the gradual removal of material from surfaces due to mechanical action, poses significant challenges in various industrial applications, including manufacturing, automotive, and biomedical sectors. Understanding the nanoscale processes underlying wear is crucial for designing durable materials and optimizing

DOI: 10.1201/9781003317272-7

tribological performance. Through experiments and simulations, researchers investigate phenomena such as abrasion, adhesion-induced wear, and surface fatigue, shedding light on the mechanisms driving material degradation and failure.

In addition to friction and wear, nanotribology encompasses the study of lubrication mechanisms at nanometer scales. Lubricants play a vital role in reducing friction and wear by forming protective layers between interacting surfaces. At the nanoscale, the behavior of lubricants becomes increasingly complex, influenced by factors such as confinement effects, surface chemistry, and molecular structure. Researchers explore various lubrication strategies, including boundary lubrication, where lubricant molecules adhere to surface asperities, and hydrodynamic lubrication, where fluid layers separate sliding surfaces. By unraveling the molecular mechanisms governing lubrication, nanotribologists aim to develop advanced lubricants with enhanced performance and durability.

Furthermore, nanotribology holds immense potential for driving innovation in nanotechnology and nanomanufacturing. As the demand for miniaturized devices and systems continues to grow, understanding and controlling friction at the nanoscale becomes paramount. Nanotribological principles guide the design and fabrication of nanoscale components with tailored frictional properties, enabling the development of novel nanomachines, sensors, and actuators. Moreover, advances in nanotribology contribute to the improvement of nanofabrication techniques, such as nanoimprint lithography and nanoassembly, by optimizing surface treatments and material selection for enhanced tribological performance.

Beyond technological applications, nanotribology also intersects with biological systems, offering insights into the frictional properties of biological interfaces. From the sliding motion of proteins within cells to the adhesion of biomolecules on surfaces, friction plays a crucial role in various biological processes. By applying nanotribological techniques, researchers can unravel the biomechanics of cellular interactions, leading to advancements in drug delivery, tissue engineering, and biomaterials design. Understanding the nanotribological behavior of biological systems holds promise for developing innovative medical therapies and diagnostics.

Thus, nanotribology represents a frontier in the study of friction, wear, and lubrication at the nanoscale. By exploring the intricacies of surface interactions at atomic and molecular levels, nanotribologists aim to unravel fundamental physical phenomena and drive technological innovation across diverse fields. From improving the durability of industrial machinery to advancing the frontiers of nanotechnology and biomedical engineering, the insights gleaned from nanotribology have the potential to revolutionize our understanding of frictional processes and shape the future of tribological science and engineering.

7.1.1 The Emergence and Significance of Micro/Nanotribology

In domains where surface interactions are prevalent, contact between surfaces often involves numerous asperities. Therefore, the recognition of investigating single asperity contacts in understanding the fundamental tribological and mechanical properties of surfaces has been a persistent pursuit. Recent progress in this field, driven by the widespread adoption of proximal probes, particularly tip-based microscopies like the scanning tunneling microscope and the atomic force microscope, coupled with computational techniques for simulating tip-surface interactions and interfacial properties, has heralded a new era of systematic exploration of interfacial challenges with unprecedented resolution. These advancements collectively gave rise to the emerging discipline known as microtribology, nanotribology, molecular tribology, or atomic-scale tribology (Bhushan, 1995; Zhang et al., 2022).

This discipline is dedicated to both experimental and theoretical investigations of processes spanning from atomic and molecular scales to microscales, encompassing phenomena encountered during adhesion, friction, wear, and thin-film lubrication at sliding interfaces. To distinguish between conventional macrotribology and micro/nanotribology, let's refer to Figure 7.1. In macrotribology, assessments are conducted on components with relatively substantial mass under heavily loaded conditions, where wear is an inevitable outcome and the bulk properties of the mating components predominantly influence tribological performance. Conversely, micro/nanotribology involves measurements on components, at least one of which possesses relatively small mass under lightly loaded conditions, resulting in minimal wear and a dominance of surface properties in governing tribological behavior.

Micro/nanotribological studies are indispensable for gaining a fundamental understanding of interfacial phenomena on a small scale. They are also crucial for examining interfacial processes involving ultrathin films, often as thin as 1–2 nanometers, and micro/nanostructures. These components find wide application in magnetic storage systems, micro/nanoelectromechanical systems (MEMS/NEMS), and various industrial sectors.

Due to their light mass (typically a few micrograms) and operation under light loads (ranging from smaller than one microgram to a few milligrams), the friction and wear operating at a nanoscale of lightly loaded micro/nanocomponents are critically dependent on surface interactions spanning only a few atomic layers. These structures are typically lubricated with molecularly thin films, making micro/nanotribological techniques ideally suited for scrutinizing friction and wear processes in the context of ultrathin films and micro/nanostructures. Notably, while micro/nanotribological studies are indispensable for investigating ultrathin films and micro/nanostructures, they also significantly contribute to a deeper understanding of interfacial phenomena in macrostructures, serving as a bridge between scientific inquiry and engineering applications (Bhushan et al., 2017). Key instruments in the field of micro/

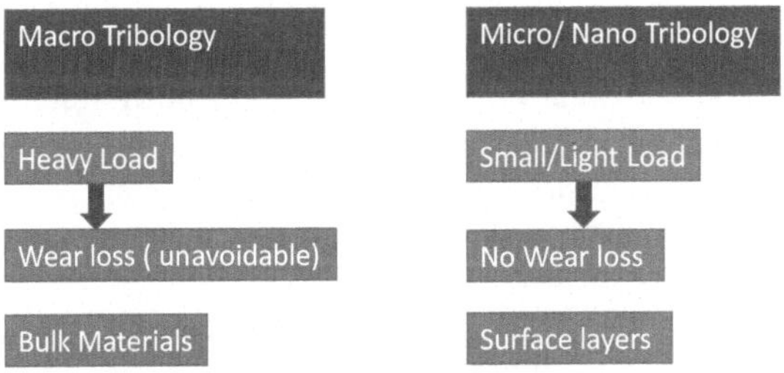

FIGURE 7.1 Differences between macrotribology and micro/nanotribology.

nanotribology include probe-based microscopes such as the scanning tunneling microscope (STM), atomic force microscope (AFM), and friction force microscope (FFM), as well as the surface force apparatus (SFA). The STM, originating from the innovative work of Binnig and Rohrer at the IBM Zurich Research Laboratory in 1981, was the first instrument capable of directly capturing three-dimensional (3D) images of solid surfaces with atomic-level precision. While STMs are suitable for studying electrically conductive surfaces, the AFM, stemming from the design of STM, emerged in 1985 and was engineered to measure ultrasmall forces (less than 1 μN) between the AFM tip and the sample surface. AFMs are versatile tools for assessing the topography of micro- to nanoscale surfaces and have been further modified into FFMs or lateral force microscopes (LFMs) for measuring friction on micro- and nanoscales. AFMs also find applications in the study of adhesion, scratching, wear, lubrication, surface temperatures, and mechanical properties such as indentation hardness and elasticity modulus (Bhushan, 2011a; Braun and Naumovets, 2006).

In addition to microscopy techniques, surface force apparatuses (SFAs), introduced in 1969, are instrumental in investigating the static and dynamic properties of molecularly thin liquid films confined between two molecularly smooth surfaces (Akbulut et al., 2005; Israelachvili et al., 2010; Ruths et al., 2004). However, SFAs necessitate the confinement of the liquid under examination between molecularly smooth surfaces, often with radii of curvature on the order of 1 mm, leading to slightly lower lateral resolution compared to AFMs. Notably, only AFMs/FFMs can be employed for the study of engineering surfaces under dry and wet conditions with atomic precision.

Parallel to experimental advancements, significant strides have been made in understanding the fundamental bonding and interactions within materials. Coupled with the evolution of computer-based modeling and simulation methods, these developments enable theoretical investigations of intricate interfacial phenomena with exceptional spatial and temporal resolution (Gusarov, 2024; Mao et al., 2023). Such simulations provide insights into atomic-scale energetics, structural dynamics, thermodynamics, transport mechanisms, and rheological aspects of tribological processes. Furthermore, these theoretical approaches serve as guides for interpreting experimental data, designing new experiments, and predicting novel phenomena grounded in atomistic principles.

7.2 FRICTION AND WEAR AT THE NANOSCALE

Friction and wear, two fundamental phenomena governing the interaction between surfaces, have intrigued scientists and engineers for centuries. From the early observations of Leonardo da Vinci to the modern advancements in nanotechnology, understanding friction and wear at the nanoscale has been a quest marked by significant discoveries and technological innovations. At the nanoscale, where surfaces exhibit unique properties and interactions, the traditional understanding of friction and wear undergoes profound transformations, opening avenues for new applications and challenges.

In this section, we delve into the intriguing realm of friction and wear at the nanoscale. We begin by defining the fundamental concepts of friction and wear, examining their significance in various fields, and elucidating their manifestations at the nanoscale. Subsequently, we delve into the mechanisms underlying nanoscale friction and wear, exploring phenomena such as adhesion, deformation, and surface interactions. Through a multidisciplinary lens, we investigate the role of materials science, physics, and engineering in unraveling the complexities of nanoscale friction and wear.

Furthermore, we discuss the technological implications of nanoscale friction and wear, ranging from the development of advanced lubricants to the design of nanomaterials with tailored tribological properties. Throughout this exploration, we highlight recent advancements, challenges, and future directions in the field, emphasizing the interdisciplinary nature of research in nanotribology.

7.2.1 Fundamental Concepts of Friction and Wear

Friction, defined as the resistance to motion between two surfaces in contact, is a ubiquitous phenomenon with far-reaching implications in everyday life and engineering. At the macroscopic scale, friction is governed by classical laws, such as Amontons' laws and Coulomb's law of friction, which describe the relationship between frictional force, normal force, and surface properties. However, at the nanoscale, these classical laws may no longer hold true, as surface interactions become increasingly dominated by atomic and molecular forces.

Wear, often described as the progressive loss of material from the surface of a solid body due to mechanical action, is intricately linked to friction. While wear manifests itself in various forms, including abrasion, adhesion, and fatigue, its underlying mechanisms are influenced by factors such as surface roughness, contact pressure, and environmental conditions. Understanding wear processes at the nanoscale is essential for improving the durability and performance of engineering components, ranging from micro-electro-mechanical systems (MEMS) to biomedical implants.

7.2.2 Manifestations of Friction and Wear at the Nanoscale

At the nanoscale, where surface-to-volume ratios are significantly amplified, friction and wear phenomena exhibit unique characteristics that defy conventional wisdom. One of the most prominent features of nanoscale friction is the transition from stick-slip motion to continuous sliding, driven by intricate interactions between surface asperities and adhesion forces. Moreover, nanoscale wear mechanisms, such as atomic-scale plowing and surface reconstruction, challenge our traditional understanding of material removal processes.

Surface chemistry and tribological properties play a crucial role in dictating nanoscale friction and wear behavior. For instance, the presence of contaminants, such as adsorbed layers or surface films, can significantly alter the frictional response of materials, leading to unexpected outcomes. Furthermore, nanomaterials, such as graphene and carbon nanotubes, exhibit remarkable tribological properties, including ultralow friction and exceptional wear resistance, making them promising candidates for various applications.

7.2.3 Mechanisms Underlying Nanoscale Friction and Wear

The mechanisms underlying nanoscale friction and wear are governed by a complex interplay of intermolecular forces, surface roughness, and contact mechanics. Adhesion, arising from Van der Waals, electrostatic, and capillary forces, plays a pivotal role in initiating frictional contact and promoting wear processes. At the same time, deformation mechanisms, such as plasticity and elastic recovery, dictate the response of materials to applied loads, influencing their tribological behavior.

Surface topography, characterized by features on the nanometer scale, profoundly affects friction and wear properties. Surface roughness, in particular, can enhance or diminish frictional forces, depending on the nature of the contact interface. Moreover, nanoscale wear mechanisms, including atomic-scale wear and tribochemical reactions, underscore the importance of understanding material behavior at the atomic level.

7.2.4 Technological Implications and Applications

The study of friction and wear at the nanoscale has significant implications for various technological applications, ranging from nanoelectromechanical systems (NEMS) to biomedical devices. By harnessing the unique properties of nanomaterials and nanotribological phenomena, researchers can develop advanced lubricants, coatings, and surface treatments with tailored friction and wear characteristics. For instance, the incorporation of nanoparticles into lubricating oils can improve their tribological performance, reducing friction and enhancing durability in mechanical systems.

Furthermore, the design of nanoscale interfaces and materials offers unprecedented opportunities for controlling friction and wear in diverse applications. Self-assembled monolayers, nanocomposites, and surface patterning techniques enable the engineering of surfaces with tailored tribological properties, catering to specific needs in automotive, aerospace, and manufacturing industries. Additionally, the development of biomimetic surfaces, inspired by nature's design principles, holds promise for creating friction-reducing surfaces with applications in medical implants and prosthetics.

7.2.5 Challenges and Future Directions

Despite significant advancements in nanotribology, numerous challenges persist in unraveling the complexities of friction and wear at the nanoscale. Characterizing nanoscale contacts with high spatial and temporal resolution remains a daunting task, requiring the integration of advanced experimental techniques, such as atomic force microscopy (AFM) and transmission electron microscopy (TEM). Moreover, understanding the influence of environmental factors, such as temperature, humidity, and surface chemistry, on nanoscale tribological behavior is essential for predicting material performance under real-world conditions.

Looking ahead, future research in nanotribology will focus on developing predictive models, multiscale simulations, and machine learning algorithms to elucidate the underlying mechanisms of friction and wear. By combining experimental insights with computational approaches, researchers aim to gain a comprehensive understanding of nanoscale tribological phenomena and pave the way for innovative solutions in materials science, engineering, and nanotechnology.

Thus, the study of friction and wear at the nanoscale represents a fascinating frontier in tribology, offering profound insights into surface interactions, material properties, and technological applications. From the exploration of fundamental concepts to the development of advanced materials and coatings, nanotribology continues to drive innovation across diverse fields. By addressing key challenges and embracing interdisciplinary approaches, researchers can unlock new opportunities for controlling friction and wear at the nanoscale, shaping the future of materials science and engineering.

7.3 MEASUREMENT TECHNIQUES IN NANOTRIBOLOGY: TOOLS FOR PROBING INTERFACIAL PHENOMENA

In the realm of nanotribology, where surface interactions occur at the nanometer scale, precise and accurate measurement techniques are indispensable for probing interfacial phenomena, understanding material behavior, and developing innovative solutions for friction and wear control. From atomic force microscopy (AFM) to surface force apparatus (SFA), a diverse array of experimental methods and instrumentation have been developed to investigate nanoscale friction, adhesion, and wear. In this section, we delve into the principles, applications, and advancements of measurement techniques in nanotribology, highlighting their crucial role in advancing our understanding of interfacial processes and guiding the development of tribological solutions.

7.3.1 Atomic Force Microscopy (AFM)

Atomic force microscopy (AFM) is a powerful and versatile tool that plays a fundamental role in the field of nanotribology, offering high-resolution imaging and precise force measurements at the nanometer scale. This technique has become a cornerstone in tribology research, providing valuable insights into the surface interactions and mechanical properties of materials. The principle behind AFM involves scanning a sharp probe tip across the surface of a sample while measuring the interaction forces between the tip and the sample. By monitoring the deflection of a cantilever-mounted tip, AFM can generate topographic images with atomic resolution and quantify mechanical properties such as adhesion, elasticity, and friction.

One of the key features of AFM is its ability to operate in various modes, each tailored to investigate different aspects of nanotribology. In contact mode AFM, the tip maintains constant contact with the sample surface, allowing for simultaneous topographic imaging and friction force measurement. This mode is particularly useful for studying nanoscale friction behavior, including stick-slip motion, lateral force modulation, and frictional anisotropy. Additionally, dynamic modes such as tapping mode AFM and non-contact mode AFM enable non-destructive imaging of delicate samples while probing surface interactions with minimal tip-sample interaction forces.

Recent advancements in AFM technology have further expanded its capabilities for nanotribological studies. High-speed AFM, for example, enables dynamic imaging of surface processes with temporal resolution down to milliseconds, unveiling transient phenomena such as wear debris formation and surface restructuring. This capability is invaluable for studying dynamic processes in tribological systems and understanding the mechanisms underlying friction and wear at the nanoscale. Figure 7.2 represents the schematic of atomic force microscope. Moreover, developments in multifunctional AFM probes have revolutionized nanotribology research. These probes are equipped with integrated force sensors and micro/nanomanipulators, allowing for simultaneous measurement of multiple tribological parameters. By integrating these functionalities into a single probe, researchers can obtain comprehensive insights into nanoscale friction and wear mechanisms, including adhesion, frictional forces, wear debris formation, and surface topography changes. This holistic approach to nanotribology enables researchers to gain a deeper understanding of the complex interplay between surface interactions and mechanical properties at the nanometer scale.

In addition to its applications in fundamental research, AFM has also found widespread use in various industrial sectors. For example, in the semiconductor industry, AFM is used for characterizing surface roughness, morphology, and defects on silicon wafers, which are critical for ensuring the quality and performance of electronic devices. In the biomedical field, AFM is employed for studying the mechanical properties of biological materials such as cells, tissues, and biomolecules, providing valuable insights into disease mechanisms and tissue engineering applications. Furthermore, AFM-based techniques have been applied in areas such as

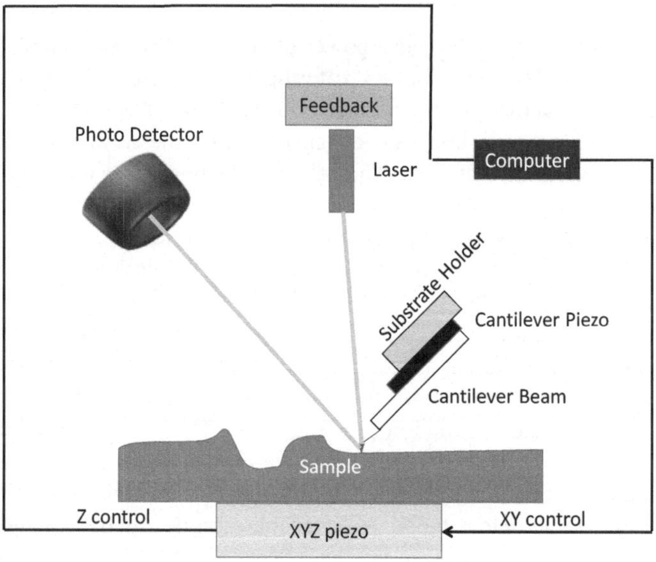

FIGURE 7.2 Schematic representation of tapping mode in AFM.

nanomanufacturing, materials science, and environmental monitoring, highlighting the versatility and importance of AFM in modern technology and research.

Looking ahead, the continued development of AFM technology is expected to further enhance its capabilities for nanotribological studies. Advances in probe design, imaging techniques, and data analysis algorithms will enable researchers to explore new frontiers in nanotribology, unraveling the mysteries of friction and wear at the atomic and molecular levels. By combining AFM with complementary techniques such as scanning electron microscopy (SEM), transmission electron microscopy (TEM), and X-ray photoelectron spectroscopy (XPS), researchers will be able to obtain a more comprehensive understanding of tribological phenomena and their impact on material performance and reliability.

Hence, atomic force microscopy is a versatile and powerful tool for investigating nanotribology, offering high-resolution imaging and precise force measurements at the nanometer scale. With its ability to operate in various modes and its recent advancements in technology, AFM continues to push the boundaries of nanotribology research, enabling researchers to gain deeper insights into the complex world of friction and wear at the nanoscale.

7.3.2 Scanning Tunneling Microscopy (STM)

Scanning tunneling microscopy (STM) has emerged as a powerful tool in the field of nanotribology, offering unparalleled capabilities for atomic-scale imaging, manipulation, and spectroscopy of surfaces. Unlike atomic force microscopy (AFM), which relies on mechanical interactions between the probe tip and the sample surface, STM operates based on the quantum mechanical phenomenon of tunneling current between the tip and the sample. This unique principle allows STM to provide atomic resolution images and detailed information about surface electronic properties and chemical reactivity, making it an invaluable tool for investigating nanoscale friction and wear phenomena.

7.3.2.1 Principles of Scanning Tunneling Microscopy (STM)

In STM, a sharp metal tip is brought into close proximity to the sample surface, typically within a few angstroms. A small bias voltage is applied between the tip and the sample, creating an electric field that facilitates the tunneling of electrons between them. The tunneling current, which exponentially depends on the tip-sample distance according to the quantum mechanical tunneling effect, is measured and used to generate images of the surface topography.

7.3.2.2 Surface Imaging With STM

One of the primary applications of STM is atomic-scale imaging of surfaces. By scanning the tip across the sample surface while maintaining a constant tunneling current, STM can generate topographic images with atomic resolution. The resulting images reveal surface features such as atomic steps, defects, and adsorbates, providing detailed insights into surface morphology and structure. STM has been instrumental in elucidating the atomic arrangements of various materials, including

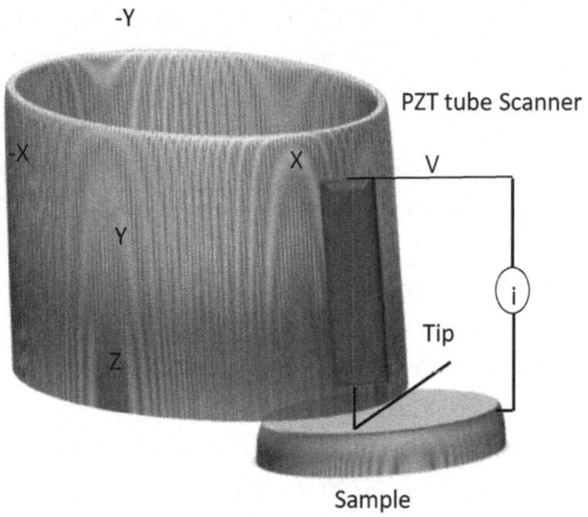

FIGURE 7.3 Schematic representation of scanning tunneling microscopes (STM).

metals, semiconductors, and insulators, and has played a crucial role in advancing our understanding of surface science. Figure 7.3 shows the schematic representation of scanning tunneling microscopes (STM).

7.3.2.3 Probing Surface Electronic Properties

In addition to surface imaging, STM can also probe surface electronic properties with high spatial resolution. By measuring the local density of states (LDOS) at the sample surface, STM can provide information about the electronic structure, band gaps, and charge distribution of materials. This capability enables researchers to investigate the electronic properties of nanostructures, quantum dots, and other nanomaterials, paving the way for the development of novel electronic devices and materials.

7.3.2.4 Scanning Tunneling Potentiometry (STP)

Scanning tunneling potentiometry (STP) is a technique based on STM that allows for the mapping of local surface potential variations. By measuring the voltage drop across the tunneling junction as the tip scans across the surface, STP can provide insights into surface charge distribution, work function variations, and electrochemical processes. Correlating surface potential maps with topographic images obtained from STM allows researchers to elucidate the relationship between surface chemistry, electronic structure, and friction behavior.

7.3.2.5 Friction Force Microscopy (FFM) With STM

Friction force microscopy (FFM) extends the capabilities of STM to directly measure lateral forces between the tip and the sample surface. By applying lateral force feedback control, FFM can maintain a constant lateral force between the tip and the sample, enabling the mapping of frictional properties with high spatial resolution. FFM has been instrumental in studying friction anisotropy, adhesion hysteresis, and nanotribological phenomena on various surfaces. By combining FFM with STM, researchers can investigate the correlation between surface topography, electronic properties, and friction behavior at the nanoscale.

7.3.2.6 Applications of STM in Nanotribology

The unique capabilities of STM have led to its widespread use in nanotribology research. By providing atomic-scale imaging and spectroscopic information, STM allows researchers to study the mechanisms of friction, wear, and lubrication at the nanoscale. STM has been applied to investigate friction phenomena on a wide range of surfaces, including metals, semiconductors, polymers, and biomaterials. Additionally, STM-based techniques such as STP and FFM have enabled researchers to explore the relationship between surface properties, electronic structure, and friction behavior, offering valuable insights into the design and optimization of nanomaterials for tribological applications.

7.3.2.7 Future Directions and Challenges

While STM has revolutionized our ability to probe surface properties and friction behavior at the nanoscale, several challenges remain to be addressed. Improving the stability, sensitivity, and resolution of STM instruments will be crucial for advancing our understanding of nanotribology. Additionally, integrating STM with complementary techniques such as AFM, transmission electron microscopy (TEM), and molecular dynamics simulations will enable researchers to gain deeper insights into the complex interplay between surface structure, electronic properties, and frictional behavior. By overcoming these challenges and continuing to innovate in the field of nanotribology, STM holds tremendous promise for addressing critical issues in materials science, nanotechnology, and engineering.

7.3.3 Surface Force Apparatus (SFA)

The surface force apparatus (SFA) represents a versatile tool for investigating interfacial interactions, including adhesion, lubrication, and deformation, at the nanometer scale. Based on the principle of measuring the force between two surfaces across a thin liquid or solid film, SFA provides quantitative data on the mechanical, chemical, and rheological properties of confined nanostructures.

In SFA, two opposing surfaces are mounted on separate force sensors and brought into close proximity within a liquid or solid medium. By monitoring the force-distance relationship between the surfaces, SFA can measure various interfacial forces, including Van der Waals forces, electrostatic forces, and steric repulsion. Moreover, SFA can characterize the tribological behavior of confined films, such as boundary lubrication, hydrodynamic lubrication, and squeeze-out phenomena, under controlled environmental conditions. SFA offers several operational modes tailored to different experimental requirements. In the normal force mode, the separation distance between the surfaces is controlled while measuring the normal force acting between them, providing insights into adhesion, deformation, and rupture processes. In the lateral force mode, lateral motion of one surface relative to the other is induced while measuring the lateral force between them, enabling the investigation of friction, shear, and wear phenomena.

Recent advancements in SFA technology have expanded its capabilities for nanotribological studies. Multi-axis SFA configurations allow for simultaneous measurement of normal and lateral forces in different directions, providing comprehensive insights into anisotropic friction behavior and surface interactions. Moreover, developments in high-pressure SFA systems enable the investigation of tribological phenomena under extreme conditions, such as high pressures and temperatures, relevant to real-world applications in engineering and materials science.

7.3.4 X-Ray Photoelectron Spectroscopy (XPS)

X-ray photoelectron spectroscopy (XPS), also known as electron spectroscopy for chemical analysis (ESCA), serves as a valuable tool for characterizing surface chemistry, composition, and bonding states, offering insights into the chemical mechanisms underlying nanotribological processes. Based on the principle of measuring the kinetic energy of photoelectrons emitted from a sample surface upon X-ray irradiation, XPS provides quantitative data on elemental composition, chemical state, and electronic structure.

In XPS, a monochromatic X-ray beam is directed onto the sample surface, causing the ejection of photoelectrons from the top few nanometers of the sample. By analyzing the energy distribution of emitted photoelectrons using a hemispherical analyzer, XPS can identify the elemental composition and chemical bonding configurations of surface species, as shown in Figure 7.4. Moreover, XPS can probe changes in surface chemistry induced by friction, wear, and surface treatments, providing insights into tribological mechanisms and material degradation processes. XPS offers several operational modes tailored to different experimental requirements. In survey mode, broad-spectrum XPS scans are performed to identify the elemental composition of the sample, enabling rapid screening of surface contaminants and impurities. In high-resolution mode, narrow-spectrum

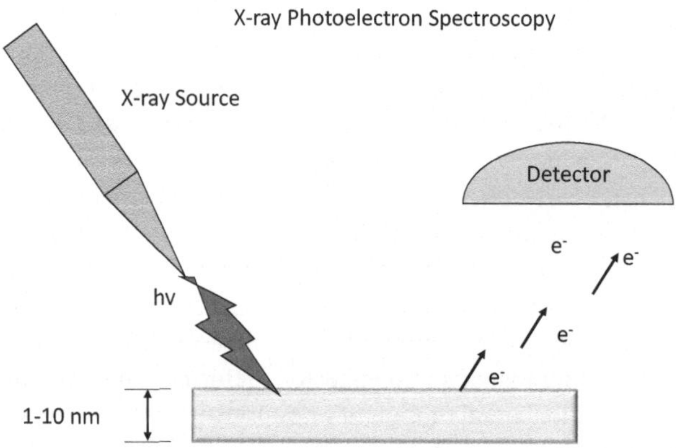

FIGURE 7.4 Schematic representation of XPS spectroscopy.

XPS scans are performed to resolve fine chemical shifts and splitting patterns, facilitating the identification of specific chemical bonds and functional groups. Moreover, depth-profiling techniques such as angle-resolved XPS (ARXPS) and sputter depth profiling, enable the characterization of compositional gradients and interface structures within multilayered samples.

Recent advancements in XPS technology have expanded its capabilities for nanotribological studies. Ultra-high vacuum (UHV) XPS systems offer enhanced sensitivity and spatial resolution, enabling the detection of trace surface species and subtle chemical changes with atomic-level precision. Moreover, developments in ambient-pressure XPS systems allow for in-situ investigation of surface reactions and dynamic processes under controlled environmental conditions, providing insights into tribological phenomena in realistic operating environments.

7.3.5 Nanoindentation

Nanoindentation serves as a powerful technique for probing mechanical properties, including hardness, modulus, and elastic/plastic deformation behavior, at the nanometer scale. Based on the principle of applying controlled load and displacement to a sharp indenter tip in contact with a sample surface, nanoindentation provides quantitative data on material response to localized deformation and stress, offering insights into nanoscale wear, fracture, and deformation mechanisms. In nanoindentation, a sharp indenter tip, typically made of diamond or other hard materials, is brought into contact with the sample surface under controlled loading conditions. By applying a precisely controlled load or displacement to the indenter tip, nanoindentation can induce localized deformation in the sample, generating a load-depth curve that reflects the material's mechanical properties. By analyzing the shape of the load-depth curve using established indentation models, nanoindentation can quantify parameters such as hardness, modulus, and elastic/plastic deformation behavior.

Nanoindentation offers several operational modes tailored to different experimental requirements. In continuous stiffness measurement (CSM) mode, a small oscillating force is superimposed on the constant loading force during indentation, enabling real-time monitoring of dynamic mechanical properties such as storage modulus, loss modulus, and damping behavior. In depth-sensing indentation (DSI) mode, the displacement of the indenter tip is controlled while measuring the load response, enabling the investigation of time-dependent mechanical properties such as creep, relaxation, and viscoelastic behavior.

Recent advancements in nanoindentation technology have expanded its capabilities for nanotribological studies. Multi-axis nanoindentation systems allow for simultaneous measurement of mechanical properties in different directions, providing insights into anisotropic material behavior and surface heterogeneity. Moreover, developments in in-situ nanoindentation techniques enable the investigation of mechanical properties under controlled environmental conditions, such as temperature, humidity, and gas atmosphere, relevant to real-world tribological applications.

Thus, measurement techniques play a pivotal role in nanotribology, enabling the precise characterization of interfacial phenomena, material properties, and mechanical behavior at the nanometer scale. From atomic force microscopy (AFM) to surface force apparatus (SFA), a diverse array of experimental methods and instrumentation offer unique capabilities for investigating nanoscale friction, adhesion, and wear. By leveraging these measurement techniques, researchers can gain fundamental insights into tribological mechanisms, guide the development of advanced materials and coatings, and pave the way for innovative solutions in engineering, materials science, and nanotechnology.

7.4 NANOSCALE INDENTATION

Nanoscale indentation is a powerful technique used to characterize the mechanical properties of materials at the nanometer scale. It involves applying controlled loads or displacements to a small region of a sample surface using a sharp indenter tip and measuring the resulting mechanical response. This method provides valuable insights into material behavior, including hardness, modulus, elastic/plastic deformation, and fracture resistance, which are essential for understanding the mechanical properties of nanomaterials and nanostructures. In this comprehensive exploration, we delve into the principles, applications, instrumentation, data analysis, and recent advancements in nanoscale indentation, highlighting its significance in materials science, nanotechnology, and engineering.

7.4.1 Principles of Nanoscale Indentation

Nanoscale indentation is based on the principle of applying controlled mechanical loads or displacements to a small area of a material surface using a sharp indenter tip while monitoring the resulting deformation response. The mechanical properties of the material, such as hardness, modulus, and elastic/plastic behavior, are then inferred from the indentation data.

The process typically involves several key steps:

Sample Preparation: The sample surface is carefully prepared to ensure flatness, cleanliness, and uniformity. Surface roughness and contaminants can affect the accuracy of indentation measurements.

Indenter Selection: A suitable indenter tip, often made of diamond or other hard materials, is chosen based on the desired measurement parameters and sample properties. The shape, size, and geometry of the indenter tip influence the indentation behavior and data interpretation.

Loading: The indenter tip is brought into contact with the sample surface, and a controlled load or displacement is applied. The loading rate and maximum load are selected based on the material properties and measurement objectives.

Deformation Monitoring: The deformation response of the sample, typically expressed as indentation depth or load, is continuously monitored throughout the indentation process. Various sensing techniques, such as force transducers or displacement sensors, are used to measure the applied load and indentation depth with high precision.

Unloading: After reaching the desired maximum load or displacement, the load is gradually removed, and the indentation depth is recorded as a function of time

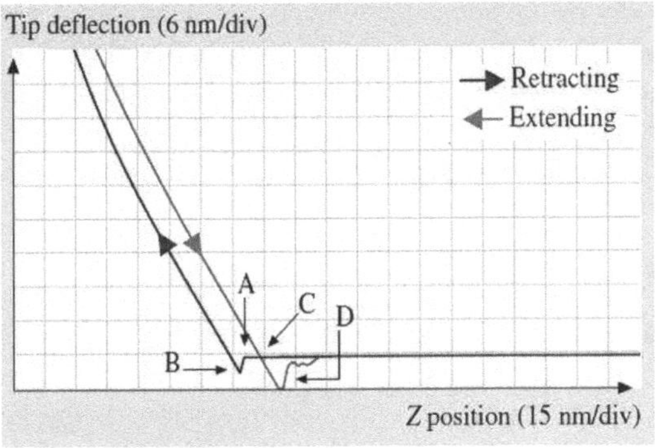

FIGURE 7.5 Schematic representation of AFM tip deflection for a polymeric magnetic tape. (Reprinted with permission from Bhushan (2011b). Copyright © 2011, Springer.)

or load. This unloading curve provides valuable information about the material's elastic recovery and plastic deformation behavior.

Data Analysis: The indentation data, including load-depth curves and indentation images, are analyzed to extract mechanical properties such as hardness, modulus, and elastic/plastic behavior, as shown in Figure 7.5. Various mathematical models, such as the Oliver-Pharr method and Hertzian contact theory, are commonly used to interpret the indentation data and calculate material properties.

7.4.2 Applications of Nanoscale Indentation

Nanoscale indentation finds widespread applications in materials science, nanotechnology, and engineering due to its ability to characterize mechanical properties at the nanometer scale. Some key applications include:

Material Characterization: Nanoscale indentation is used to measure fundamental mechanical properties such as hardness, modulus, and elastic/plastic behavior of materials. This information is crucial for understanding material performance, designing new materials, and optimizing manufacturing processes.

Thin Film Analysis: Nanoscale indentation is employed to characterize the mechanical properties of thin films and coatings, including adhesion strength, fracture resistance, and residual stress. This is essential for evaluating the quality and reliability of thin film-based devices and structures in microelectronics, optics, and coatings industries.

Biomaterials and Biomechanics: Nanoscale indentation is used to study the mechanical properties of biological materials such as cells, tissues, and biomolecules. This includes measuring the stiffness of cells, the elasticity of tissues, and the adhesion strength of biomaterials, which are essential for understanding biological processes, disease mechanisms, and tissue engineering applications.

Nanomaterials and Nanocomposites: Nanoscale indentation is employed to characterize the mechanical properties of nanomaterials such as nanoparticles, nanotubes, and nanowires, as well as nanocomposite materials. This includes measuring the strength, stiffness, and fracture toughness of nanomaterials, which are critical for developing advanced nanomaterial-based devices and structures.

7.4.3 Instrumentation for Nanoscale Indentation

Nanoscale indentation requires specialized instrumentation capable of applying controlled loads or displacements to a small area of a sample surface while measuring the resulting deformation response with high precision. Some key components of nanoscale indentation instrumentation include:

Indenter System: The indenter system consists of a sharp indenter tip, often made of diamond or other hard materials, mounted on a precision positioning mechanism. The indenter tip shape, size, and geometry can vary depending on the measurement requirements and sample properties.

Load and Displacement Control: The indentation system includes mechanisms for applying controlled loads or displacements to the sample surface. This typically involves a load cell or force transducer for measuring the applied load and a piezoelectric actuator or nanopositioning stage for controlling the displacement of the indenter tip.

Sensing and Measurement: The indentation system is equipped with sensors and measurement devices for monitoring the deformation response of the sample during indentation. This may include displacement sensors, force transducers, piezoelectric actuators, and optical or laser-based detection systems for measuring indentation depth, applied load, and other relevant parameters.

Environmental Control: Some advanced indentation systems incorporate environmental control features such as temperature and humidity control chambers, vacuum chambers, and inert gas environments. This enables characterization of mechanical properties under specific environmental conditions relevant to real-world applications.

Data Acquisition and Analysis: The indentation system is coupled with data acquisition and analysis software for recording and analyzing indentation data in real-time. This includes processing load-depth curves, calculating mechanical properties, and generating graphical representations of the indentation results.

7.4.4 Recent Advancements in Nanoscale Indentation

Recent advancements in nanoscale indentation have focused on enhancing measurement accuracy, expanding measurement capabilities, and improving instrumentation performance. Some notable advancements include:

High-Resolution Imaging: Advanced indentation systems incorporate high-resolution imaging techniques such as atomic force microscopy (AFM) or scanning electron microscopy (SEM) to visualize indentation events with nanometer-scale resolution. This enables precise positioning of the indenter tip, accurate measurement of indentation depth, and characterization of surface topography before and after indentation.

In-Situ Characterization: Newer indentation systems offer in-situ characterization capabilities, allowing real-time observation of mechanical processes such as elastic deformation, plastic deformation, and fracture. This enables researchers to study dynamic mechanical behavior and transient phenomena during indentation, providing valuable insights into material response under varying loading conditions.

Multi-Parameter Measurement: Advanced indentation techniques enable simultaneous measurement of multiple mechanical parameters, including hardness, modulus, adhesion strength, and frictional behavior. This multi-parameter approach provides a comprehensive characterization of material properties and surface interactions, enhancing the understanding of complex mechanical phenomena at the nanometer scale.

Environmental Control: Recent advancements in nanoscale indentation instrumentation include improved environmental control features, such as temperature stability, humidity control, and gas atmosphere regulation. This enables researchers to study mechanical properties under controlled environmental conditions relevant to specific applications, such as microelectronics, biomedicine, and environmental sensing.

Nanoscale indentation is a versatile and powerful technique for characterizing the mechanical properties of materials at the nanometer scale. By applying controlled loads or displacements to a small area of a sample surface using a sharp indenter tip, nanoscale indentation provides valuable insights into material behavior, including hardness, modulus, elastic/plastic deformation, and fracture resistance. This technique finds widespread applications in materials science, nanotechnology, and engineering, enabling researchers to understand the mechanical properties of nanomaterials, thin films, biomaterials, and nanocomposites. Recent advancements in nanoscale indentation instrumentation have further expanded its capabilities, offering enhanced measurement accuracy, multi-parameter characterization, and in-situ observation of mechanical processes. As nanotechnology continues to advance, nanoscale indentation will remain a critical tool for characterizing and optimizing the mechanical properties of nanomaterials and nanostructures for various applications in science and technology.

7.5 NANOSCALE LUBRICATION

Nanoscale lubrication refers to the study and application of lubrication principles at the nanometer scale, where surfaces interact with extremely high precision. Lubrication plays a crucial role in minimizing friction and wear between moving surfaces, thereby enhancing the efficiency, durability, and performance of mechanical systems and devices. At the nanoscale, understanding and controlling lubrication phenomena become even more critical due to the increased influence of surface forces and the intricate interactions between lubricants and solid surfaces.

7.5.1 Importance of Nanoscale Lubrication

Nanoscale lubrication is of paramount importance across various industries and applications for the following reasons:

Friction Reduction: Proper lubrication reduces friction between surfaces, leading to energy savings, improved efficiency, and reduced wear and tear on mechanical components.

Wear Protection: Lubricants form a protective layer between moving surfaces, preventing direct contact and minimizing wear, thereby extending the lifespan of components and machinery.

Surface Protection: Lubricants can protect surfaces from corrosion, oxidation, and chemical reactions, preserving the integrity and functionality of materials in diverse environments.

Enhanced Performance: Optimized lubrication at the nanoscale results in smoother operation, increased reliability, and enhanced performance of mechanical systems, contributing to overall productivity and cost-effectiveness.

7.5.2 Fundamentals of Nanoscale Lubrication

Surface Forces: Nanoscale lubrication involves understanding various surface forces, including Van der Waals forces, electrostatic interactions, and capillary forces, which influence lubricant behavior and interactions between surfaces.

Boundary Lubrication: Lubricants at the nanoscale form a boundary layer between surfaces, preventing direct contact and reducing friction. This boundary lubrication is essential for minimizing wear and maintaining smooth operation.

Surface Roughness: Surface roughness plays a crucial role in nanotribology, affecting lubricant film formation, adhesion, and friction behavior. Understanding and controlling surface roughness are key to optimizing lubrication performance.

7.5.3 Nanolubricants and Lubricant Additives

Types of Nanolubricants: Nanolubricants come in various forms, including solid nanoparticles, nanofluids, and graphene-based materials. Each type offers unique properties and advantages for specific applications.

Synthesis Methods: Nanolubricants are synthesized using bottom-up or top-down approaches, involving techniques such as chemical vapor deposition, sol-gel synthesis, and mechanical milling. Synthesis methods influence the properties and performance of nanolubricants.

Performance Enhancement: Nanolubricants enhance friction reduction, wear resistance, and load-carrying capacity compared to traditional lubricants. They can improve the performance of mechanical systems in diverse applications.

7.5.4 Characterization Techniques for Nanoscale Lubrication

Atomic Force Microscopy (AFM): AFM is a powerful tool for studying surface topography, friction behavior, and lubricant film properties at the nanoscale. It provides high-resolution imaging and precise force measurements, enabling detailed analysis of lubricant-surface interactions.

Scanning Tunneling Microscopy (STM): STM offers atomic-scale imaging and spectroscopy capabilities, allowing researchers to study surface topography, electronic properties, and chemical reactivity. STM techniques provide insights into nanoscale friction phenomena.

Surface Force Apparatus (SFA): SFA is used to probe intermolecular forces and study lubricant behavior in confined geometries at the nanoscale. It enables researchers to investigate lubricant film formation and mechanical properties under controlled conditions.

7.5.5 Molecular Dynamics Simulations in Nanolubrication

Overview of Molecular Dynamics (MD) Simulations: MD simulations model lubricant behavior at the molecular level, providing insights into lubricant-surface interactions, film formation, and friction mechanisms.

Lubricant Film Formation: MD simulations elucidate the processes involved in lubricant film formation, including molecular adsorption, desorption, and diffusion at solid-liquid interfaces. They help understand the factors influencing lubricant film stability and effectiveness.

Boundary Lubrication Mechanisms: MD simulations reveal the molecular mechanisms underlying boundary lubrication, such as molecular layering, shear-induced ordering, and load-bearing capacity of lubricant films. They provide valuable information for designing effective lubricants and optimizing lubrication performance.

7.5.6 Applications of Nanoscale Lubrication

Automotive Industry: Nanolubricants improve engine efficiency, reduce wear, and enhance fuel economy in automotive applications, leading to better performance and reduced environmental impact.

Aerospace Industry: Nanolubricants are used in aircraft engines, space exploration vehicles, and satellite mechanisms to improve reliability, durability, and performance under extreme conditions.

Manufacturing Processes: Nanolubrication techniques improve precision machining, metal forming, and microfabrication processes, enhancing surface finish, tool life, and dimensional accuracy in manufacturing.

Biomedical Engineering: Nanolubricants minimize wear, friction, and tissue damage in medical implants and prosthetics, improving the performance and longevity of biomedical devices.

7.5.7 Challenges and Future Directions in Nanoscale Lubrication

Scalability and Durability: Challenges include scaling up nanolubrication techniques for practical applications and ensuring long-term stability and durability under operating conditions.

Environmental Sustainability: Future research focuses on developing environmentally friendly lubricants and lubrication techniques to minimize environmental impact and promote sustainability.

Emerging Trends and Innovations: Emerging trends include self-healing lubricants, bio-inspired lubrication strategies, and stimuli-responsive materials, offering novel solutions for nanoscale lubrication challenges.

Future Research Directions: Future research aims to address current challenges and advance the field of nanoscale lubrication through interdisciplinary collaboration, innovation, and the development of new materials and technologies.

Thus, nanoscale lubrication is a multidisciplinary field that plays a crucial role in various industries, offering opportunities for improving efficiency, reliability, and sustainability of mechanical systems and devices. Continued research and innovation in nanoscale lubrication are essential for addressing current challenges and unlocking new opportunities for technological advancement.

7.6 HYDROPHILIC AND HYDROPHOBIC SURFACES

7.6.1 Introduction to Hydrophilic and Hydrophobic Surfaces

Hydrophilic and hydrophobic surfaces represent two fundamental categories of surface properties that dictate how materials interact with water. Understanding the behavior of these surfaces is crucial in various fields, including materials science, chemistry, biology, and engineering. In this comprehensive exploration, we delve into the definitions, properties, characteristics, applications, and underlying principles of hydrophilic and hydrophobic surfaces.

7.6.2 Definitions and Characteristics

Hydrophilic Surfaces

Hydrophilic surfaces are those that have an affinity for water. The term "hydrophilic" originates from the Greek words "hydro" (water) and "philos" (loving). On hydrophilic surfaces, water molecules exhibit high surface tension, causing them to spread out and form a thin, uniform film rather than bead up. Hydrophilic surfaces are characterized by their ability to readily absorb or adsorb water molecules, allowing them to maintain a wetted state. Examples of hydrophilic materials include metals, metal oxides, certain polymers, and surfaces treated with hydrophilic coatings.

Hydrophobic Surfaces

Conversely, hydrophobic surfaces repel water and have an aversion to water molecules. The term "hydrophobic" comes from the Greek words "hydro" (water) and "phobos" (fear). On hydrophobic surfaces, water droplets exhibit low surface tension, causing them to bead up and form spherical shapes, minimizing contact with the surface. Hydrophobic surfaces are characterized by their ability to repel water and resist wetting. Common examples of hydrophobic materials include non-polar substances such as oils, waxes, fluoropolymers, and certain types of plastics.

7.6.3 Surface Energy and Wetting Behavior

Surface energy plays a central role in determining the wetting behavior of surfaces. Surface energy is a measure of the energy required to create a unit area of new surface and is influenced by the chemical composition and structure of the surface. Hydrophilic surfaces typically have high surface energy due to the presence of polar functional groups that interact favorably with water molecules. In contrast, hydrophobic surfaces have low surface energy, primarily consisting of non-polar molecules that repel water.

Wetting behavior refers to how a liquid spreads or beads up on a solid surface. It is governed by the balance between adhesive forces, which attract the liquid to the surface, and cohesive forces, which hold the liquid molecules together. On hydrophilic surfaces, adhesive forces dominate, leading to wetting and spreading of the liquid. On hydrophobic surfaces, cohesive forces are stronger, causing the liquid to bead up and minimize contact with the surface.

7.6.4 Mechanisms of Wetting and Contact Angle

The wetting behavior of a surface is quantitatively described by the contact angle, which measures the angle formed between the tangent to the liquid surface and the solid surface at the point of contact. A contact angle of 0° indicates complete wetting, while a contact angle of 180° indicates complete non wetting. Hydrophilic surfaces typically exhibit low contact angles (< 90°), indicating good wetting, while hydrophobic surfaces have high contact angles (> 90°), indicating poor wetting, as shown in Figure 7.6.

The contact angle is determined by the interplay of surface tension forces at the solid-liquid interface. Figure 7.7 illustrates water droplets on surfaces with varying

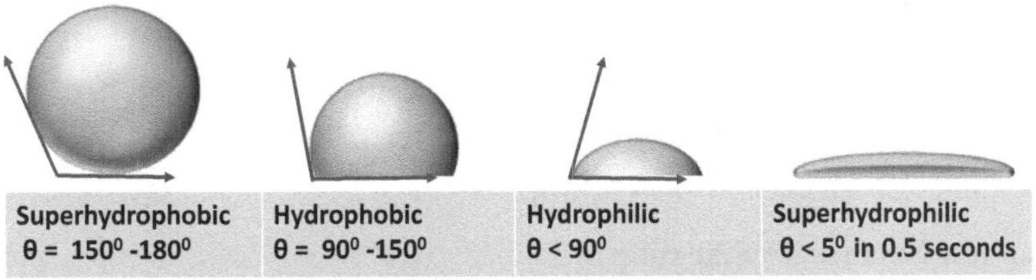

FIGURE 7.6 Schematic representation of different type of water droplet interactions (super hydrophobic, hydrophobic superhydrophilic, and hydrophilic surfaces).

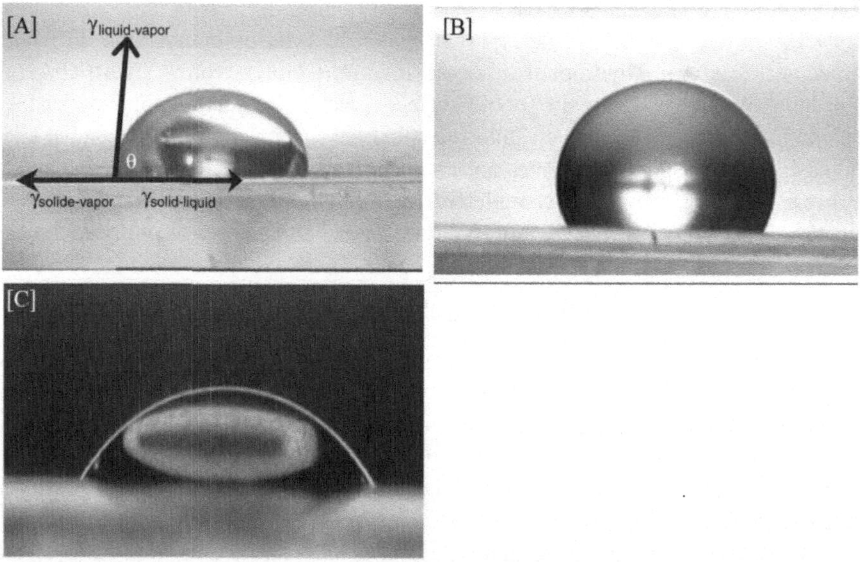

FIGURE 7.7 (a) Schematic representation of water droplet interaction (contact angle), (b) Water droplet interaction (contact angle) with Teflon surfaces, (c) contact angle (θ) with glass substrate. (Reprinted with permission from Elkhyat et al. (2004). Copyright ©2005 John Wiley & Sons.)

characteristics. In Figure 7.7a, the intersection between the liquid-solid and liquid-vapor phases is depicted. As the water droplet contact angle increases, indicating a larger angle, the material's hydrophobicity intensifies, as shown in Figure 7.7b. Conversely, a decreased contact angle results in a less hydrophobic or more hydrophilic material, as seen in Figure 7.7c (Elkhyat et al., 2004).

Young's equation relates the contact angle (θ) to the surface tensions of the solid (γ_s), liquid (γ_l), and vapor (γ_v) phases:

$$\gamma sv = \gamma sl + \gamma lv \cos(\theta) \tag{7.1}$$

Where,

γsv is the solid-vapor surface tension.
γsl is the solid-liquid surface tension.
γlv is the liquid-vapor surface tension.
θ is the contact angle.

7.6.5 Surface Modification Techniques

Hydrophilic Surface Modification

Chemical Functionalization: Hydrophilic functional groups such as hydroxyl (-OH), carboxyl (-COOH), and amine ($-NH_2$) groups can be introduced onto the surface through chemical reactions, enhancing its affinity for water.

Plasma Treatment: Plasma treatment introduces polar functional groups onto the surface by bombarding it with ionized gas, increasing its surface energy and hydrophilicity.

Surface Coatings: Hydrophilic coatings containing polar molecules or polymers can be applied to the surface to enhance its hydrophilicity and promote wetting.

Hydrophobic Surface Modification

Surface Coatings: Hydrophobic coatings composed of low-surface-energy materials such as fluoropolymers, silanes, or organosiloxanes can be applied to the surface to reduce its surface energy and promote water repellency.

Surface Roughening: Introducing micro or nanoscale roughness to the surface can increase its apparent contact angle by minimizing the contact area between the liquid and the surface, enhancing its hydrophobicity. Laser processing has emerged as a promising fabrication technique due to several key advantages: (I) precise control of surface roughness at the nano to microscale, (II) single-step processing under suitable conditions, and (III) compatibility with various materials (Baldacchini et al., 2006). Laser treatment can enhance the static water droplet contact angle by increasing surface roughness, typically followed by coating the material with a fluorine-based substance to create a superhydrophobic (SHO) surface. This combination of the fluorine-based material and surface roughness results in highly hydrophobic properties (Chun et al., 2016, 2014). Figure 7.8 represents the distinct SEM images illustrating laser-textured surfaces subjected to varying laser powers. Figure 7.8a exhibits the lowest laser power, which yields a less distinct grid pattern compared to Figures 7.8b, c, and d.

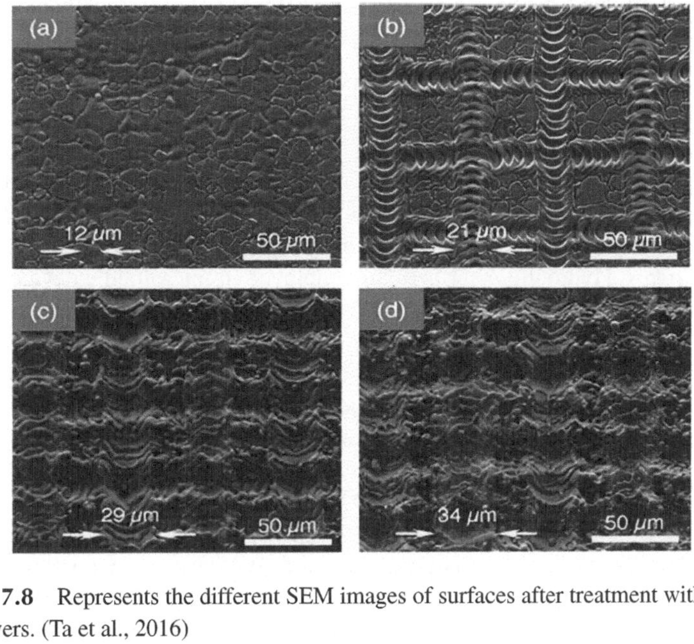

FIGURE 7.8 Represents the different SEM images of surfaces after treatment with different laser powers. (Ta et al., 2016)

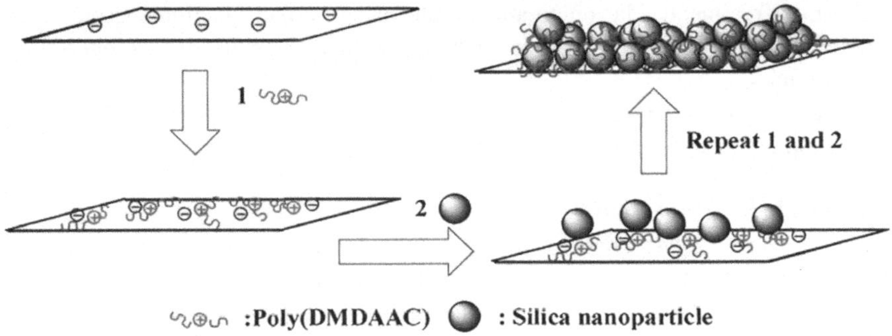

FIGURE 7.9 Schematic representation of layer-by-layer assembly process. (Reprinted with permission from Zhang et al. (2012), Copyright ©2012 Elsevier.)

Self-Assembled Monolayers (SAMs): SAMs formed by the spontaneous adsorption of long-chain hydrophobic molecules onto the surface can modify its chemistry and structure, imparting hydrophobic properties. Layer-by-layer (LBL) assembly stands out as an attractive fabrication method due to its ability to create multifunctional films on various materials while preserving their bulk properties (Buck et al., 2010). LBL can be applied to a wide range of substrates, including noble metals, oxides, and synthetic polymers. However, applying LBL layers to synthetic polymers can be challenging, often requiring aggressive pre-treatment such as plasma treatment, oxidative chemical reactions, or polymeric adsorption. The LBL deposition typically involves four steps: the adsorption of a positive component, washing, adsorption of a negative component, and final washing. These steps can be repeated to build thicker films (Michel et al., 2012). Figure 7.9 shows the schematic representation of layer-by-layer assembly method.

7.6.6 Applications of Hydrophilic and Hydrophobic Surfaces

Hydrophilic Surfaces

Biomedical Devices: Hydrophilic surfaces are used in medical implants, drug delivery systems, and diagnostic devices to promote cell adhesion, minimize protein adsorption, and enhance biocompatibility.

Microfluidics: Hydrophilic surfaces are employed in microfluidic devices for precise control and manipulation of small volumes of liquid, facilitating applications in lab-on-a-chip systems, chemical analysis, and biosensing.

Optics and Photovoltaics: Hydrophilic coatings are utilized in optics and photovoltaic devices to prevent fogging, reduce reflection, and improve light transmission, enhancing performance and durability.

Hydrophobic Surfaces

Water Repellent Coatings: Hydrophobic surfaces are used in textiles, coatings, and consumer products to repel water, prevent staining, and enhance durability, imparting properties such as self-cleaning and water resistance.

Oil and Gas Industry: Hydrophobic surfaces are employed in oil and gas pipelines, tanks, and equipment to inhibit corrosion, prevent fouling, and improve fluid flow, reducing maintenance and operational costs.

Electronics and Electrical Engineering: Hydrophobic coatings are utilized in electronic devices, printed circuit boards, and electrical insulation to protect against moisture ingress, improve reliability, and prolong lifespan.

7.6.7 Challenges and Future Directions

While hydrophilic and hydrophobic surfaces offer numerous benefits and applications, several challenges and opportunities for improvement remain:

Durability: Ensuring the long-term stability and durability of surface modifications is essential for practical applications, especially in harsh environments or high-wear conditions.

Scalability: Developing scalable and cost-effective fabrication methods for producing hydrophilic and hydrophobic surfaces on large scales is critical for widespread adoption across industries.

Multifunctionality: Integrating multiple functionalities, such as anti-fouling, anti-corrosion, and self-healing, into single surface coatings can enhance performance and versatility in various applications.

Biocompatibility: Improving the biocompatibility of hydrophobic surfaces is essential for biomedical applications, minimizing adverse effects such as inflammation, thrombosis, and tissue rejection.

Environmental Impact: Addressing the environmental impact of surface treatments, including the use of hazardous chemicals, waste generation, and end-of-life disposal, is important for sustainable manufacturing and product lifecycle management.

Thus, hydrophilic and hydrophobic surfaces play critical roles in a wide range of applications, from biomedical devices and microfluidic systems to textiles and electronics. Understanding the principles, properties, and applications of these surfaces is essential for designing functional materials, optimizing processes, and advancing technology. By addressing challenges and exploring new opportunities, researchers and engineers can continue to innovate in the field of surface science, paving the way for enhanced performance, sustainability, and functionality in diverse applications.

7.7 SURFACE FORCES AND NANORHEOLOGY

Surface forces and nanorheology are essential concepts in the field of nanoscience and nanotechnology, playing a crucial role in understanding the behavior of materials at the nanoscale. Surface forces govern the interactions between surfaces at the molecular and atomic levels, influencing phenomena such as adhesion, friction, and lubrication. Nanorheology, on the other hand, focuses on the study of mechanical properties and flow behavior of materials on nanometer length scales, providing insights into their rheological properties and behavior under confinement. In this comprehensive exploration, we delve into the principles, measurement techniques, applications, and advancements in surface forces and nanorheology.

7.7.1 Surface Forces: Fundamentals and Mechanisms

Surface forces arise from the interactions between molecules and atoms on the surfaces of materials, leading to attractive or repulsive forces between them. These forces can be classified into several categories based on their origin and mechanism, including Van der Waals forces, electrostatic forces, hydration forces, and steric forces.

7.7.1.1 Van der Waals Forces

Van der Waals forces are weak, short-range forces that result from fluctuations in the electron distribution within molecules and atoms. They include London dispersion forces, which arise from temporary dipoles induced by instantaneous fluctuations in electron density, and Kccsom and Debye forces, which arise from permanent or induced dipoles in polar molecules. Van der Waals forces play a significant role in determining the adhesive and cohesive properties of materials, influencing phenomena such as adhesion, friction, and colloidal stability.

Van der Waals interactions encompass three distinct and additive forces responsible for short-range attraction (cohesion) between atoms and molecules, whether they are separated by a vacuum or a medium. All three types of Van der Waals interactions exhibit interaction energies that are directly proportional to r^{-6}, where r represents the separation distance. These interactions can occur both in a vacuum and within a medium.

7.7.1.2 Van der Waals Interactions Among Molecules in a Vacuum (and Non-Reactive Gases)

The dipole-dipole interaction, also known as the orientational or Keesom interaction, arises from electrostatic interactions between polar molecules, which possess permanent dipoles. In liquids and vapors, these dipoles can freely rotate but tend to orient themselves in such a way that, on average, an attractive force emerges. The angle-averaged interaction free energy between two freely rotating permanent dipoles in a vacuum is described by Equation 7.2:

$$w(r) = -\frac{\mu_1^2 \mu_2^2}{3kT\left(4\pi\varepsilon_0\right)^2 r^6} \tag{7.2}$$

Here, "r" represents the distance between the molecules, "μ_i" denotes their permanent dipole moments, and Equation 7.2 accounts for the influence of thermal motion, which diminishes the alignment of dipoles (Israelachvili, 2011).

The dipole-induced dipole interaction, also referred to as the induction or Debye interaction, arises from induced dipole moments in atoms or in polar/non-polar molecules due to the presence of an electric field emanating from permanent dipoles. This results in an attractive force between the induced and permanent dipoles.

The interaction free energy between two freely rotating permanent dipoles, each possessing an induced dipole moment "α_i" is expressed by Equation 7.3:

$$w(r) = -\frac{\alpha_1\mu_2^2 + \alpha_2\mu_1^2}{(4\pi\varepsilon_0)^2 r^6} \tag{7.3}$$

In both cases, "r" represents the separation distance, "μ_i" signifies permanent dipole moments, "α_i" represents induced dipole moments, and "ε_0" is the vacuum permittivity constant. This interaction remains temperature-independent because the induced dipole moment aligns itself with the permanent dipole moment as the molecule undergoes reorientation due to thermal motion (Israelachvili, 2011).

The induced-dipole-induced-dipole interaction, also known as the fluctuation or London dispersion interaction, exists between all atoms and molecules, regardless of their polarity. It arises from the spontaneous polarization of one atom or molecule in response to fluctuations in the charge distribution of a neighboring atom or molecule. This interaction is orientation-independent but becomes stronger with increasing polarizability. This leads to the initial molecule experiencing more significant fluctuations in its charge distribution and, in turn, induces a larger dipole moment in the responding molecule. An approximate equation describing the interaction energy, assuming a single characteristic vibration frequency of electrons for each molecule, is given by Equation 7.4:

$$w(r) = -\frac{3}{2} \times \frac{\alpha_1\alpha_2}{(4\pi\varepsilon_0)^2 r^6} \frac{hn_1n_2}{(n_1 + n_2)} \tag{7.4}$$

Here, "r" represents the separation distance, α_1 and α_2 are the polarizabilities of the interacting molecules, "ε_0" is the vacuum permittivity constant, "h" is Planck's constant, and "n_1" and "n_2" are the characteristic vibration frequencies of the electrons for each molecule. Here, "h" represents Planck's constant, and "n_i" denotes the frequency of the lowest electron transition, often approximated by the ionization frequency of an electron in one of the outer orbitals. The dominant contribution (70–100%) to Van der Waals interactions typically arises from the London interaction, except in the case of highly polar substances like water. The rapid fluctuations in charge distributions lead to a weakening of this interaction at larger separation distances (> 10 nm), a phenomenon referred to as "retardation" (Casimir and Polder, 1948; Israelachvili, 2011).

However, at extremely small separation distances, when the electron clouds of molecules begin to overlap, all three of these interactions ultimately become repulsive due to the Pauli exclusion principle. Therefore, the overall Van der Waals interaction is a combination of very short-range repulsion and longer-range attraction.

7.7.1.3 Van der Waals Interactions Between Molecules in a Liquid

When molecules interact across another material or a medium, the Van der Waals forces tend to be significantly weaker compared to interactions in a vacuum. This reduction in strength arises because the interaction between two solute molecules within a medium (solvent) involves the reorientation and displacement of nearby solvent molecules. These solvent molecules interact with both the solute and other solvent molecules. To partially account for these solvent-related effects, the symbol ε_0 in the equations mentioned earlier is replaced with $\varepsilon_0\varepsilon_r$, where represents the relative permittivity of the bulk medium, and "α_i" is replaced by the excess polarizability of the solute molecule.

While the surrounding medium (solvent) may also alter the dipole moment and polarizability of the solute molecule from their values in a vacuum, the interaction between two identical molecules continues to be attractive. The theory describing Van der Waals interactions has been extended to accommodate scenarios involving more than one absorption frequency in an atom or molecule and the influence of a solvent (Fennell Evans and Wennerström, 1999 n.d.; Israelachvili, 2011).

7.7.1.4 Van der Waals Interactions Between Macroscopic Bodies

The contribution of Van der Waals interactions to the overall interaction energy between two macroscopic bodies, such as colloidal particles or a nanometer-sized asperity and a flat surface, can be determined by summing the pair-wise interactions (7.2)–(7.4) among all the molecules in one body with those in the other. The interaction energy for various geometries, including the case of two spheres with identical radii, R, interacting across a vacuum, was originally formulated by Hamaker (Hamaker, 1937; Israelachvili, 2011). Similar expressions have been derived for other geometries frequently encountered in nanotribological measurements, like two flat plates or a sphere interacting with a flat plate (equivalently, two crossed cylinders):

Interaction energy for two spheres:

$$W = -\frac{A}{12h} \tag{7.5}$$

Interaction energy for two flat plates:

$$W = -\frac{A}{12\pi h^2} \tag{7.6}$$

Interaction energy for a sphere and a flat plate:

$$W = -\frac{AR}{6h} \tag{7.7}$$

Here, "A" represents the Hamaker constant (see section "The Hamaker Constant"), and "h" denotes the closest separation distance (with R being the radius). However, it's essential to note that these equations may overestimate dispersion forces at large separations (when bodies interact across a medium with h > 10 nm) because they do not account for the time required for the propagation of induced dipole moments due to fluctuations, which is known as the retardation effect.

Upon comparing the dependence on separation distance in equations (7.2)–(7.4) with that in equations (7.5)–(7.7), it becomes evident that the interaction energy between macroscopic bodies exhibits significantly longer-range characteristics than the interaction between two individual molecules.

7.7.1.5 The Hamaker Constant

The non-retarded Hamaker constant can be determined through two distinct methods: the pair-wise interactions, known as the original "London-Hamaker microscopic approach," evaluated for a single oscillation frequency; or the "macroscopic" Lifshitz approach (Israelachvili, 2011), where the interacting bodies and the intervening medium are treated as continuous phases. In the Lifshitz approach, the polarizabilities and dipole moments are derived from bulk dielectric properties.

For solids interacting across a vacuum (or air), Hamaker constants typically range from 10^{-20} to 10^{-19} J, with higher values observed for metals. However, when these interacting bodies are separated by a medium, such as a liquid, the Van der Waals interactions are significantly diminished, necessitating the use of an effective Hamaker constant. A further approximation of the Hamaker constant expression, derived from the Lifshitz theory for a single vibration frequency and interactions between two different media (denoted as 1 and 3) separated by medium 2, is represented by Equation 7.8:

$$A_{123} = A_{n=0} + A_{n>0} \approx \frac{3kT}{4}\left(\frac{\varepsilon_1 - \varepsilon_2}{\varepsilon_1 + \varepsilon_2}\right)\frac{\varepsilon_3 - \varepsilon_2}{\varepsilon_3 + \varepsilon_2} + \frac{3hn_e}{8\sqrt{2}} \frac{\left(n_1^2 - n_2^2\right)\left(n_3^2 - n_2^2\right)}{\sqrt{n_1^2 + n_2^2}\sqrt{n_3^2 + n_2^2}\left\{\sqrt{n_1^2 + n_2^2} + \sqrt{n_3^2 + n_2^2}\right\}} \tag{7.8}$$

In this equation, the first term ($n = 0$) accounts for the Keesom and Debye interactions, while the second term ($n > 0$) represents the London (dispersion) interaction. Here, ε_i and n_i ($i = 1, 2, 3$) denote the static dielectric constants and refractive indexes of the respective media, and n_e stands for the frequency of the lowest electron transition, typically around 3 x 10^{15} s^{-1}

If the dielectric properties of the intervening medium 2 in Equation 7.8 fall between those of media 1 and 3, the interactions between 1 and 3 across 2 will have an opposite sign (repulsive, $A < 0$ in Equation 7.8) compared to the interactions between 1 and 1 across 3, and 2 and 2 across 3. In the symmetrical scenario where two bodies of medium 1 interact across medium 2 (i.e., $3 \rightarrow 1$ in Equation 7.8), A is always positive. Therefore, the Van der Waals interaction energy between bodies of the same material remains attractive when interacting with another material, regardless of its dielectric properties, following the principle of "like attracts like." This principle is akin to the interactions observed between identical solute molecules in a binary mixture (see section "Van der Waals Interactions Between Molecules in a Liquid").

7.7.1.6 Electrostatic Forces

Electrostatic forces arise from the interactions between charged particles on the surfaces of materials. They can be attractive or repulsive depending on the sign and magnitude of the charges involved. Electrostatic forces play a crucial role in determining the stability and behavior of colloidal suspensions, as well as the adhesion and friction between charged surfaces.

7.7.1.7 Electrostatic Double-Layer Interactions in a Liquid

When surfaces come into contact with highly polar liquids like water, they can become charged. This charge can originate from the dissolution of ions that become solvated and distribute into the solution due to entropy or from the preferential adsorption of ions from the solution. The surface's charge is balanced by a decreasing concentration of counterions found within a thin layer of solution close to the surface. Together, the surface charges and these counterions form what is known as a "diffuse double layer." In dilute solutions (i.e., those with low ionic strength), the thickness of this layer can be measured by the Debye length (also known as the Debye screening length), denoted as k^{-1}. One can draw an analogy between the diffuse double layer and a charged capacitor, where k^{-1} represents the distance between plates with equal charge density, mimicking that of the surface within the double layer. Essentially, the Debye length serves as a measure of the thickness of the "atmosphere of counterions" near the surface, exceeding the concentration found in the bulk solution. The Debye length's value is influenced by the ionic strength of the solution.

The Debye length (k^{-1}) can be calculated using Equation 7.9:

$$k^{-1} = \sqrt{\frac{\varepsilon_0 \varepsilon_{rk_B} T}{\left(e^2 N_{A \sum_i z_i^2 c_i}\right)}} \tag{7.9}$$

In this equation, ε_0 represents the vacuum permittivity, ε_r is the relative permittivity of the medium, $k_B T$ denotes thermal energy, e is the electron charge, N_A stands for Avogadro's constant, z_i represents the ionic charge, and c_i represents the solution concentration.

The Debye length is an inherent property of the solution and is relatively large in dilute solutions, for instance, approximately 30 nm in a 10^{-4} M 1:1 electrolyte solution and roughly 1 mm in pure water. The thickness of the diffuse double layer determines the range of the interaction forces between two surfaces. As the surfaces get closer together, counterions remain trapped between them due to their attraction to the surface charges. This results in repulsion caused by increased osmotic pressure within the confined thin film. This electrostatic "double-layer force" primarily originates from entropy considerations. At large separation distances (where there is weak overlap between ion number densities and potentials), the interaction energy between two similarly charged surfaces is generally repulsive and decays exponentially with separation distance (h), with the decay length being the Debye length (k^{-1}). This relationship can be expressed as Equation 7.10:

$$W(h) = Constant \times e^{-kh} \tag{7.10}$$

The constant in this equation depends on various factors, including the geometry of the interacting surfaces, the surface charge density, and the solution conditions (temperature and ionic strength). Solving the Poisson–Boltzmann equation under different constraints and approximations provides values for this constant.

It's important to note that the equations mentioned above are accurate primarily at large separation distances, specifically those greater than approximately one Debye length. At very small separations, the interaction potential energy is influenced by whether the surfaces are kept at a constant potential, at constant charge density (resulting in stronger repulsion than the equations above), or somewhere in between these limits. Under "constant charge" conditions, the total number of counterions in the confined solution remains unchanged, while under "constant potential," the concentration remains constant. The limiting force per unit area, known as the "disjoining pressure," at constant charge is essentially the osmotic pressure of the confined ions.

In certain cases, even when the surfaces possess similar charges, the double-layer force can become attractive at small separation distances. This can occur when the surface contains chargeable chemical groups that can undergo dissociation (charge regulation) or when ions condense from the solution onto the surfaces, effectively reducing the effective surface charge density. Particularly at high charge density, the mobility and polarizability of the numerous counterions near the surface can decrease the repulsion, a phenomenon known as ion correlation or charge fluctuation (Evans and Wennerström 1999 n.d.; Israelachvili 2011).

7.7.1.8 Hydration Forces

Hydration forces arise from the interactions between water molecules and charged or polar groups on the surfaces of materials. They result from the restructuring of the water molecules near the surface, leading to attractive or repulsive forces between hydrated surfaces. Hydration forces can significantly influence the adhesion, friction, and swelling behavior of materials in aqueous environments.

7.7.1.9 Steric Forces

Steric forces arise from the repulsion between molecules or atoms due to their geometric and steric constraints. They result from the overlap of electron clouds or the exclusion of solvent molecules from confined spaces, leading to repulsive forces between surfaces. Steric forces play a critical role in determining the stability and behavior of colloidal suspensions, as well as the lubrication and adhesion properties of materials under confinement.

7.8 MEASUREMENT TECHNIQUES FOR SURFACE FORCES

Measuring surface forces at the nanoscale requires precise and sensitive experimental techniques capable of quantifying interactions between surfaces with high resolution. Several techniques have been developed for this purpose, including atomic force microscopy (AFM), surface force apparatus (SFA), and quartz crystal microbalance (QCM).

7.8.1 Atomic Force Microscopy (AFM)

AFM is a versatile and powerful technique for imaging and measuring surface forces at the nanoscale. In AFM, a sharp probe tip mounted on a cantilever is brought into close proximity to the sample surface, allowing for the measurement of forces between them. By monitoring the deflection of the cantilever as it interacts with the surface, AFM can generate high-resolution images of surface topography and quantify surface forces such as adhesion, friction, and elasticity.

7.8.2 Surface Force Apparatus (SFA)

SFA is a specialized experimental technique for measuring the interaction forces between two surfaces across a liquid or vacuum gap. In SFA, two parallel surfaces are mounted on separate force-sensitive cantilevers and brought into close proximity to each other. By monitoring the deflection of the cantilevers as they interact with each other, SFA can measure the forces between the surfaces as a function of separation distance. SFA is particularly well-suited for studying the fundamental mechanisms of surface forces and their dependence on environmental factors such as temperature, humidity, and solvent composition.

7.8.3 Quartz Crystal Microbalance (QCM)

QCM is a sensitive and versatile technique for measuring changes in mass or viscoelastic properties of thin films or surfaces. In QCM, a quartz crystal resonator coated with the sample material is excited at its resonance frequency, and changes in frequency or damping are monitored as the sample interacts with its environment. By measuring the frequency shift or dissipation of the resonator, QCM can quantify surface mass changes, adsorption, desorption, and viscoelastic properties of thin films or surface layers.

7.9 APPLICATIONS OF SURFACE FORCES

Surface forces play a critical role in numerous natural and technological processes, influencing phenomena such as adhesion, friction, lubrication, colloidal stability, and self-assembly. Understanding and controlling surface forces is essential for a wide range of applications, including:

Adhesion: Surface forces determine the adhesion between surfaces in various contexts, including biological adhesion, adhesion in microelectronic devices, and adhesion between solid surfaces and coatings or adhesives.

Friction and Lubrication: Surface forces govern the frictional behavior and lubrication properties of materials, influencing phenomena such as boundary lubrication, hydrodynamic lubrication, and wear resistance in mechanical systems.

Colloidal Stability: Surface forces are crucial for the stability and behavior of colloidal suspensions, influencing phenomena such as particle aggregation, sedimentation, and stability against flocculation or coagulation.

Self-Assembly: Surface forces drive the self-assembly of molecules and nanoparticles into ordered structures, enabling the fabrication of functional materials and devices with tailored properties and functionalities.

7.10 NANORHEOLOGY: PRINCIPLES AND MEASUREMENT TECHNIQUES

Nanorheology is the study of mechanical properties and flow behavior of materials on nanometer length scales, providing insights into their rheological properties and behavior under confinement. Nanorheological measurements are typically performed using techniques such as AFM-based force spectroscopy, microfluidic devices, and particle tracking methods.

7.11 AFM-BASED FORCE SPECTROSCOPY

AFM-based force spectroscopy is a powerful technique for measuring the mechanical properties of materials at the nanoscale. In force spectroscopy experiments, the AFM probe tip is brought into contact with the sample surface, and the resulting force-distance curve is recorded as the tip is retracted. By analyzing the force-distance curve, parameters such as adhesion, elasticity, and viscosity of the sample material can be quantified with high spatial resolution.

7.12 MICROFLUIDIC DEVICES

Microfluidic devices are miniature fluidic systems that allow for precise control and manipulation of small volumes of liquid on the micrometer or nanometer scale. By incorporating features such as channels, valves, and pumps into microfabricated structures, microfluidic devices can be used to study the rheological properties of fluids, suspensions, and soft materials under controlled flow conditions.

7.13 PARTICLE TRACKING METHODS

Particle tracking methods involve tracking the motion of individual particles or molecules suspended in a fluid using microscopy techniques. By analyzing the trajectories of tracked particles, parameters such as diffusion coefficients, viscoelastic moduli, and flow profiles can be determined, providing insights into the rheological properties of the surrounding medium.

7.14 APPLICATIONS OF NANORHEOLOGY

Nanorheology has diverse applications in fields such as materials science, biophysics, and soft matter research, enabling the characterization of mechanical properties and flow behavior of complex fluids, biomolecules, and soft materials on the nanometer scale. Key applications include:

Biological Systems: Nanorheology is used to study the mechanical properties of biological materials such as cells, tissues, and biomolecules, providing insights into their structure, function, and disease states.

Polymer Nanocomposites: Nanorheology is employed to characterize the rheological properties of polymer nanocomposites, which consist of polymer matrices reinforced with nanoscale fillers such as nanoparticles or nanofibers. Understanding the rheological behavior of these materials is essential for optimizing their mechanical properties and processing conditions.

Drug Delivery Systems: Nanorheology is used to study the rheological properties of drug delivery systems such as liposomes, micelles, and hydrogels, providing insights into their stability, release kinetics, and transport mechanisms in biological tissues.

Soft Matter Physics: Nanorheology is employed to investigate the rheological properties of soft materials such as colloidal suspensions, emulsions, and gels, providing insights into their flow behavior, structure, and phase transitions under controlled conditions.

7.15 CHALLENGES AND FUTURE DIRECTIONS

Despite significant advancements in surface forces and nanorheology, several challenges and opportunities for improvement remain:

Complexity of Interactions: Understanding the complex interplay between different surface forces and their effects on material properties and behavior remains a major challenge, requiring interdisciplinary approaches and advanced theoretical and computational models.

Measurement Techniques: Developing new measurement techniques and instrumentation with higher sensitivity, spatial resolution, and throughput is essential for advancing our understanding of surface forces and nanorheology and their applications in diverse fields.

Biological and Environmental Applications: Extending the principles and techniques of surface forces and nanorheology to biological and environmental applications presents exciting opportunities for addressing pressing challenges such as drug delivery, tissue engineering, and environmental remediation.

Multiscale Modeling and Simulation: Integrating experimental data with multiscale modeling and simulation approaches can provide deeper insights into the underlying mechanisms of surface forces and nanorheology, enabling predictive modeling and design of materials with tailored properties and functionalities.

Thus, surface forces and nanorheology are fundamental concepts in nanoscience and nanotechnology, playing a crucial role in understanding the behavior of materials at the nanoscale. By investigating the mechanisms, measurement techniques, applications, and challenges of surface forces and nanorheology, researchers can continue to advance our understanding of material properties and behavior and develop new technologies with transformative applications in diverse fields.

REFERENCES

Akbulut, M., Chen, N., Maeda, N., Israelachvili, J., Grunewald, T., Helm, C.A., 2005. Crystallization in Thin Liquid Films Induced by Shear. The Journal of Physical Chemistry B 109, 12509–12514. https://doi.org/10.1021/jp0448200

Baldacchini, T., Carey, J.E., Zhou, M., Mazur, E., 2006. Superhydrophobic Surfaces Prepared by Microstructuring of Silicon Using a Femtosecond Laser. Langmuir 22, 4917–4919. https://doi.org/10.1021/la053374k

Bhushan, B., 2011a. Introduction – Measurement Techniques and Applications, in: Bhushan, B. (Ed.), Nanotribology and Nanomechanics I: Measurement Techniques and Nanomechanics. Springer, Berlin; Heidelberg, pp. 1–34. https://doi.org/10.1007/978-3-642-15283-2_1

Bhushan, B., 2011b. Nanotribology, Nanomechanics, and Materials Characterization, in: Bhushan, B. (Ed.), Nanotribology and Nanomechanics II: Nanotribology, Biomimetics, and Industrial Applications. Springer, Berlin; Heidelberg, pp. 3–106. https://doi.org/10.1007/978-3-642-15263-4_12

Bhushan, B., 1995. Micro/nanotribology and Its Applications to Magnetic Storage Devices and MEMS. Tribology International 28, 85–96. https://doi.org/10.1016/0301-679X(95)92698-5

Bhushan, B., Eminent, O., Winbigler, H.D., 2017. Nanotribology, Nanomechanics and Materials Characterization Studies Using Scanning Probe Microscopy, in: Bhushan, B. (Ed.), Nanotribology and Nanomechanics: An Introduction. Springer International Publishing, Cham, pp. 373–455. https://doi.org/10.1007/978-3-319-51433-8_8

Braun, O.M., Naumovets, A.G., 2006. Nanotribology: Microscopic mechanisms of friction. Surface Science Reports 60, 79–158. https://doi.org/10.1016/j.surfrep.2005.10.004

Buck, M.E., Schwartz, S.C., Lynn, D.M., 2010. Superhydrophobic Thin Films Fabricated by Reactive Layer-by-Layer Assembly of Azlactone-Functionalized Polymers. Chemistry of Materials 22, 6319–6327. https://doi.org/10.1021/cm102115e

Casimir, H.B.G., Polder, D., 1948. The Influence of Retardation on the London-van der Waals Forces. Physical Review 73, 360–372. https://doi.org/10.1103/PhysRev.73.360

Chun, D.-M., Davaasuren, G., Ngo, C.-V., Kim, C.-S., Lee, G.-Y., Ahn, S.-H., 2014. Fabrication of Transparent Superhydrophobic Surface on Thermoplastic Polymer Using Laser Beam Machining and Compression Molding for Mass Production. CIRP Annals 63, 525–528. https://doi.org/10.1016/j.cirp.2014.03.085

Chun, D.-M., Ngo, C.-V., Lee, K.-M., 2016. Fast Fabrication of Superhydrophobic Metallic Surface Using Nanosecond Laser Texturing and Low-temperature Annealing. CIRP Annals 65, 519–522. https://doi.org/10.1016/j.cirp.2016.04.019

Elkhyat, A., Courderot-Masuyer, C., Gharbi, T., Humbert, P., 2004. Influence of the hydrophobic and hydrophilic characteristics of sliding and slider surfaces on friction coefficient: in vivo human skin friction comparison. Skin Research and Technology 10, 215–221. https://doi.org/10.1111/j.1600-0846.2004.00085.x

Fennell Evans, D., Wennerström, H., 1999. The Colloidal Domain: Where Physics, Chemistry, Biology, and Technology Meet, 2nd ed., Advances in Interfacial Engineering. Wiley, Germany.

Gusarov, S., 2024. Advances in Computational Methods for Modeling Photocatalytic Reactions: A Review of Recent Developments. Materials 17, 2119. https://doi.org/10.3390/ma17092119

Hamaker, H.C., 1937. The London—van der Waals attraction between spherical particles. Physica 4, 1058–1072. https://doi.org/10.1016/S0031-8914(37)80203-7

Israelachvili, J., Min, Y., Akbulut, M., Alig, A., Carver, G., Greene, W., Kristiansen, K., Meyer, E., Pesika, N., Rosenberg, K., Zeng, H., 2010. Recent Advances in the Surface Forces Apparatus (SFA) Technique. Reports on Progress in Physics 73, 036601. https://doi.org/10.1088/0034-4885/73/3/036601

Israelachvili, J.N., 2011. Chapter 13 – Van der Waals Forces between Particles and Surfaces, in: Israelachvili, J.N. (Ed.), Intermolecular and Surface Forces (Third Edition). Academic Press, San Diego, pp. 253–289. https://doi.org/10.1016/B978-0-12-375182-9.10013-2

Mao, Q., Feng, M., Jiang, X.Z., Ren, Y., Luo, K.H., Duin, A.C.T. van, 2023. Classical and Reactive Molecular Dynamics: Principles and Applications in Combustion and Energy Systems. Progress in Energy and Combustion Science 97, 101084. https://doi.org/10.1016/j.pecs.2023.101084

Michel, M., Toniazzo, V., Ruch, D., Ball, V., 2012. Deposition Mechanisms in Layer-by-Layer or Step-by-Step Deposition Methods: From Elastic and Impermeable Films to Soft Membranes with Ion Exchange Properties. ISRN Materials Science 2012, 701695. https://doi.org/10.5402/2012/701695

Ruths, M., Berman, A.D., Israelachvili, J.N., 2004. Surface Forces and Nanorheology of Molecularly Thin Films, in: Bhushan, B. (Ed.), Springer Handbook of Nanotechnology. Springer, Berlin; Heidelberg, pp. 543–603. https://doi.org/10.1007/3-540-29838-X_18

Ta, V.D., Dunn, A., Wasley, T.J., Li, J., Kay, R.W., Stringer, J., Smith, P.J., Esenturk, E., Connaughton, C., Shephard, J.D., 2016. Laser Textured Superhydrophobic Surfaces and Their Applications for Homogeneous Spot Deposition. Applied Surface Science 365, 153–159. https://doi.org/10.1016/j.apsusc.2016.01.019

Zhang, M., Wang, S., Wang, C., Li, J., 2012. A Facile Method to Fabricate Superhydrophobic Cotton Fabrics. Applied Surface Science 261, 561–566. https://doi.org/10.1016/j.apsusc.2012.08.055

Zhang, S., Li, D., Liu, Y., 2022. Friction Behavior of Rough Surfaces on the Basis of Contact Mechanics: A Review and Prospects. Micromachines 13, 1907. https://doi.org/10.3390/mi13111907

8 Structural Application

8.1 INTRODUCTION

Nanotechnology, a rapidly advancing field encompassing the manipulation of matter at the nanoscale, has emerged as a transformative force across various sectors, including structural engineering. This introduction will explore the multifaceted applications of nanotechnology in structural engineering, elucidating its profound impact on material properties, construction methodologies, and structural performance. With the ability to engineer materials and structures at the atomic and molecular levels, nanotechnology offers unprecedented opportunities to enhance the strength, durability, sustainability, and functionality of infrastructure systems (Nasrollahzadeh et al., 2019).

Nanomaterials, characterized by their unique properties arising from quantum effects and increased surface area-to-volume ratios, represent a cornerstone of nanotechnology-enabled advancements in structural engineering. Carbon nanotubes (CNTs), for instance, exhibit exceptional mechanical properties, such as high tensile strength and stiffness, surpassing those of conventional structural materials like steel and concrete. Similarly, graphene, a two-dimensional material composed of a single layer of carbon atoms, demonstrates remarkable electrical conductivity, thermal conductivity, and mechanical strength, making it an ideal candidate for reinforcing and protecting structural elements. By harnessing the exceptional properties of nanomaterials, engineers can develop innovative construction materials with enhanced performance characteristics, paving the way for stronger, lighter, and more durable structures. The integration of nanotechnology in structural engineering extends beyond material enhancement to encompass novel construction techniques and methodologies. Advanced manufacturing processes, such as 3D printing and additive manufacturing, leverage nanomaterials to fabricate intricate structural components with unparalleled precision and complexity. These additive techniques enable the customization of designs and the rapid prototyping of structural elements, revolutionizing traditional construction practices. Furthermore, self-healing materials, embedded with nanocapsules containing healing agents, offer autonomous repair capabilities, mitigating damage and prolonging the lifespan of structures. Nanotechnology-enabled construction methodologies not only improve efficiency and cost-effectiveness but also contribute to the development of sustainable and resilient infrastructure systems (Paras et al., 2023).

DOI: 10.1201/9781003317272-8

In addition to material enhancement and construction methodologies, nanotechnology facilitates the development of advanced structural monitoring, sensing, and diagnostics technologies. Nanoscale sensors, capable of detecting changes in temperature, strain, humidity, and chemical composition, can be embedded within structural elements to provide real-time feedback on performance and health status. These sensors enable predictive maintenance strategies, allowing engineers to anticipate and mitigate potential structural failures before they occur. Moreover, smart materials, responsive to external stimuli such as temperature, pressure, and light, enable adaptive structures capable of self-monitoring and self-regulation. By integrating sensing and actuating functionalities, nanotechnology enables the creation of intelligent infrastructure systems that optimize performance, maximize efficiency, and enhance safety (Tovar-Lopez, 2023).

Furthermore, nanotechnology plays a pivotal role in advancing the sustainability and environmental impact of structural engineering practices. Nanomaterials derived from abundant and renewable sources offer sustainable alternatives to conventional construction materials, reducing reliance on finite resources and minimizing carbon footprints. Lightweight nanocomposites enable the construction of energy-efficient structures with reduced material consumption and transportation costs. Additionally, nanotechnology facilitates the recycling and upcycling of construction waste, transforming it into value-added materials for future building projects. By promoting sustainable construction practices and eco-friendly materials, nanotechnology contributes to the development of resilient infrastructure systems that mitigate environmental impact and adapt to evolving societal needs. Thus, nanotechnology represents a paradigm shift in structural engineering, offering unprecedented opportunities to enhance the performance, sustainability, and resilience of infrastructure systems. From material enhancement to construction methodologies to structural monitoring and sustainability, nanotechnology permeates every aspect of the built environment. As research and development efforts continue to push the boundaries of nanoscience and engineering, the integration of nanotechnology into structural applications holds immense potential for innovation and transformation. By harnessing the unique properties of nanomaterials and leveraging advanced manufacturing techniques and sensing technologies, engineers can develop infrastructure systems that are stronger, smarter, and more sustainable, shaping a resilient built environment for generations to come (Khan, 2020).

8.2 ADDITIVE MANUFACTURING

Additive Manufacturing (AM), commonly known as 3D printing, has emerged as a disruptive technology that revolutionizes the way we design, prototype, and produce objects. Unlike traditional subtractive manufacturing methods, which involve cutting away material from a solid block, AM builds objects layer by layer from digital 3D models. This flexibility and precision have opened up new possibilities across various industries, from aerospace and automotive to healthcare and consumer goods. Additive manufacturing encompasses a diverse range of processes that share the fundamental principle of adding material layer by layer to create a three-dimensional object. These processes offer several advantages over conventional manufacturing techniques, including:

Design Flexibility: AM allows for the creation of complex geometries and intricate designs that would be difficult or impossible to produce using traditional methods.
Rapid Prototyping: AM enables the quick and cost-effective production of prototypes, allowing designers and engineers to iterate and refine their designs more efficiently.
Customization: AM facilitates the customization of products, allowing for personalized items tailored to individual preferences and requirements.
Material Efficiency: AM minimizes material waste by only using the necessary amount of material for each part, reducing overall material consumption.

8.2.1 Types of Additive Manufacturing Processes

There are different types of additive manufacturing processes such as, Fused Deposition Modeling (FDM), Stereolithography (SLA), Selective Laser Sintering (SLS), Binder Jetting, Direct Metal Laser Sintering (DMLS), Selective Laser Melting (SLM), Material Jetting, Laminated Object Manufacturing (LOM), Directed Energy Deposition (DED) and Drop on Demand (DOD). Each of these additive manufacturing processes utilizes different technologies, materials, and techniques to build three-dimensional objects layer by layer. They vary in terms of their applications, materials compatibility, resolution, speed, and cost-effectiveness, allowing for a diverse range of manufacturing solutions across industries.

8.2.1.1 Fused Deposition Modeling (FDM)

Fused Deposition Modeling (FDM) stands as one of the most popular and widely used additive manufacturing techniques worldwide. This method, often referred to as Fused Filament Fabrication (FFF) in some contexts, has reshaped prototyping, manufacturing, and even hobbyist creation due to its accessibility, versatility, and relative affordability. In this detailed exploration, we will delve into the intricacies of FDM, understanding its underlying principles, applications across industries, advantages, limitations, and ongoing advancements. Fused Deposition Modeling, developed by Scott Crump in the late 1980s, represents a significant breakthrough in additive manufacturing. The process involves the layer-by-layer deposition of melted thermoplastic filament to construct three-dimensional objects based on digital designs. FDM systems consist of three primary components: a filament spool, a heated extrusion nozzle, and a build platform. The FDM process begins with the creation or acquisition of a digital 3D model using Computer-Aided Design (CAD) software. This digital model serves as the blueprint for the physical object to be produced. The digital model is then sliced into thin horizontal layers using slicing software. Each layer is converted into a series of cross-sectional 2D slices that the FDM printer will replicate layer by layer. A thermoplastic filament, typically made of materials like PLA (polylactic acid) or ABS (acrylonitrile butadiene styrene), is fed from a spool into the printer's extrusion

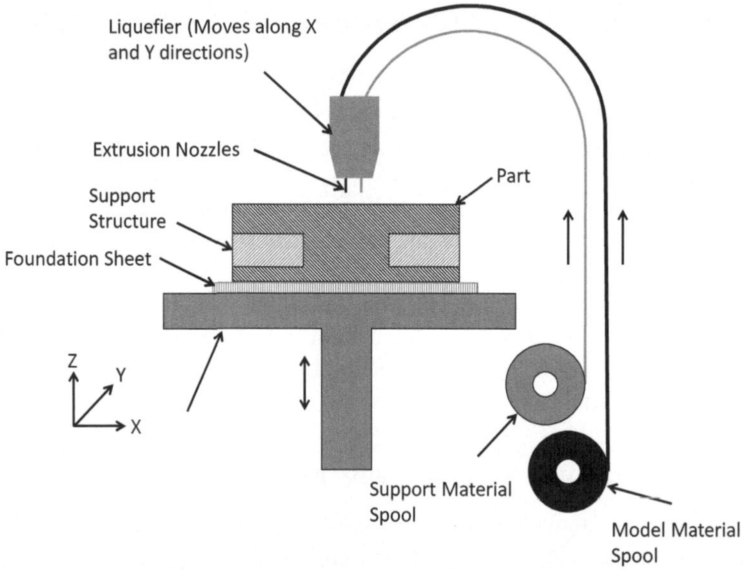

FIGURE 8.1 Schematic representation of FDM process. (Gebisa and Lemu, 2018)

system. The filament passes through a heated extrusion nozzle, where it is melted to a semi-liquid state. The extrusion nozzle moves along the X and Y axes, depositing the melted filament in a precise pattern determined by the sliced layers. The printer builds up the object layer by layer, with each new layer fusing to the previous one as it cools and solidifies. This process continues until the entire object is complete. For objects with overhangs or unsupported features, the printer may use a secondary, water-soluble material or the same material to create temporary support structures. These supports are removed after printing is complete. Once printing is finished, the object may require post-processing steps such as removing support structures, sanding rough surfaces, or applying surface finishes like painting or polishing (Masood, 2014). Figure 8.1 shows the schematic of the FDM process.

Applications of FDM

Fused Deposition Modeling finds extensive applications across various industries due to its versatility and cost-effectiveness. Some notable applications include:

1. Prototyping:
   FDM is widely used for rapid prototyping in industries ranging from automotive and aerospace to consumer goods and electronics. Its ability to quickly produce functional prototypes allows for iterative design improvements and accelerated product development cycles.
2. Manufacturing Tooling:
   Manufacturing companies utilize FDM to produce custom jigs, fixtures, and tooling for assembly lines and production processes. These bespoke tools can improve efficiency, reduce costs, and streamline manufacturing operations.

3. End-Use Parts:
While traditionally used for prototyping and tooling, FDM is increasingly being employed to produce end-use parts in low-volume manufacturing scenarios. Industries such as healthcare, where custom medical devices and prosthetics are required, benefit from the ability to produce small batches of parts on-demand.
4. Education and Research:
FDM has become a valuable tool in educational settings, allowing students to gain hands-on experience in design and manufacturing. Additionally, researchers utilize FDM to fabricate custom lab equipment, prototypes, and research models across various scientific disciplines.

Advantages of FDM

Fused Deposition Modeling offers several advantages that contribute to its widespread adoption,

1. Cost-Effectiveness:
FDM is relatively affordable compared to other additive manufacturing methods, making it accessible to small businesses, educational institutions, and hobbyists.
2. Material Variety:
FDM supports a wide range of thermoplastic materials with different properties, including ABS, PLA, PETG, and TPU. This material versatility allows for the production of objects with varying mechanical, thermal, and aesthetic properties.
3. Design Flexibility:
With FDM, designers have the freedom to create complex geometries, organic shapes, and lightweight structures that would be challenging or impossible to achieve with traditional manufacturing methods.
4. Rapid Prototyping:
FDM excels in rapid prototyping, enabling designers and engineers to quickly iterate on designs, test concepts, and validate ideas before committing to full-scale production.
5. Minimal Waste:
Since FDM only uses the amount of material required to build the object, it generates minimal waste compared to subtractive manufacturing methods, where excess material is often discarded.

Limitations of FDM

Despite its numerous advantages, Fused Deposition Modeling also has some limitations:

1. Surface Finish:
FDM parts may exhibit visible layer lines and rough surface finishes, especially on curved or sloped surfaces. Post-processing techniques such as sanding, painting, or vapor smoothing may be required to achieve a smoother finish.

2. Dimensional Accuracy:
   FDM parts may experience dimensional inaccuracies, particularly in features with tight tolerances or small details. Factors such as material shrinkage and thermal expansion can affect the overall accuracy of printed parts.
3. Support Structures:
   Objects with overhangs or complex geometries may require support structures, which add time and material to the printing process. Removing these supports can be labor intensive and may leave behind surface blemishes.
4. Material Limitations:
   While FDM supports a wide range of thermoplastic materials, it may not be suitable for applications requiring high temperature resistance, chemical resistance, or biocompatibility.

Ongoing Developments and Future Outlook

Fused Deposition Modeling continues to evolve with ongoing advancements in materials, hardware, and software. Manufacturers are developing new filaments with enhanced properties, such as increased strength, flexibility, and heat resistance. Additionally, improvements in printer technology, including higher resolution, faster print speeds, and larger build volumes, are expanding the capabilities of FDM. As the technology matures, FDM is expected to play an increasingly significant role in manufacturing, prototyping, and customization across diverse industries. Fused Deposition Modeling stands as a cornerstone of additive manufacturing, offering a cost-effective, versatile, and accessible solution for producing complex three-dimensional objects. With its wide range of applications, from prototyping and tooling to end-use parts production, FDM continues to drive innovation and enable creativity in design and manufacturing. While it may have some limitations, ongoing developments and advancements in FDM technology are poised to further expand its capabilities and applications in the years to come.

8.2.1.2 Stereolithography (SLA)

Stereolithography (SLA) stands as one of the earliest and most influential additive manufacturing processes, paving the way for the 3D printing revolution. Developed by Chuck Hull in the 1980s, SLA introduced a groundbreaking method for creating three-dimensional objects from digital designs. In this comprehensive exploration, we delve into the intricacies of Stereolithography, understanding its principles, working mechanism, materials, applications across industries, advantages, limitations, and ongoing advancements. Stereolithography represents the first commercially available 3D printing technology, marking a significant milestone in the evolution of additive manufacturing. This process enables the creation of highly detailed, complex objects by selectively curing liquid photopolymer resin layer by layer using a precisely controlled UV laser. SLA is widely utilized for prototyping, product development, and low-volume production in industries ranging from aerospace and automotive to healthcare and consumer goods.

Steps in Stereolithography

The Stereolithography process begins with the creation or acquisition of a digital 3D model using Computer-Aided Design (CAD) software. This digital model serves as

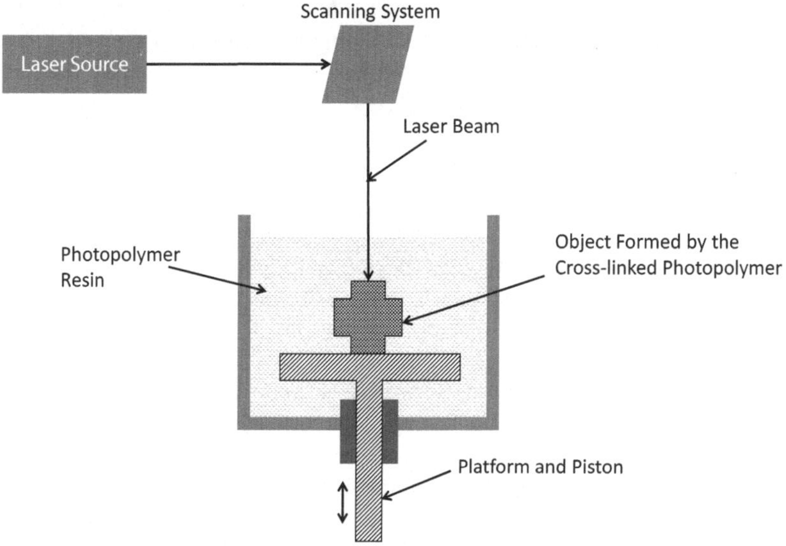

FIGURE 8.2 Schematic representation of SLA process. (Konta et al., 2017)

the blueprint for the physical object to be produced. The digital model is sliced into thin horizontal layers using slicing software. Each layer is converted into a series of cross-sectional 2D slices that the SLA printer will replicate layer by layer. The SLA printer contains a vat filled with liquid photopolymer resin, as shown in Figure 8.2. The build platform is submerged in the resin, with a thin layer of resin covering the surface. A UV laser is directed onto the surface of the liquid resin, selectively curing it according to the pattern of each layer. The laser beam follows the contour of the sliced cross-section, solidifying the resin wherever it strikes. After each layer is cured, the build platform moves incrementally downward, allowing the next layer of liquid resin to flow over the cured layer. The process repeats, with the laser curing each new layer onto the previous one, until the entire object is complete. Once printing is finished, the object is removed from the SLA printer and cleaned to remove excess resin. Depending on the desired finish, the object may undergo additional post-processing steps such as support removal, sanding, and surface finishing (Konta et al., 2017).

Materials Used in SLA

SLA utilizes liquid photopolymer resins as the raw material for printing. These resins typically consist of monomers, oligomers, photoinitiators, and other additives. Photopolymer resins come in a variety of formulations with different properties, including standard, flexible, transparent, and high-temperature-resistant resins. Each type of resin offers unique mechanical, thermal, and aesthetic characteristics, allowing for a wide range of applications.

Applications of Stereolithography

1. Stereolithography is widely used for rapid prototyping and product development across industries. Designers and engineers utilize SLA to create

functional prototypes, concept models, and validation prototypes for testing and evaluation. The ability to produce highly detailed, accurate parts quickly and cost-effectively accelerates the product development cycle and reduces time to market.

2. Manufacturing companies leverage SLA to produce custom tooling, jigs, fixtures, and molds for use in production processes. SLA-produced tooling offers advantages such as reduced lead times, lower costs, and enhanced customization compared to traditional manufacturing methods. These bespoke tools improve efficiency, precision, and consistency in manufacturing operations.
3. While traditionally used for prototyping and tooling, SLA is increasingly being employed to produce end-use parts in low-volume production scenarios. Industries such as healthcare, aerospace, and consumer goods benefit from the ability to manufacture small batches of production-quality parts with intricate geometries and fine details.
4. SLA is popular in the jewelry industry for creating highly detailed, intricate designs with precious metals and resins. Jewelry designers use SLA to produce prototypes, master patterns, and custom pieces with precise dimensions and surface finishes. Additionally, artists and sculptors utilize SLA to create sculptures, figurines, and art installations with complex shapes and textures.

Advantages of Stereolithography

1. SLA offers exceptional resolution and surface finish, allowing for the production of highly detailed, intricate parts with fine features and smooth surfaces. This makes SLA ideal for applications requiring precision and aesthetics.
2. SLA supports a variety of photopolymer resins with different properties, including standard, flexible, transparent, and high-temperature-resistant resins. This material versatility enables the production of parts with diverse mechanical, thermal, and optical characteristics.
3. SLA is a fast and efficient additive manufacturing process, capable of producing parts quickly and cost-effectively. The layer-by-layer build approach allows for the simultaneous production of multiple parts in a single print job, further increasing efficiency.
4. SLA offers designers unparalleled freedom to create complex geometries, intricate details, and organic shapes that would be difficult or impossible to achieve with traditional manufacturing methods. This design flexibility encourages innovation and creativity in product development.

Limitations of Stereolithography

1. SLA is limited to photopolymer resins, which may not possess the mechanical, thermal, or chemical properties required for certain applications. Additionally, the range of available resins may be limited compared to other additive manufacturing processes.
2. SLA parts require post-processing to remove excess resin, support structures, and achieve the desired surface finish. Depending on the application,

post-processing steps such as cleaning, curing, sanding, and painting may be necessary, adding time and labor to the manufacturing process.

3. SLA printers typically have smaller build volumes compared to other additive manufacturing technologies, limiting the size of parts that can be produced in a single print job. Large or oversized parts may need to be printed in multiple pieces and assembled post-printing.
4. Photopolymer resins used in SLA are sensitive to UV light and may degrade over time when exposed to sunlight or other sources of UV radiation. Proper storage and handling are required to maintain resin quality and prevent premature curing.

Ongoing Developments and Future Outlook

Stereolithography is continuously evolving with advancements in materials and methods utilized and also in terms of software. Manufacturers are developing new photopolymer resins with enhanced properties like increased strength, flexibility, and temperature resistance. Improvements in SLA printer technology, including higher resolution, faster print speeds, and larger build volumes, are broadening the process's capabilities. As the technology matures, Stereolithography is expected to remain a leading additive manufacturing process, driving innovation and enabling new applications across various industries.

Stereolithography stands as a pioneering additive manufacturing process that has revolutionized product development, prototyping, and low-volume production. With its high resolution, diverse material options, design flexibility, and efficiency, SLA continues to be a preferred choice for producing highly detailed, complex parts across numerous industries. Despite some limitations, ongoing advancements and innovations in Stereolithography technology are set to expand its capabilities and applications, solidifying its legacy as a cornerstone of additive manufacturing.

8.2.1.3 Selective Laser Sintering (SLS)

Selective Laser Sintering (SLS) stands at the forefront of additive manufacturing technologies, offering unparalleled versatility, precision, and efficiency in producing complex three-dimensional objects. Developed in the mid-1980s, SLS has since evolved into a widely adopted manufacturing method across various industries, from aerospace and automotive to healthcare and consumer goods. In this section, we will delve into the principles, process, materials, applications, advantages, limitations, and future prospects of Selective Laser Sintering. Selective Laser Sintering (SLS) is an additive manufacturing technique that fabricates objects by selectively fusing powdered materials, typically polymers, metals, or ceramics, using a high-powered laser beam. Unlike traditional subtractive manufacturing methods, which involve cutting away material from a solid block, SLS builds parts layer by layer from a bed of powdered material. This layer-by-layer approach enables the creation of intricate, geometrically complex parts with high precision and accuracy.

Working Principle of Selective Laser Sintering

The SLS process begins with the creation or acquisition of a digital 3D model using Computer-Aided Design (CAD) software. This digital model serves as the blueprint

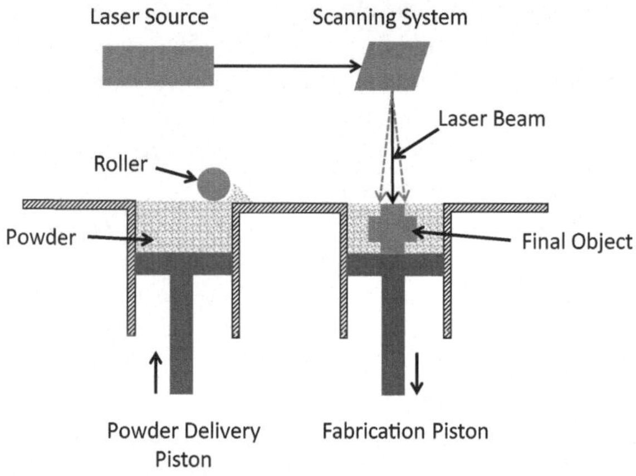

FIGURE 8.3 Schematic representation of selective laser sintering process. (Sanna M. Peltola and Kellomäki, 2008)

for the physical object to be manufactured. A thin layer of powdered material, such as nylon, polyamide, or metal alloy, is evenly spread across the build platform of the SLS machine. The powdered material serves as the raw feedstock for the manufacturing process. A high-powered laser beam is directed onto the surface of the powdered material, selectively sintering or fusing the particles together in a controlled manner, as shown in Figure 8.3. The laser follows the contour of each cross-section of the object as dictated by the digital model, solidifying the material where the laser strikes. After each layer is sintered, the build platform descends by a fraction of the layer thickness, and a new layer of powdered material is spread over the previously sintered layer. The process repeats iteratively, with the laser sintering each new layer onto the previous one, gradually building up the desired object. Once the entire object is sintered layer by layer, it is allowed to cool inside the SLS machine. The gradual solidification of the material results in a fully formed, three-dimensional object embedded within the unsintered powder bed. After printing is complete, the fabricated object is carefully removed from the SLS machine. Excess powder clinging to the surface of the part is removed using compressed air or vacuum systems. Depending on the application, additional post-processing steps such as sanding, surface finishing, or heat treatment may be performed to achieve the desired final properties (Peltola and Kellomäki, 2008). Selective Laser Sintering supports a diverse range of powdered materials, each with its unique properties and applications. Common materials used in SLS include:

1. Nylon (Polyamide):
   Nylon-based powders, such as PA11 and PA12, are widely used in SLS for their excellent mechanical properties, including strength, flexibility, and impact resistance. Nylon parts produced via SLS exhibit high dimensional accuracy and surface quality, making them suitable for functional prototypes, end-use parts, and customized components.

2. Thermoplastic Elastomers (TPE):
   Thermoplastic elastomers, such as TPU (thermoplastic polyurethane), are used in SLS to produce flexible, rubber-like parts with elastomeric properties. TPE materials are often employed in applications requiring shock absorption, vibration damping, or tactile characteristics, such as gaskets, seals, and cushioning components.
3. Metals:
   Selective Laser Sintering can also fabricate metal parts using powdered metal alloys, such as stainless steel, aluminum, titanium, and cobalt-chromium. Metal SLS parts exhibit high strength, corrosion resistance, and thermal conductivity, making them suitable for demanding applications in aerospace, automotive, and medical industries.
4. Ceramics:
   SLS supports ceramic powders, such as alumina, zirconia, and silicon carbide, for producing parts with exceptional hardness, wear resistance, and thermal stability. Ceramic SLS parts find applications in electronics, semiconductor, and aerospace industries, where high-temperature resistance and electrical insulation properties are essential.

Applications of Selective Laser Sintering

1. Selective Laser Sintering plays a critical role in the aerospace and defense industries, where lightweight, high-strength components are required. SLS is used to manufacture structural aircraft components, engine parts, UAV (unmanned aerial vehicle) components, and missile components with complex geometries and optimized designs.
2. In the automotive sector, SLS is employed for rapid prototyping, tooling, and low-volume production of customized parts. Automotive manufacturers use SLS to produce interior components, dashboard prototypes, functional prototypes of engine components, and customized parts for concept cars and specialty vehicles.
3. Selective Laser Sintering is widely utilized in the healthcare and medical device industries for producing patient-specific implants, surgical instruments, orthopedic implants, and dental prosthetics. SLS enables the fabrication of anatomically accurate medical models, personalized prostheses, and custom implants tailored to individual patient requirements.
4. SLS finds applications in the consumer products and electronics industries for producing consumer electronics, wearables, consumer goods, and electronic enclosures. Manufacturers leverage SLS for rapid prototyping of product designs, customization of consumer products, and low-volume production of specialty items.
5. SLS is also used for producing manufacturing tooling, jigs, fixtures, and assembly aids used in production processes. SLS-produced tooling offers advantages such as reduced lead times, lower costs, and enhanced customization compared to traditional machining methods, improving efficiency and productivity in manufacturing operations.

Advantages of Selective Laser Sintering

1. Selective Laser Sintering offers unparalleled design freedom, allowing designers to create complex geometries, intricate features, and lightweight structures that would be difficult or impossible to achieve with traditional manufacturing methods. SLS eliminates many of the design constraints associated with subtractive manufacturing, enabling innovative and optimized product designs.
2. SLS supports a wide range of materials, including polymers, metals, ceramics, and composites, offering flexibility in material selection based on application requirements. This versatility allows manufacturers to choose materials with specific mechanical, thermal, electrical, or chemical properties tailored to the intended use of the fabricated parts.
3. SLS is capable of producing parts with exceptional precision, dimensional accuracy, and surface finish. The layer-by-layer build process and fine resolution of the laser beam result in parts with tight tolerances and fine details, meeting the stringent requirements of various industries, including aerospace, medical, and automotive.
4. Selective Laser Sintering enables rapid prototyping and low-volume production of parts with quick turnaround times. The additive nature of SLS eliminates the need for costly tooling and setup, allowing for on-demand manufacturing of customized parts without long lead times or minimum order quantities.
5. SLS generates minimal material waste compared to traditional manufacturing methods, as only the necessary amount of powder is used to build each part. Unused powder can be recycled and reused in subsequent printing jobs, minimizing environmental impact and material costs.

Limitations of Selective Laser Sintering

1. SLS parts may exhibit rough surface finishes and visible layer lines, especially on curved or sloped surfaces. Post-processing techniques, such as sanding, polishing, or surface coating, may be required to achieve the desired surface finish.
2. SLS parts may experience dimensional inaccuracies, particularly in features with tight tolerances or small details. Factors such as powder bed density, laser energy, and thermal effects can affect the overall accuracy of printed parts.
3. SLS parts with overhangs or complex geometries may require support structures to prevent warping and distortion during printing. These support structures add material and time to the printing process and may require post-processing removal.
4. SLS machines and materials can be costly compared to other additive manufacturing technologies, limiting accessibility for small businesses, startups, and educational institutions.

Ongoing Developments and Future Outlook

Selective Laser Sintering (SLS) is advancing rapidly due to continuous progress in material science, engineering, and digital innovation. Researchers are creating new high-performance materials with greater durability, flexibility, and thermal resistance, which greatly expand the possible uses for SLS. Enhancements in SLS machinery, such as more powerful lasers, quicker scanning speeds, and larger build platforms, are boosting efficiency and productivity. As these technological improvements make SLS more accessible, its role in the manufacturing sector is expected to grow, fostering greater innovation, customization, and sustainability across various industries.

SLS has transformed the process of designing, prototyping, and producing complex components. Its remarkable design flexibility, wide range of material options, precision, and efficiency have made it essential in sectors like aerospace, automotive, healthcare, and consumer goods. Despite some challenges, ongoing advancements in SLS technology are likely to further extend its capabilities and applications, paving the way for a future where customized, on-demand manufacturing becomes standard practice.

8.2.1.4 Binder Jetting

Binder Jetting stands as a revolutionary additive manufacturing process that has gained significant traction in recent years due to its versatility, speed, and cost-effectiveness. Developed as an alternative to traditional manufacturing methods, binder jetting offers a unique approach to fabricating complex three-dimensional objects with a wide range of materials. In this comprehensive exploration, we will delve into the principles, working mechanism, materials, applications, advantages, limitations, and ongoing advancements of binder jetting.

Binder Jetting is an additive manufacturing technique that builds parts layer by layer by selectively depositing a binding agent onto a powdered material bed. The binding agent selectively bonds the powder particles together, forming a solid object according to the digital design. Unlike other additive manufacturing processes that utilize lasers or heat to fuse materials, binder jetting relies on the precise application of liquid binders to create parts with intricate geometries and fine details.

Working Mechanism of Binder Jetting

The binder jetting process begins with the creation or acquisition of a digital 3D model using Computer-Aided Design (CAD) software. This digital model serves as the blueprint for the physical object to be fabricated. A thin layer of powdered material, such as polymer, metal, ceramic, or sand, is spread evenly across the build platform of the binder jetting machine. The powdered material serves as the raw feedstock for the manufacturing process. An inkjet printhead moves across the surface of the powdered material bed, depositing droplets of liquid binding agent onto specific locations according to the digital design. The binding agent selectively binds the powder particles together, forming solidified layers corresponding to the cross-sections of the object being fabricated. After each layer is printed with the binding agent, the build platform descends by a fraction of the layer thickness, and a new

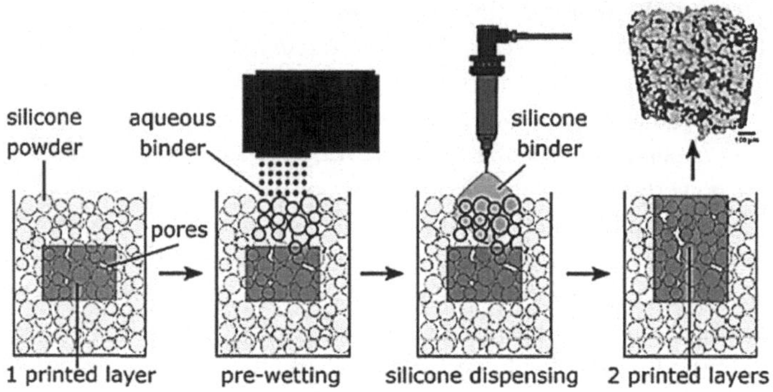

FIGURE 8.4 Schematic representation of Binder Jetting. (Reprinted with permission from Liravi and Vlasea (2018), Copyright © 2018, Elsevier.)

layer of powdered material is spread over the previously printed layer. The process repeats iteratively, with the inkjet printhead depositing binder droplets onto each layer until the entire object is fully formed. Once printing is complete, the fabricated object is carefully removed from the binder jetting machine, as shown in Figure 8.4. Excess powder clinging to the surface of the part is removed using compressed air or vacuum systems. Depending on the application, additional post-processing steps such as curing, infiltration, or surface finishing may be performed to enhance the mechanical properties and surface quality of the part. Binder Jetting supports a wide range of powdered materials, each with its unique properties and applications. Common materials used in binder jetting includes, polymers such as polyamide, nylon, polypropylene, metals and ceramics (Du et al., 2020; Gokuldoss et al., 2017; Liravi and Vlasea, 2018).

Applications of Binder Jetting

1. Binder jetting is widely used for rapid prototyping and product development across industries. Designers and engineers utilize binder jetting to create functional prototypes, concept models, and validation prototypes for testing and evaluation. The ability to produce parts quickly and cost-effectively accelerates the product development cycle and reduces time to market.
2. Binder Jetting enables low-volume production of customized parts, small batches, and on-demand manufacturing. Industries such as aerospace, automotive, and consumer goods leverage binder jetting for producing end-use parts, tooling, and manufacturing aids with reduced lead times and production costs compared to traditional manufacturing methods.
3. Manufacturing companies utilize binder jetting to produce custom tooling, jigs, fixtures, and assembly aids for use in production processes. Binder Jetting-produced tooling offers advantages such as reduced lead times, lower costs, and enhanced customization compared to conventional machining methods, improving efficiency and productivity in manufacturing operations.

4. Binder Jetting is used to fabricate sand molds and cores for metal casting applications in foundries and manufacturing industries. Sand binder jetting enables the rapid production of complex, intricate molds and cores with precise geometries, reducing lead times and costs associated with traditional mold-making methods.
5. Binder Jetting is employed in the dental industry for producing dental crowns, bridges, and prosthetics with high precision and accuracy. Dental labs utilize binder jetting to create customized restorations tailored to individual patient requirements, ensuring optimal fit, aesthetics, and functionality.

Advantages of Binder Jetting

1. Binder Jetting is a fast and efficient additive manufacturing process, capable of producing parts quickly and cost-effectively. The layer-by-layer build approach allows for the simultaneous production of multiple parts in a single print job, increasing efficiency and productivity.
2. Binder Jetting offers unparalleled design freedom, allowing for the creation of complex geometries, intricate details, and lightweight structures that would be difficult or impossible to achieve with traditional manufacturing methods. This design flexibility encourages innovation and creativity in product development.
3. Binder Jetting supports a wide range of powdered materials, including polymers, metals, ceramics, and composites, offering flexibility in material selection based on application requirements. This material versatility enables the production of parts with diverse mechanical, thermal, and chemical properties.
4. Binder Jetting is a cost-effective manufacturing solution, particularly for low-volume production and customized parts. The additive nature of binder jetting eliminates the need for expensive tooling and setup, reducing production costs and lead times compared to traditional manufacturing methods.
5. Binder Jetting generates minimal material waste compared to subtractive manufacturing methods, as only the necessary amount of powder is used to build each part. Unused powder can be recycled and reused in subsequent printing jobs, minimizing environmental impact and material costs.

Limitations of Binder Jetting

1. Binder Jetting parts may exhibit rough surface finishes and visible layer lines, especially on curved or sloped surfaces. Post-processing techniques, such as sanding, polishing, or surface coating, may be required to achieve the desired surface finish.
2. Binder Jetting parts may experience dimensional inaccuracies, particularly in features with tight tolerances or small details. Factors such as powder bed density, binder penetration, and curing process can affect the overall accuracy of printed parts.

3. Binder Jetting parts may have inferior mechanical properties compared to parts produced by other additive manufacturing processes such as Selective Laser Sintering (SLS) or Fused Deposition Modeling (FDM). The binding agent used in binder jetting may result in reduced strength, stiffness, and durability of the final parts.
4. Binder Jetting parts with overhangs or complex geometries may require support structures to prevent sagging and deformation during printing. These support structures add material and time to the printing process and may require post-processing removal.

Ongoing Developments and Future Outlook

Binder jetting technology is experiencing significant advancements driven by continuous innovations in materials, machinery, and digital processes. New binding agents and powders with enhanced properties, such as improved strength, density, and precision, are expanding the potential applications of binder jetting. These advancements are making it possible to produce parts with higher structural integrity and finer detail. In terms of machinery, advancements such as faster printing speeds, larger build volumes, and more advanced control systems are boosting the efficiency and scalability of binder jetting. Additionally, the integration of AI and machine learning into binder jetting processes is optimizing production parameters, resulting in higher quality outputs and reduced waste. As binder jetting technology becomes more accessible and cost-effective, its impact on various industries—ranging from aerospace and automotive to healthcare and consumer goods—is expected to grow. This growth will drive innovation in product design and manufacturing, enabling more sustainable and customized production methods. The future of binder jetting looks promising, with ongoing advancements poised to further enhance its capabilities and applications, solidifying its role in the next generation of manufacturing technologies.

8.2.1.5 Electron Beam Melting (EBM)

Electron Beam Melting (EBM) represents a cutting-edge additive manufacturing technology that has transformed the landscape of metal fabrication. Developed by Arcam AB in the 1990s, EBM offers unparalleled precision, material properties, and design flexibility in producing complex metal components. In this comprehensive overview, we will delve into the principles, working mechanism, materials, applications, advantages, limitations, and ongoing advancements of electron beam melting. Electron Beam Melting (EBM) is an additive manufacturing technique that utilizes an electron beam to selectively melt layers of metal powder, building up parts layer by layer. Unlike traditional manufacturing methods that involve subtractive processes, EBM creates parts from the ground up, offering unmatched design freedom and versatility. EBM is particularly well-suited for producing high-quality, fully dense metal components with intricate geometries and superior mechanical properties.

Working Mechanism of Electron Beam Melting

The EBM process begins with the creation or acquisition of a digital 3D model using Computer-Aided Design (CAD) software. This digital model serves as the

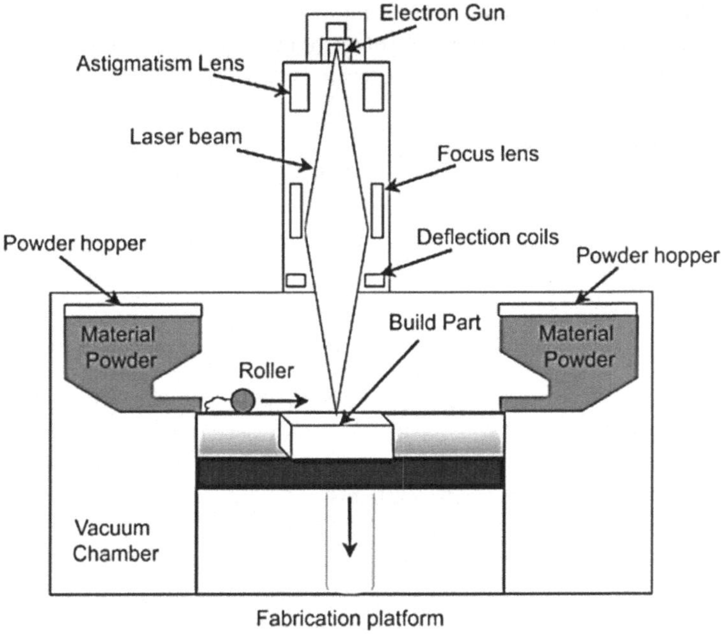

FIGURE 8.5 Schematic representation of Electron Beam Melting (EBM) process. (Ameen et al., 2022)

blueprint for the physical object to be fabricated. A thin layer of metal powder, typically titanium, titanium alloys, or other high-performance metals, is spread evenly across the build platform of the EBM machine. The powdered metal serves as the raw material for the manufacturing process.

An electron beam is generated within the vacuum chamber of the EBM machine and directed onto the surface of the powdered metal bed as shown in Figure 8.5. The electron beam selectively melts the metal powder particles in specific locations according to the digital design, fusing them together to form solidified layers. After each layer is melted by the electron beam, the build platform descends by a fraction of the layer thickness, and a new layer of powdered metal is spread over the previously melted layer. The process repeats iteratively, with the electron beam melting each new layer onto the previous one until the entire object is fully formed. Once printing is complete, the fabricated metal object is carefully removed from the EBM machine. Excess powder clinging to the surface of the part is removed using compressed air or vacuum systems. Depending on the application, additional post-processing steps such as heat treatment, surface finishing, or machining may be performed to achieve the desired final properties (Galati and Iuliano, 2018; Gokuldoss et al., 2017).

Materials Used in Electron Beam Melting

Electron Beam Melting supports a wide range of metal powders, each with its unique properties and applications. Common materials used in EBM include:

1. Titanium and Titanium Alloys:
   Titanium and its alloys, such as Ti6Al4V (Titanium 6% Aluminum 4% Vanadium), are widely used in EBM for their excellent strength-to-weight ratio, corrosion resistance, and biocompatibility. Titanium EBM parts find applications in aerospace, medical, and automotive industries, where high-performance, lightweight components are required.
2. Nickel-Based Superalloys:
   Nickel-based superalloys, such as Inconel 718 and Hastelloy X, are utilized in EBM for their exceptional heat resistance, creep strength, and oxidation resistance. Nickel EBM parts are commonly used in aerospace, power generation, and chemical processing industries, where components operate in high-temperature, high-stress environments.
3. Stainless Steel:
   Stainless steel powders, such as 316L and 17-4 PH, are employed in EBM for their corrosion resistance, mechanical properties, and affordability. Stainless steel EBM parts find applications in aerospace, automotive, and consumer goods industries, where durability, hygiene, and aesthetics are essential.
4. Cobalt-Chromium Alloys:
   Cobalt-chromium alloys, such as CoCrMo and CoCrW, are used in EBM for their high strength, wear resistance, and biocompatibility. Cobalt-chromium EBM parts are commonly used in medical and dental applications, such as orthopedic implants, dental prosthetics, and surgical instruments.

Applications of Electron Beam Melting

1. Electron Beam Melting plays a critical role in the aerospace industry for producing lightweight, high-strength components such as turbine blades, structural brackets, and engine components. EBM parts offer the performance characteristics required for aerospace applications, including fatigue resistance, thermal stability, and corrosion resistance.
2. In the medical sector, electron beam melting is utilized for manufacturing patient-specific implants, surgical instruments, and medical devices with complex geometries and tailored properties. EBM enables the fabrication of orthopedic implants, cranial implants, and dental prosthetics with precise fit, biocompatibility, and osseointegration.
3. Electron Beam Melting is employed in the automotive industry for producing lightweight, high-performance components such as engine parts, suspension components, and transmission components. EBM enables automotive manufacturers to reduce vehicle weight, improve fuel efficiency, and enhance overall performance.
4. In the energy sector, electron beam melting is used for manufacturing components for power generation, including gas turbine blades, steam turbine

blades, and heat exchanger components. EBM-produced parts offer superior mechanical properties, thermal stability, and corrosion resistance, ensuring reliable performance in demanding operating conditions.

Advantages of Electron Beam Melting

1. Electron Beam Melting produces parts with exceptional mechanical properties, including high strength, ductility, and fatigue resistance. EBM parts exhibit isotropic properties, meaning they have consistent material properties in all directions, making them ideal for structural applications.
2. Electron Beam Melting offers unparalleled design freedom, allowing designers to create complex geometries, internal features, and lightweight structures that would be difficult or impossible to achieve with traditional manufacturing methods. EBM eliminates many of the design constraints associated with subtractive manufacturing, enabling innovative and optimized product designs.
3. Electron Beam Melting is a highly efficient manufacturing process that minimizes material waste. The additive nature of EBM ensures that only the necessary amount of material is used to build each part, reducing material costs and environmental impact compared to traditional manufacturing methods.
4. Electron Beam Melting enables rapid prototyping and production of metal components with quick turnaround times. The layer-by-layer build approach allows for the simultaneous production of multiple parts in a single print job, increasing efficiency and productivity.
5. EBM parts exhibit excellent high-temperature capability, making them suitable for applications in aerospace, energy, and automotive industries where components operate in extreme temperature environments. EBM-produced parts maintain their structural integrity and mechanical properties at elevated temperatures, ensuring reliable performance under thermal stress.

Limitations of Electron Beam Melting

1. Electron Beam Melting parts may exhibit rough surface finishes and visible layer lines, especially on curved or sloped surfaces. Post-processing techniques, such as machining, grinding, or polishing, may be required to achieve the desired surface finish.
2. Electron Beam Melting machines typically have limited build volumes, restricting the size of parts that can be produced in a single print job. Large components may need to be divided into multiple segments and assembled post-production, increasing complexity and lead times.
3. Electron Beam Melting machines and materials can be costly compared to other additive manufacturing technologies, limiting accessibility for small businesses, startups, and educational institutions.
4. Electron Beam Melting supports a limited range of metal powders compared to other additive manufacturing processes such as Powder Bed Fusion (PBF).

While EBM offers superior mechanical properties and material efficiency, the selection of suitable materials may be restricted based on application requirements.

Ongoing Developments and Future Outlook

Electron Beam Melting technology continues to advance with ongoing research and development efforts aimed at improving process capabilities, material options, and machine performance. Manufacturers are developing new metal alloys, optimizing process parameters, and enhancing post-processing techniques to address existing limitations and expand the range of applications for EBM technology. As EBM technology matures and becomes more widely adopted, it is expected to play an increasingly significant role in the future of metal additive manufacturing, driving innovation, customization, and sustainability across industries. Thus, electron beam melting (EBM) stands as a transformative manufacturing technology that has revolutionized the way complex metal parts are designed, prototyped, and produced. With its superior mechanical properties, design freedom, and material efficiency, EBM has become an indispensable tool for industries ranging from aerospace and automotive to medical and energy. Despite some limitations, ongoing advancements in EBM technology are poised to further expand its capabilities and applications, driving innovation and shaping the future of metal additive manufacturing.

8.2.1.6 Direct Metal Laser Sintering (DMLS)

Direct Metal Laser Sintering (DMLS) stands at the forefront of metal additive manufacturing technologies, offering unprecedented precision, complexity, and material diversity in the production of metal parts. Developed as a derivative of Selective Laser Sintering (SLS), DMLS utilizes a high-powered laser to selectively fuse metal powders, layer by layer, into fully functional components. In this comprehensive exploration, we will explore into the principles, process, materials, applications, advantages, limitations, and ongoing advancements of Direct Metal Laser Sintering.

DMLS is an advanced additive manufacturing process that fabricates metal parts by selectively sintering or melting metal powder particles using a high-powered laser beam. Unlike traditional subtractive manufacturing methods, which involve cutting away material from a solid block, DMLS builds parts layer by layer from a bed of powdered metal. This layer-by-layer approach enables the creation of complex geometries, internal structures, and customized components with high precision and accuracy.

Working Principle of Direct Metal Laser Sintering

The DMLS process begins with the creation or acquisition of a digital 3D model using Computer-Aided Design (CAD) software. This digital model serves as the blueprint for the physical object to be manufactured. A thin layer of metal powder, such as stainless steel, aluminum, titanium, or cobalt-chromium alloy, is spread evenly across the build platform of the DMLS machine. The powdered metal serves as the raw feedstock for the manufacturing process. A high-powered laser beam is directed onto the surface of the powdered metal bed, selectively sintering

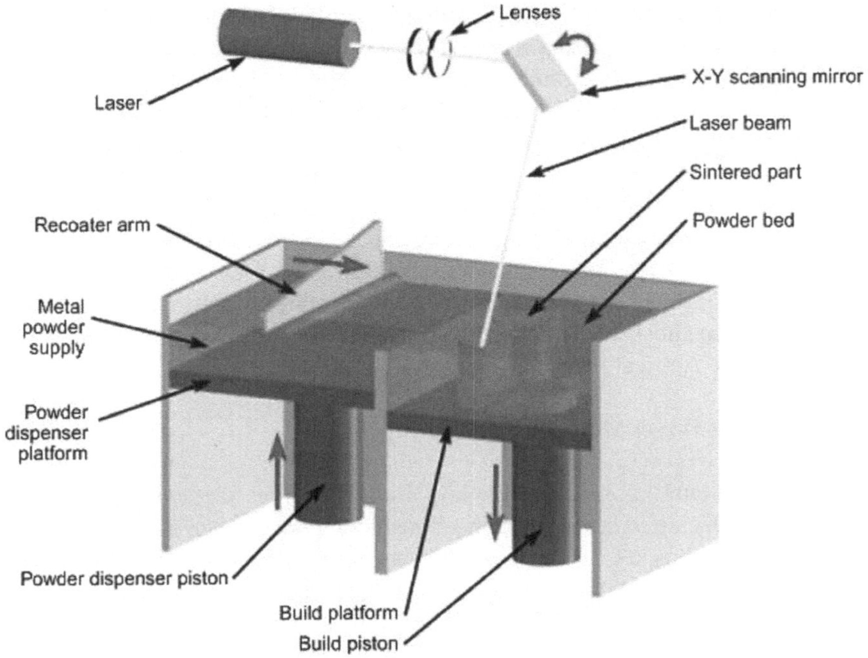

FIGURE 8.6 Schematic representation of Direct Metal Laser Sintering (Reprinted with permission from Criales et al. (2017), Copyright © 2017, Elsevier.)

or melting the metal powder particles in specific locations according to the digital design. The laser follows the contour of each cross-section of the object as dictated by the digital model, solidifying the material where the laser strikes. After each layer is sintered or melted by the laser, the build platform descends by a fraction of the layer thickness, and a new layer of powdered metal is spread over the previously sintered or melted layer. The process repeats iteratively, with the laser sintering or melting each new layer onto the previous one until the entire object is fully formed. Once the entire object is sintered or melted layer by layer, it is allowed to cool inside the DMLS machine, as shown in Figure 8.6. The gradual solidification of the metal results in a fully formed, three-dimensional object embedded within the unsintered or unmelted powder bed. After printing is complete, the fabricated metal object is carefully removed from the DMLS machine. Excess powder clinging to the surface of the part is removed using compressed air or vacuum systems. Depending on the application, additional post-processing steps such as heat treatment, surface finishing, or machining may be performed to achieve the desired final properties. Direct Metal Laser Sintering supports a wide range of metal powders, each with its unique properties and applications. Stainless steel powders, such as 316L and 17-4 PH, are widely used in DMLS for their corrosion resistance, mechanical properties, and affordability. Stainless steel DMLS parts find applications in aerospace, automotive, and consumer goods industries, where durability, hygiene, and aesthetics are essential. Aluminum powders are utilized in DMLS for their lightweight, high strength-to-weight ratio, and thermal conductivity. Aluminum DMLS parts are commonly used in aerospace, automotive, and electronics industries, where weight reduction and thermal management are critical. Titanium and its alloys, such as Ti6Al4V (Titanium 6% Aluminum 4% Vanadium), are frequently used in DMLS for their excellent strength-to-weight ratio, corrosion resistance, and biocompatibility. Titanium DMLS parts find applications in aerospace, medical, and automotive industries, where high-performance, lightweight components are required. Cobalt-chromium alloys, such as CoCrMo and CoCrW, are used in DMLS for their high strength, wear resistance, and biocompatibility. Cobalt-chromium DMLS parts are commonly used in medical and dental applications, such as orthopedic implants, dental prosthetics, and surgical instruments (Martinho, 2021).

Applications of Direct Metal Laser Sintering

1. Direct Metal Laser Sintering plays a critical role in the aerospace industry for producing lightweight, high-strength components such as turbine blades, structural brackets, and engine components. DMLS parts offer the performance characteristics required for aerospace applications, including fatigue resistance, thermal stability, and corrosion resistance.
2. In the medical sector, Direct Metal Laser Sintering is utilized for manufacturing patient-specific implants, surgical instruments, and medical devices with complex geometries and tailored properties. DMLS enables the fabrication of orthopedic implants, cranial implants, and dental prosthetics with precise fit, biocompatibility, and osseointegration.
3. Direct Metal Laser Sintering is employed in the automotive industry for producing lightweight, high-performance components such as engine parts, suspension components, and transmission components. DMLS enables automotive manufacturers to reduce vehicle weight, improve fuel efficiency, and enhance overall performance.
4. Manufacturing companies utilize Direct Metal Laser Sintering to produce custom tooling, jigs, fixtures, and assembly aids for use in production processes. DMLS-produced tooling offers advantages such as reduced lead times, lower costs, and enhanced customization compared to conventional machining methods, improving efficiency and productivity in manufacturing operations.

Advantages of Direct Metal Laser Sintering

1. Direct Metal Laser Sintering offers unparalleled precision and accuracy in producing complex metal parts with tight tolerances and fine features. The layer-by-layer build approach allows for the creation of intricate geometries, internal structures, and customized components with minimal post-processing requirements.
2. Direct Metal Laser Sintering provides designers with unprecedented design freedom, allowing for the creation of complex shapes, organic forms, and lightweight structures that would be challenging or impossible to achieve with

traditional manufacturing methods. DMLS eliminates many of the design constraints associated with subtractive manufacturing, enabling innovative and optimized product designs.

3. Direct Metal Laser Sintering is a highly efficient manufacturing process that minimizes material waste. The additive nature of DMLS ensures that only the necessary amount of material is used to build each part, reducing material costs and environmental impact compared to traditional manufacturing methods.
4. Direct Metal Laser Sintering enables rapid prototyping and production of metal components with quick turnaround times. The layer-by-layer build approach allows for the simultaneous production of multiple parts in a single print job, increasing efficiency and productivity.
5. Direct Metal Laser Sintering produces parts with exceptional mechanical properties, including high strength, ductility, and fatigue resistance. DMLS parts exhibit isotropic properties, meaning they have consistent material properties in all directions, making them ideal for structural applications.

Limitations of Direct Metal Laser Sintering

1. Direct Metal Laser Sintering parts may exhibit rough surface finishes and visible layer lines, especially on curved or sloped surfaces. Post-processing techniques, such as machining, grinding, or polishing, may be required to achieve the desired surface finish.
2. Direct Metal Laser Sintering machines typically have limited build volumes, restricting the size of parts that can be produced in a single print job. Large components may need to be divided into multiple segments and assembled post-production, increasing complexity and lead times.
3. Direct Metal Laser Sintering machines and materials can be costly compared to other additive manufacturing technologies, limiting accessibility for small businesses, startups, and educational institutions.
4. Direct Metal Laser Sintering supports a limited range of metal powders compared to other additive manufacturing processes such as Powder Bed Fusion (PBF). While DMLS offers superior mechanical properties and material efficiency, the selection of suitable materials may be restricted based on application requirements.

Ongoing Developments and Future Outlook

Direct Metal Laser Sintering (DMLS) technology continues to progress through ongoing research and development efforts focused on enhancing process capabilities, expanding material options, and improving machine performance. Manufacturers are innovating with new metal alloys, refining process parameters, and advancing post-processing techniques to overcome current limitations and broaden the scope of applications for DMLS. As DMLS matures and gains wider acceptance, it is poised to play an increasingly pivotal role in the future of metal additive manufacturing, fostering innovation, customization, and sustainability across various sectors.

DMLS represents a transformative manufacturing technology that has revolutionized the design, prototyping, and production of intricate metal parts. Its exceptional precision, design flexibility, and material efficiency have made it indispensable in industries such as aerospace, automotive, medical, and tooling. Despite inherent challenges, ongoing advancements in DMLS technology are expected to significantly enhance its capabilities and applications, driving continued innovation and shaping the future of metal additive manufacturing.

8.3 NANO COMPOSITES IN STRUCTURAL APPLICATIONS

Nanocomposites represent a class of materials that integrate nanoparticles into a matrix material, resulting in enhanced mechanical, thermal, electrical, and optical properties. In structural applications, nanocomposites offer significant advantages such as improved strength-to-weight ratio, durability, corrosion resistance, and thermal stability. This comprehensive exploration will delve into the principles, fabrication methods, properties, applications, advantages, challenges, and future prospects of nanocomposites in structural engineering.

8.3.1 Introduction to Nanocomposites in Structural Applications

Nanocomposites have emerged as promising materials for structural engineering due to their unique properties and potential to overcome limitations of conventional materials. By incorporating nanoscale reinforcements, such as nanoparticles, nanotubes, or nanofibers, into a matrix material, nanocomposites can achieve superior mechanical, thermal, and electrical performance compared to traditional composites. In structural applications, nanocomposites offer opportunities for lightweight construction, improved energy efficiency, and enhanced sustainability (Sen, 2020).

8.3.2 Fabrication Methods of Nanocomposites

1. Polymer Nanocomposites:
   Polymer nanocomposites are fabricated by dispersing nanoparticles, such as clay platelets or carbon nanotubes, into a polymer matrix using various processing techniques. Common methods include melt blending, solution mixing, in-situ polymerization, and electrospinning. These techniques allow for precise control over nanoparticle dispersion, orientation, and interfacial bonding, resulting in tailored properties for specific structural applications.
2. Metal Matrix Nanocomposites:
   Metal matrix nanocomposites are produced by incorporating nanoparticles, such as carbon nanotubes or graphene, into a metal matrix via powder metallurgy, mechanical alloying, or electrodeposition. These methods enable the homogeneous distribution of nanoparticles within the metal matrix, enhancing mechanical strength, stiffness, and thermal conductivity. Metal matrix nanocomposites find applications in aerospace, automotive, and electronics industries for lightweight structural components and heat management systems.

3. Ceramic Matrix Nanocomposites:
   Ceramic matrix nanocomposites are synthesized by dispersing nanoparticles, such as alumina or silicon carbide, into a ceramic matrix through techniques like sol-gel processing, chemical vapor deposition, or spark plasma sintering. These methods facilitate the incorporation of nanoparticles with controlled morphology and distribution, resulting in enhanced mechanical toughness, wear resistance, and thermal stability. Ceramic matrix nanocomposites are used in high-temperature structural applications, such as aerospace components, turbine blades, and thermal barrier coatings (Sen, 2020).

8.3.3 Properties of Nanocomposites

1. Mechanical Properties:
   Nanocomposites exhibit superior mechanical properties compared to conventional materials due to the reinforcement provided by nanoparticles. Enhanced tensile strength, modulus of elasticity, and fracture toughness are observed in nanocomposites, making them ideal for structural applications requiring high strength-to-weight ratios and impact resistance.
2. Thermal Properties:
   Nanocomposites demonstrate improved thermal conductivity, heat resistance, and thermal stability attributed to the high aspect ratio and surface-to-volume ratio of nanoparticles. These properties make nanocomposites suitable for thermal management systems, fire-resistant materials, and high-temperature structural components.
3. Electrical Properties:
   Certain nanocomposites exhibit enhanced electrical conductivity, dielectric strength, and electromagnetic interference shielding capabilities due to the conductive nature of nanoparticles. These properties enable applications in electrical and electronic devices, smart structures, and electromagnetic shielding materials.
4. Optical Properties:
   Nanocomposites can possess tailored optical properties, including transparency, opacity, and light absorption, depending on the composition and morphology of nanoparticles. These properties find applications in optical coatings, sensors, and photovoltaic devices.

8.3.4 Applications of Nanocomposites in Structural Engineering

1. Nanocomposites are increasingly used in aerospace structures for lightweight construction, improved fuel efficiency, and enhanced mechanical performance. Carbon nanotube-reinforced polymer composites are employed in aircraft components, such as fuselage panels, wings, and interior fittings, to reduce weight and increase strength.
2. In the automotive industry, nanocomposites are utilized for manufacturing lightweight structural components, such as body panels, chassis parts, and engine components. Polymer nanocomposites reinforced with carbon

nanotubes or graphene offer superior strength, stiffness, and impact resistance, contributing to fuel economy and vehicle safety.

3. Nanocomposites are incorporated into building materials, such as concrete, steel, and polymer composites, to enhance structural performance, durability, and sustainability. Nanoparticles, such as nanosilica or carbon nanofibers, improve the strength, ductility, and corrosion resistance of construction materials, leading to longer service life and reduced maintenance costs.
4. Nanocomposites are employed in the rehabilitation and strengthening of civil infrastructure, such as bridges, tunnels, and pipelines. Carbon nanotube-reinforced polymer composites are used for structural reinforcement, crack repair, and corrosion protection, extending the lifespan of aging infrastructure and reducing the need for costly replacements.

8.3.5 Advantages of Nanocomposites in Structural Engineering

1. Nanocomposites offer significant weight savings compared to traditional materials, making them ideal for lightweight structural components in aerospace, automotive, and marine applications. The high strength-to-weight ratio of nanocomposites enables fuel efficiency, payload capacity, and overall performance improvements.
2. Nanocomposites exhibit superior mechanical properties, including strength, stiffness, toughness, and fatigue resistance, resulting in durable and reliable structural components. The reinforcement provided by nanoparticles enhances load-bearing capacity, impact resistance, and structural integrity, reducing the risk of failure under various operating conditions.
3. Nanocomposites offer enhanced durability, corrosion resistance, and weatherability compared to conventional materials, leading to longer service life and reduced maintenance requirements. The barrier properties of nanoparticles inhibit moisture ingress, chemical degradation, and environmental degradation, ensuring structural integrity and performance over time.
4. Nanocomposites can be tailored to meet specific performance requirements for diverse structural applications by adjusting the type, size, shape, and concentration of nanoparticles. Customized nanocomposite formulations enable engineers to optimize properties such as strength, stiffness, thermal conductivity, and electrical conductivity for optimal performance in various environments and operating conditions.

8.3.6 Challenges and Limitations of Nanocomposites in Structural Engineering

1. The production cost of nanocomposites remains relatively high compared to traditional materials due to the expense of nanoparticles, fabrication techniques, and quality control measures. Scaling up nanocomposite manufacturing processes to industrial levels while maintaining cost-effectiveness poses challenges for widespread adoption in structural engineering applications.

2. Achieving uniform dispersion of nanoparticles within the matrix material is essential for maximizing the performance of nanocomposites. However, nanoparticles tend to agglomerate or cluster together, leading to uneven distribution, weak interfacial bonding, and degradation of mechanical properties. Developing effective dispersion techniques and surface modification strategies is crucial for overcoming this challenge.
3. The interface between nanoparticles and the matrix material plays a critical role in determining the mechanical properties and performance of nanocomposites. Poor interfacial bonding, incompatible materials, or mismatched properties can result in reduced strength, delamination, or premature failure of nanocomposite structures. Enhancing interfacial adhesion and compatibility through surface treatments, functionalization, or interlayers is essential for optimizing structural performance.
4. The potential toxicity and environmental impact of nanoparticles pose health and safety concerns during nanocomposite fabrication, handling, and disposal. Inhalation or skin exposure to nanoparticles may cause adverse health effects, necessitating proper safety protocols and protective measures in nanocomposite manufacturing facilities. Additionally, the release of nanoparticles into the environment during production, use, or end-of-life disposal raises environmental concerns and regulatory considerations.

8.3.7 Ongoing Developments and Future Prospects

Despite the challenges and limitations, ongoing research and development efforts are advancing the field of nanocomposites in structural engineering. Innovations in nanoparticle synthesis, dispersion techniques, manufacturing processes, and characterization methods are driving progress toward the commercialization of nanocomposite materials for a wide range of applications. As nanocomposite technology matures and becomes more cost-effective, sustainable, and scalable, it is expected to play an increasingly significant role in enhancing the performance, durability, and sustainability of structural systems worldwide. Thus Nanocomposites represent a transformative class of materials that hold great promise for revolutionizing structural engineering. By harnessing the unique properties of nanoparticles, nanocomposites offer lightweight construction, enhanced mechanical performance, improved durability, and tailored properties for diverse structural applications. Despite the challenges and limitations, ongoing advancements in nanocomposite fabrication, characterization, and application are paving the way for widespread adoption in aerospace, automotive, construction, and infrastructure sectors. With continued research, innovation, and collaboration across disciplines, nanocomposites are poised to shape the future of structural engineering, offering sustainable solutions for a rapidly evolving built environment.

8.4 SELF-HEALING MATERIALS

In the realm of material science and engineering, the quest for resilient and durable materials has been an ongoing pursuit. One of the most intriguing developments in this field is the emergence of self-healing materials. These materials possess the remarkable ability to autonomously repair damage, akin to the regenerative capabilities found in biological systems. This section explores the fascinating world of self-healing materials, delving into their mechanisms, types, applications, challenges, and future prospects.

8.4.1 Inspiration

The concept of self-healing materials is deeply rooted in nature's ability to heal and regenerate. Drawing inspiration from biological systems such as the human body's wound healing process, researchers have sought to mimic these mechanisms in synthetic materials. The idea of self-healing materials is not only to repair damage but also to extend the lifespan of materials and reduce the need for frequent maintenance and replacements.

8.4.2 Mechanisms of Self-Healing

Self-healing materials employ various mechanisms to achieve autonomous repair. These mechanisms include:

8.4.2.1 Microcapsule-Based Systems

Microcapsule-based systems represent a fascinating approach to achieving self-healing properties in materials. These systems involve the incorporation of tiny capsules, typically ranging from a few micrometers to a few hundred micrometers in size, throughout the material matrix. Each capsule contains a healing agent, which can be a liquid, solid, or gel-like substance capable of repairing the material upon activation. The process begins with the fabrication of the microcapsules, which are often made from polymers or other materials that can encapsulate the healing agent effectively. These capsules are then dispersed homogeneously within the bulk material during the manufacturing process. The distribution of the capsules throughout the material ensures that the healing agent is readily available to repair any damage that may occur. When the material experiences external damage, such as cracks or fractures due to mechanical stress, the microcapsules located near the damaged area are ruptured. This rupture can occur due to the pressure exerted by the propagating crack or through the application of external stimuli, such as heat or pressure. Once the capsules rupture, the healing agent is released and flows into the crack or fracture site. The healing agent then comes into contact with any reactive species present in the damaged area, initiating a chemical reaction or phase change that leads to the formation of a solid and durable filler material. This filler material effectively seals the crack, restoring the material's integrity and preventing further propagation of the damage. The choice of healing agent depends on the specific application requirements and the type of damage expected to occur. Common healing agents include epoxy resins, polyurethanes, and other polymers that can undergo rapid polymerization or cross-linking reactions to form a strong bond with the surrounding material as shown in Figure 8.7.

One of the key advantages of microcapsule-based systems is their ability to provide localized healing precisely where it is needed. Since the capsules are distributed

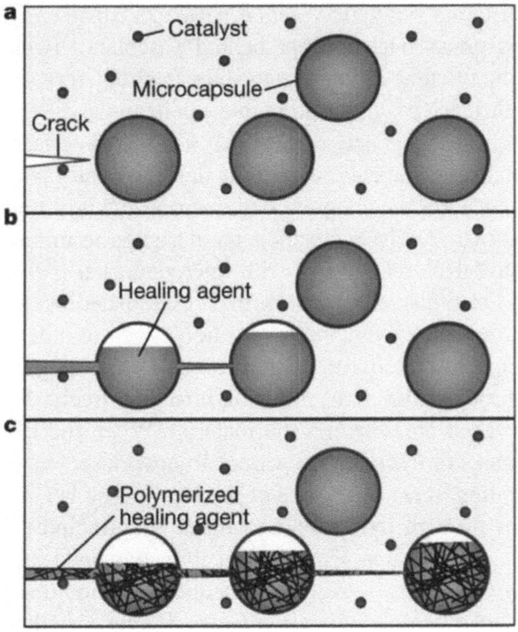

FIGURE 8.7 Schematic representation of microcapsule based self-healing system. (Reprinted with permission from White et al. (2001). Copyright © 2001, Springer.)

throughout the material, they can respond to damage occurring at any location within the material, ensuring comprehensive and effective repair. Additionally, the encapsulation of the healing agent protects it from degradation or premature activation, preserving its efficacy until it is needed.

However, microcapsule-based systems also face certain challenges. The size and distribution of the capsules can affect the mechanical properties of the material, potentially compromising its overall performance. Moreover, the healing process may not be instantaneous, particularly if the reaction kinetics of the healing agent are slow. Researchers continue to explore ways to optimize the design and performance of microcapsule-based systems, addressing these challenges to unlock their full potential in a wide range of applications. Thus, microcapsule-based systems offer a promising avenue for achieving self-healing properties in materials, allowing them to repair damage autonomously and prolong their lifespan. By encapsulating healing agents within tiny capsules dispersed throughout the material, these systems provide a localized and efficient means of repair, offering significant benefits for various engineering applications.

8.4.2.2 Vascular Networks

Vascular networks in self-healing materials represent a remarkable bioinspired approach to achieving autonomous repair. Inspired by the intricate network of blood vessels in the human circulatory system, researchers have developed innovative materials that incorporate synthetic vascular networks capable of delivering healing agents to damaged areas. The concept behind vascular networks in self-healing materials is to mimic the functionality of the human circulatory system, where blood vessels transport nutrients, oxygen, and immune cells to tissues throughout the body. In a similar manner, vascular networks in self-healing materials serve as conduits for the delivery of healing agents to sites of damage. The fabrication of vascular networks typically involves the integration of microchannels or microfluidic channels within the material matrix. These channels are designed to mimic the structure and functionality of natural blood vessels, with interconnected pathways that span the material's volume. Once the vascular network is established within the material, it is infused with a healing agent, which may be a liquid, gel, or other reactive substance capable of repairing the material upon activation. The healing agent is introduced into the network through inlet ports, allowing it to flow freely through the channels and reach any damaged areas within the material. When the material experiences damage, such as cracks or fractures, the vascular network serves as a pathway for the transport of the healing agent to the site of the injury. The healing agent is actively pumped or diffuses through the channels, guided by principles of fluid dynamics, until it reaches the damaged area. Upon arrival at the damaged site, the healing agent interacts with reactive species or triggers a chemical reaction that initiates the repair process. This may involve polymerization, cross-linking, or other mechanisms that result in the formation of a solid and durable filler material, effectively sealing the crack or fracture and restoring the material's integrity. One of the key advantages of vascular networks in self-healing materials is their ability to provide on-demand and localized delivery of healing agents precisely where they are needed. Unlike other self-healing mechanisms that rely on passive diffusion or diffusion from embedded reservoirs, vascular networks offer active and controlled transport of healing agents, ensuring rapid and efficient repair. Furthermore, vascular networks can be designed to respond dynamically to changes in the material's condition or external stimuli. For example, the flow rate of the healing agent through the channels can be modulated in response to changes in temperature, pressure, or mechanical stress, allowing for adaptive and autonomous repair in real-time.

However, the design and fabrication of vascular networks in self-healing materials present several challenges. Achieving the desired channel geometry, distribution, and connectivity within the material matrix requires precise fabrication techniques, such as 3D printing or microfluidics. Moreover, ensuring the long-term stability and functionality of the vascular network in diverse operating conditions remains a topic of ongoing research. Thus, vascular networks in self-healing materials offer a promising approach to achieving autonomous repair, inspired by the remarkable efficiency and adaptability of the human circulatory system. By mimicking nature's design principles, these innovative materials hold great potential for enhancing the durability, reliability, and sustainability of engineered systems in various applications.

8.4.2.3 Chemical Reactions

Chemical reactions are a fundamental mechanism employed by certain self-healing materials to repair damage and restore their integrity. These materials undergo reversible chemical reactions when subjected to external stress or damage, enabling them to reform bonds and heal the inflicted injury. The concept of chemical reactions in self-healing materials revolves around the dynamic nature of chemical bonds. When these materials experience mechanical deformation or fracture, the chemical bonds within the material are disrupted. However, instead of irreversibly breaking apart, some of these bonds can undergo reversible reactions under specific conditions. Upon damage, the material's chemical structure undergoes a transformation as the broken bonds are exposed to reactive species or trigger stimuli. These reactive species can be intrinsic to the material or introduced externally, depending on the design and composition of the self-healing system.

The reversible nature of these chemical reactions allows the material to self-heal by reforming the broken bonds and restoring its original structure. This process typically involves the migration of molecular components to the damaged area, where they react with the exposed functional groups or reactants, forming new chemical bonds. The choice of chemical reactions employed in self-healing materials depends on several factors, including the material's composition, environmental conditions, and desired healing kinetics. Common examples of reversible chemical reactions used in self-healing materials include:

Dynamic Covalent Bonds: These bonds undergo reversible reactions under specific conditions, such as changes in temperature, pH, or concentration. Examples include imine bonds, disulfide bonds, and boronic ester bonds.

Supramolecular Interactions: Non-covalent interactions, such as hydrogen bonding, π-π stacking, and host-guest interactions, can facilitate self-healing by promoting molecular reorganization and bond formation.

Metal-Ligand Coordination: Coordination complexes formed between metal ions and ligands can undergo reversible coordination-decoordination reactions, allowing for dynamic exchange and rearrangement of chemical bonds.

The effectiveness of chemical reactions in self-healing materials depends on various factors, including the accessibility of reactive sites, the kinetics of the reaction, and the stability of the healed bonds. Optimization of these parameters is essential to ensure rapid and efficient healing while maintaining the material's mechanical properties and performance. One of the key advantages of chemical reactions in self-healing materials is their inherent versatility and adaptability. Unlike other self-healing mechanisms that rely on encapsulated healing agents or vascular networks, chemical reactions occur intrinsically within the material, allowing for autonomous and continuous repair without the need for external intervention.

However, the implementation of chemical reactions in self-healing materials also presents challenges, such as controlling the reaction kinetics, preventing premature activation, and ensuring compatibility with the material's matrix. Additionally, the long-term stability and durability of the healed bonds need to be carefully evaluated to assess the material's reliability over time. In summary, chemical reactions represent a powerful mechanism for achieving self-healing properties in materials, allowing them to repair damage autonomously and restore their integrity. By harnessing the reversible nature of chemical bonds, self-healing materials offer a promising solution for enhancing the durability, resilience, and sustainability of engineered systems in diverse applications. Continued research and innovation in this field hold the potential to unlock new opportunities for advancing self-healing technology and addressing real-world challenges.

8.4.2.4 Shape Memory Polymers

Shape memory polymers (SMPs) are a fascinating class of materials that exhibit the remarkable ability to recover their original shape after undergoing significant deformation. This unique property allows SMPs to effectively "heal" themselves by returning to their initial configuration, even after experiencing substantial mechanical stress or deformation. The concept of shape memory in polymers stems from the reversible nature of certain polymer chains and their ability to undergo structural transitions in response to external stimuli. SMPs are typically composed of polymer chains with two distinct states: a temporary, or "deformed," state and a permanent, or "original," state. The process of shape memory in SMPs can be understood through the following sequence of events:

Programming: SMPs are initially synthesized or processed into a specific shape, which represents their permanent state. This shape is then "programmed" into the material through a series of heating and cooling cycles or mechanical deformation processes.

Deformation: When subjected to external stress or deformation, such as bending, stretching, or compression, SMPs transition from their permanent state to a temporary, deformed state. During this deformation, the polymer chains rearrange and adopt a new configuration, allowing the material to accommodate the applied stress.

Recovery: Upon removal of the external stress or application of a specific stimulus, such as heat, light, or a change in pH, SMPs undergo a phase transition, reverting from their temporary, deformed state back to their permanent, original state. This transition triggers the polymer chains to return to their initial configuration, causing the material to recover its original shape.

The ability of SMPs to recover their original shape after deformation is governed by the unique properties of the polymer chains and their response to external stimuli. Depending on the specific composition and design of the SMP, the transition between the temporary and permanent states can be triggered by different stimuli, allowing for tailored responses and applications.

One of the key advantages of SMPs is their ability to undergo repeated cycles of deformation and recovery without significant degradation in performance. This makes them ideal candidates for applications requiring dynamic shape changes or adaptive responses to changing environmental conditions. SMPs find a wide range of applications across various industries, including:

Biomedical Devices: SMPs are used in medical implants, such as stents and sutures, where they can adapt to the body's natural movements and provide targeted therapy.

Aerospace: SMPs are employed in aerospace components, such as morphing aircraft wings and deployable structures, to achieve aerodynamic efficiency and reduce weight.

Textiles: SMPs are incorporated into smart textiles and fabrics for applications such as shape-changing garments, self-adjusting apparel, and responsive textiles.

However, the design and implementation of SMPs also pose certain challenges, such as controlling the transition temperatures, optimizing mechanical properties, and ensuring long-term stability and reliability. Ongoing research in material science and engineering is focused on addressing these challenges and unlocking the full potential of SMPs for advanced applications. Thus, shape memory polymers represent a versatile and innovative class of materials with the ability to "heal" themselves by recovering their original shape after deformation. By harnessing the principles of polymer chemistry and materials science, SMPs offer unique opportunities for engineering solutions that are adaptive, responsive, and resilient in diverse environments and applications.

8.4.3 Types of Self-Healing Materials

Self-healing materials encompass a wide range of compositions and structures:

Polymer-Based: These materials are versatile and can be engineered to exhibit self-healing properties through various mechanisms.

Metallic: Self-healing metals repair damage caused by fatigue or deformation, often relying on mechanisms like grain boundary migration.

Ceramic: Self-healing ceramics are suitable for high-temperature applications and typically rely on chemical reactions for repair.

8.4.4 Applications of Self-Healing Materials

Self-healing materials find applications across diverse industries:

Consumer Electronics: Scratch-resistant coatings for smartphones and tablets improve durability and aesthetics.

Automotive Industry: Self-healing polymers in car paints repair minor scratches, reducing maintenance costs.

Infrastructure: Self-healing concrete and asphalt repair cracks caused by environmental factors, extending the lifespan of roads and bridges.

Biomedical Devices: Self-healing materials in biomedical implants improve longevity and biocompatibility.

8.5 SELF-MONITORING MATERIALS

Self-monitoring materials represent an innovative paradigm shift in materials science and engineering, offering real-time monitoring and feedback capabilities directly integrated into the material matrix. This chapter provides a comprehensive review of self-monitoring materials, exploring their principles, design strategies, applications, challenges, and future prospects. By seamlessly integrating sensing functionalities within the material structure, self-monitoring materials hold the potential to revolutionize various industries and enable advanced applications in structural health monitoring, aerospace, automotive, healthcare, and beyond.

8.5.1 Principles of Self-Monitoring Materials

Self-monitoring materials leverage the principles of sensor technology and materials science to enable autonomous sensing and feedback capabilities. These materials incorporate sensing elements or functionalities within their structure, allowing them to detect changes in parameters such as temperature, strain, stress, humidity, pH, or chemical composition. The integration of sensing elements, such as nanoparticles, nanofibers, or functionalized polymers, enables self-monitoring materials to respond to external stimuli and provide real-time feedback on the material's condition.

8.5.2 Design Strategies for Self-Monitoring Materials

Designing self-monitoring materials requires careful consideration of the material's composition, structure, and sensing mechanisms. Various design strategies have been explored to achieve robust and reliable sensing capabilities within the material matrix. These strategies include:

Incorporation of conductive or piezoresistive fillers to enable sensing of mechanical deformation or strain.

Integration of optical fibers or sensors for direct measurement of temperature, chemical composition, or other parameters.

Functionalization of polymers with responsive groups or molecules to enable sensing of specific environmental conditions.

Utilization of micro- or nanostructured materials to enhance sensitivity and spatial resolution of sensors.

8.5.3 Types of Self-Monitoring Materials

Self-monitoring materials encompass a diverse range of compositions and structures, each tailored to specific applications and sensing requirements. The main types of self-monitoring materials include:

1. Polymer-based self-monitoring materials, which leverage the versatility and tunability of polymers to enable sensing of mechanical, thermal, or chemical stimuli.
2. Metal-based self-monitoring materials, which utilize the conductivity or piezoresistivity of metals to enable sensing of mechanical strain or stress.
3. Ceramic-based self-monitoring materials, which leverage the stability and durability of ceramics to enable sensing in high-temperature or harsh environments.

8.5.4 Applications of Self-Monitoring Materials

1. Self-monitoring materials find applications across diverse industries and sectors, offering benefits such as real-time monitoring, predictive maintenance, and enhanced safety and reliability. Some key applications include:
2. Structural health monitoring in civil engineering, aerospace, and automotive industries.
3. Condition monitoring of machinery and equipment in manufacturing and industrial settings.
4. Health monitoring in wearable devices, medical implants, and prosthetics.
5. Environmental monitoring in pollution detection, water quality assessment, and agriculture.

8.5.5 Challenges and Limitations

Despite their potential, self-monitoring materials face several challenges and limitations that must be addressed to realize their full capabilities. These include:

1. Reliability and robustness of sensing elements under varying environmental conditions.
2. Compatibility with the material's mechanical, thermal, and chemical properties.
3. Integration into existing manufacturing processes and applications.
4. Cost-effectiveness and scalability of fabrication methods.
5. Data management, analysis, and interpretation for effective decision-making.

REFERENCES

Ameen, W., Al-Ahmari, A., Mian, S.H., Mohammed, M.K., Kaid, H., Abdulhameed, O., 2022. Optimization of Tree-like Support for Titanium Overhang Structures Produced via Electron Beam Melting. Sustainability 14, 13017. https://doi.org/10.3390/su142013017

Criales, L.E., Arısoy, Y.M., Lane, B., Moylan, S., Donmez, A., Özel, T., 2017. Laser Powder Bed Fusion of Nickel Alloy 625: Experimental Investigations of Effects of Process Parameters on Melt Pool Size and Shape with Spatter Analysis. Int. J. Mach. Tools Manuf. 121, 22–36. https://doi.org/10.1016/j.ijmachtools.2017.03.004

Du, W., Ren, X., Pei, Z., Ma, C., 2020. Ceramic Binder Jetting Additive Manufacturing: A Literature Review on Density. J. Manuf. Sci. Eng. 142. 040801. https://doi.org/10.1115/1.4046248

Galati, M., Iuliano, L., 2018. A Literature Review of Powder-Based Electron Beam Melting Focusing on Numerical Simulations. Addit. Manuf. 19, 1–20. https://doi.org/10.1016/j.addma.2017.11.001

Gebisa, A.W., Lemu, H.G., 2018. Investigating Effects of Fused-Deposition Modeling (FDM) Processing Parameters on Flexural Properties of ULTEM 9085 Using Designed Experiment. Materials 11, 500. https://doi.org/10.3390/ma11040500

Gokuldoss, P.K., Kolla, S., Eckert, J., 2017. Additive Manufacturing Processes: Selective Laser Melting, Electron Beam Melting and Binder Jetting—Selection Guidelines. Materials 10, 672. https://doi.org/10.3390/ma10060672

Khan, S.H., 2020. Green Nanotechnology for the Environment and Sustainable Development, in: Naushad, Mu., Lichtfouse, E. (Eds.), Green Materials for Wastewater Treatment.

Springer International Publishing, Cham, pp. 13–46. https://doi.org/10.1007/978-3-030-17724-9_2

Konta, A.A., García-Piña, M., Serrano, D.R., 2017. Personalised 3D Printed Medicines: Which Techniques and Polymers Are More Successful? Bioengineering 4, 79. https://doi.org/10.3390/bioengineering4040079

Liravi, F., Vlasea, M., 2018. Powder Bed Binder Jetting Additive Manufacturing of Silicone Structures. Addit. Manuf. 21, 112–124. https://doi.org/10.1016/j.addma.2018.02.017

Martinho, P.G., 2021. Chapter 9 – Rapid Manufacturing and Tooling, in: Pouzada, A.S. (Ed.), Design and Manufacturing of Plastics Products, Plastics Design Library. William Andrew Publishing, Norwich, NY, pp. 381–456. https://doi.org/10.1016/B978-0-12-819775-2.00008-5

Masood, S.H., 2014. 10.04 – Advances in Fused Deposition Modeling, in: Hashmi, S., Batalha, G.F., Tyne, C.J.V., Yilbas, B. (Eds.), Comprehensive Materials Processing. Elsevier, Oxford, pp. 69–91. https://doi.org/10.1016/B978-0-08-096532-1.01002-5

Nasrollahzadeh, M., Sajadi, S.M., Sajjadi, M., Issaabadi, Z., 2019. Chapter 1 – An Introduction to Nanotechnology, in: Nasrollahzadeh, M., Sajadi, S.M., Sajjadi, M., Issaabadi, Z., Atarod, M. (Eds.), An Introduction to Green Nanotechnology, Interface Science and Technology. Elsevier, London, pp. 1–27. https://doi.org/10.1016/B978-0-12-813586-0.00001-8

Paras, Yadav, K., Kumar, P., Teja, D.R., Chakraborty, S., Chakraborty, M., Mohapatra, S.S., Sahoo, A., Chou, M.M.C., Liang, C.-T., Hang, D.-R., 2023. A Review on Low-Dimensional Nanomaterials: Nanofabrication, Characterization and Applications. Nanomaterials 13, 160. https://doi.org/10.3390/nano13010160

Peltola, S.M., Grijpma, D.W., Melchels, F.P.W., Kellomäki, M., 2008. A Review of Rapid Prototyping Techniques for Tissue Engineering Purposes. Ann. Med. 40, 268–280. https://doi.org/10.1080/07853890701881788

Sen, M., 2020. Nanocomposite Materials, in: Sen, M. (Ed.), Nanotechnology and the Environment. IntechOpen, Rijeka. https://doi.org/10.5772/intechopen.93047

Tovar-Lopez, F.J., 2023. Recent Progress in Micro- and Nanotechnology-Enabled Sensors for Biomedical and Environmental Challenges. Sensors 23, 5406. https://doi.org/10.3390/s23125406

White, S.R., Sottos, N.R., Geubelle, P.H., Moore, J.S., Kessler, M.R., Sriram, S.R., Brown, E.N., Viswanathan, S., 2001. Autonomic Healing of Polymer Composites. Nature 409, 794–797. https://doi.org/10.1038/35057232

9 Computational Nanotechnology

9.1 INTRODUCTION

The rapid advancements in nanotechnology demand innovative approaches for imaging, measuring, modeling, and manipulating matter at the nanoscale. Previously, we relied mainly on two methods for studying various phenomena: experimentation and theoretical analysis. In an experiment, researchers manipulate one or more variables under controlled conditions to observe the effects on other variables. The purpose of an experiment is to gather empirical evidence and draw conclusions based on the observed results. In theory, models of systems are developed and typically expressed through mathematical equations. These models are then validated based on their ability to accurately describe system behavior, often focusing on simpler instances where solutions can be computed from the equations. However, this process usually involves significant simplification to address the complexities inherent in real-world scenarios, thus facilitating problem resolution. With the advances in computers, computational methods have emerged as a third paradigm of scientific investigation. They act more like virtual experiments. In a computer experiment, models are derived from theory, but machines perform computations, incorporating a slightly higher level of complexity to enhance realism in the solutions obtained. In nanotechnology, virtual experiments hold significant potential for driving progress. Computer simulations offer time and spatial resolutions that are often unattainable through physical experiments (Merkle, 1991). Furthermore, they can aid in the interpretation of experimental data. Given the scarcity of theoretical models applicable to nanoscale systems, computer simulations can assist in modeling these systems and deriving governing equations to describe them.

Several computational techniques, such as density functional theory (DFT), molecular dynamics (MD), and the lattice Boltzmann method (LBM), are utilized to explore diverse physical phenomena. Each technique has its validity within specific lengths and timescales. Continuum modeling relies on employing numerical techniques such as the finite element method (FEM) and finite volume method (FVM) to solve governing equations or phenomenological laws, enabling a deeper comprehension of the physical behavior exhibited by the system under scrutiny. These methodologies operate on the premise of the continuum hypothesis, which treats a

DOI: 10.1201/9781003317272-9

bulk system as continuous, disregarding the discreteness of atoms. This approach applies only to systems where the characteristic length scale significantly surpasses the atomic scale, allowing for averaging and neglecting atomic properties and fluctuations over time. Mesoscale modeling is a potent tool for examining systems between atomic and macroscopic scales operating within intermediate time scales. It ensures computational precision and is adept at analyzing and modeling a spectrum broader than the atomic scale. Examples encompass coarse-grained molecular dynamics (CGMD), dissipative particle dynamics (DPD), and phase-field methods. Atomistic and molecular modeling strategies are typically employed to elucidate atomic and molecular interactions within a sample. These methods are well-suited for analyzing chemical interactions involving bond formation and breakage between atoms and molecules, making them particularly suitable for scrutinizing material properties at the nanoscale. When the length or time scales relevant to the problem are small, we must investigate the behavior of sub-atomic particles. Ab initio methods, a Latin term for "from the beginning," refers to principle-based techniques that will help us explore the wave nature of the sub-atomic particles. Various methods include DFT, quantum Monte Carlo, ab initio MD, etc.

Users must choose the most suitable method or combination thereof, based on the spatial and temporal scales relevant to their system. Other factors, such as the available computational resources and time, should also be considered when choosing an approach to solving the problem. It is a balance between the model accuracy and the computational cost. Atomistic or ab initio calculations can provide detailed insight, but they are computationally intensive and may need to be more practical for large systems. Mesoscopic or macroscopic models are generally less computationally demanding but may oversimplify a complex phenomenon. Last but not least, the computational results must be appropriately validated with the help of suitable experimental data. We also need methods to couple continuum/molecular/quantum mechanics techniques to investigate the physical behavior of specific systems. For example, most research questions relevant to nanotechnology demand using atomistic or ab initio methods for scientific investigation. This is because continuum laws, which form the backbone of macroscopic computational techniques, are no longer valid at the length and time scales relevant to nanotechnology. But, simultaneously, systems of interest to us will have components with different sizes and response times, thus operating at various lengths and time scales. Such approaches are called multiscale modeling techniques. These coupling techniques are also crucial in the context of the time and length scale limitations faced by methods like ab initio quantum mechanics or molecular dynamics (Marx and Hutter, 2009; Merkle, 1991; Ohno et al., 2018; Srivastava and Atluri, 2002).

The basic steps involved in performing a virtual experiment are as follows:

1. Identify the problem.
2. Identify a suitable computational technique relevant to the problem.
3. Write the simulation program/Choose the appropriate software package.
4. Perform the computer experiment.

The outline of the chapter is as follows: in the subsequent sections, we will discuss one example from each atomistic and ab initio computational technique, as they are widely employed to model nanoscale systems.

9.2 AB INITIO METHODS

Here, we are transferring from a particle-like description of the constituent bodies to a wave-like description, i.e., rather than just describing the state of a system using the position and velocity of a particle, in wave description, we need to specify the amplitude of the wave everywhere in space. The wave is an excitation (vibration); we need to know the amplitude of the excitation at every point and every instant. The wavefunction is represented as $\Psi(r,t)$.

The Schrodinger wave equation is the equivalent of Newton's equation of motion for a quantum particle (Equation 9.1). Hence, solving Schrodinger's equation can give us information about the electron and nuclei dynamics.

$$H\Psi = E\Psi \tag{9.1}$$

The formulation aims to find the wave function. $\Psi(r,t)$ Which will minimize the system's total energy, as expressed in Equation 9.2.

$$E = min_{\Psi}\langle\Psi|H|\Psi\rangle \tag{9.2}$$

However, obtaining a complete solution to Schrodinger's equation is very difficult, both analytically and numerically. Hence, we go for various assumptions and try to obtain numerical solutions to the approximate representations of Schrodinger's equation. Ab initio methods involve the numerical solving of a relative of the Schrodinger equation to get information about the wave function. $\Psi(r,t)$. Born-Oppenheimer approximation helps us to decouple the dynamics of the electrons and nuclei as the mass of the nuclei is very large compared to the electrons. This allows us to initially assume an equilibrium state to the nuclei, solve for the electron dynamics, and then perform calculations to update the nuclei state. Equation 9.3 represents the time-independent form of the Schrodinger equation for electrons.

$$H\Psi = E\Psi = \sum_{i=j}^{N}\left(\frac{\bar{h}^{2}}{2m}\nabla_i^2\Psi - Ze^2\sum_{R}\frac{1}{|r_i - R|}\Psi\right) + \frac{1}{2}\sum_{i\neq j}\frac{e^2}{|r_i - r_j|}\Psi \tag{9.3}$$

The Hamiltonian (H) consists of the kinetic energy, electron interaction with nuclei, and mutual interaction energy of electrons. The solution of this second-order differential equation gives us the electron dynamics. But, a complete solution to this form of Schrodinger's equation is also computationally intensive. For example, consider a benzene molecule having 42 electrons. To obtain the electronic structure information of benzene, we have to solve for the wave function $\Psi(r)$, which is an antisymmetric function of 126 coordinates and 42 electronic spin components. As per Equation 9.3, this will involve highly intensive calculations to obtain the electron mutual interaction energy. Hence, a complete solution of Equation 9.3 may be possible only for single-electron systems.

In DFT, we consider the electron density. $n(r)$, a physical observable, rather than solving for the complicated $\Psi(r)$. The $n(r)$ contains all the information about the electronic structure, and thus, we are able to reduce a multidimensional problem to just three coordinates. Here, we try to find the $n(r)$. That will minimize the total energy as per Equation 9.4.

$$E = min_n \left\{ \int V_{nuclei}(r) n(r) d^3 r + F[n(r)] \right\} \tag{9.4}$$

Here, $F[n(r)]$ is known as a universal functional. A functional is defined as a function of a function. $F[n(r)]$ is a function of the electron density $n(r)$, which is a function of position r. The first term on the right of Equation 9.4 represents the external potential due to the nuclei and the universal functional, which includes the kinetic energy and internal energy contributions to the total energy. The main roadblock in performing a DFT calculation is that there is no accurate mathematical model for $F[n(r)]$. Hence, we have to go for simplifying approximations to include the contribution of the universal functional towards total energy. The universal functional can be approximated as

$$F[n(r)] = T_s[n(r)] + E_H[n(r)] + E_{xc}[n(r)] \tag{9.5}$$

Where T_s is the kinetic energy term introduced by assuming a fictitious system of N non-interacting electrons (Kohn–Sham) electrons, the kinetic energy of non-interacting systems is well defined. E_H represents the Hatree energy (classical Coulombic energy), which can be approximated using the following equation.

$$E_H[n(r)] = \frac{1}{2} \iint \frac{n(r_1) n(r_2)}{|r_1 - r_2|} \tag{9.6}$$

Hence, the kinetic energy and Coulombic contributions to total energy can be represented by using the above assumptions (T_s and E_H). The third term on the right-hand side of Equation 9.5 is the exchange-correlation functional (E_{xc}). It considers all the unknown quantum effects, like the exchange and correlation effects of the electrons in your system. This term also includes corrections for the self-interaction of electrons. Hence, the only term that does not have an accurate mathematical representation in Equation 9.5 is the exchange-correlation functional. In DFT, various types of functionals, like local density approximation (LDA), generalized gradient approximation, etc., have been developed to approximate the exchange and correlation effects among electrons.

The Schrodinger's equation (Equation 9.3) is a many-body equation, i.e., the $\Psi(r)=\Psi(r_1,r_2,\ldots\ldots,r_m)$, where m is the total number of electrons, the Hatree approximation helps simplify this equation to many single-electron Schrodinger's equations. The Hartree-Fock approximation helps to include the antisymmetry property of the electrons into Schrodinger's equation. With the help of Equations 9.4 and 9.5 and using the Euler-Lagrange method, the single-electron Schrodinger's equation can be approximated to the form mentioned in Equation 9.7. These are known as the Kohn–Sham equations. In DFT, we will be using this form of Schrodinger's equation to investigate the dynamics of the electrons.

$$\left[-\frac{1}{2}\nabla^2+v_H(r)+v_{xc}(r)+v_{ext}(r)\right]\Psi_i(r)=\hat{H}_{KS}\Psi_i(r)=\epsilon_i\Psi_i(r) \qquad (9.7)$$

Where,

$$v_H(r)=\int\frac{n(r')}{|r-r'|}dr',$$

$$v_{xc}(r)=\frac{\delta E_{xc}}{\delta n(r)},$$

$$n(r)=\sum_{i=1}^{N}|\Psi_i(r)|^2.$$

Now, the next question is how to solve these one-particle differential equations to obtain the electron dynamics. The method followed is to expand the wave function using a set of basis functions (Equation 9.8) and substituting in the Kohn–Sham equations to obtain the solution.

$$\Psi(r)=\sum_{n=1,k}c_n\varphi_n(r) \qquad (9.8)$$

This will help us express the Kohn–Sham equations in a matrix form and use the principles of Eigen decomposition to solve them, as seen from Equations 9.9 and 9.10.

$$\sum_{n=1,k}c_n\left\langle\varphi_m\middle|\hat{H}\middle|\varphi_n\right\rangle=Ec_m \qquad (9.9)$$

$$\sum_{n=1,k}H_{mn}c_n=Ec_m \qquad (9.10)$$

The choice of basis functions is very important for the calculation: for molecules, we usually go for atomic orbitals or localized functions like Gaussians, and for solids, we use the plane wave approach, where periodic functions like sines and cosines are used.

Another simplifying assumption used in DFT is the pseudopotential approach, where we exclude the inactive atomic core electrons from an explicit treatment in our calculations. This is based on the observation that the electrons in the inner shells are tightly bound to the nuclei and do not contribute to the valence bonding (Ohno et al., 2018; Sholl and Steckel, 2009),

The following are the steps to perform DFT calculations for the systems of our interest.

1. Remove tightly bound core electrons: the pseudopotential approach.
2. Represent the orbitals with a basis: Gaussian and plane wave.
3. Calculate total energy for trial wavefunctions.
4. Minimize energy and iterate charge density to self-consistency.

DFT has emerged as an important tool for investigating systems at the quantum level and is extensively used for various applications like material design, targeted drug delivery, etc. There are many open-source and commercial software packages, like Quantum ESPRESSO, VASP, ADF, CASTEP, ONETEP, Gaussian, GAMESS, CP2K, etc, available for performing DFT calculations.

9.3 ATOMISTIC SIMULATION METHODS

When dealing with spatial and temporal scales of approximately a few nanometers and nanoseconds or when atom-level resolution is necessary for computations, techniques such as MD, Monte Carlo (MC), and similar methods are employed. Atomistic simulations have huge applications in various fields, such as analyzing the behavior of fluids confined in nanochannels or calculating the thermal properties of a nanomaterial.

In MD, the fundamental premise is that all components are comprised of atoms; in other words, the behavior of each atom within the system is studied to comprehend its properties. Here, Newton's law of motion is employed to compute the trajectory of an atom. By aggregating the trajectories of all atoms within the system, we can comprehend the system's evolution over time (Frenkel and Smit, 2001; Rapaport, 2004). The major steps involved in performing an MD simulation are as follows:

1. Define the Initial and Boundary Conditions
   To perform an MD simulation, we have to solve Newton's law of motion (Equation 9.11) for each and every atom present in the system.

$$F = ma = m\frac{d^2r}{dt^2} \tag{9.11}$$

   Since it's a second-order differential equation, we require two initial conditions, position r and velocity v, to solve the equation. This is achieved by defining the initial position and velocity of each and every atom present in the system. In short, we define a simulation box (region of study) and arrange the constituent atoms inside the box. For example, assume that you want to calculate the thermal conductivity of copper at 400 K. First, define a region in space (like a cube of 1 nm in size) and then arrange the atoms of copper according to the crystal structure inside the cube. In cases where the crystal structure is unknown, for example, in the case of liquids, we assume a crystal structure, like a simple cubic unit cell, and arrange the atoms accordingly. Also, when simulating materials like polymers, carbon nanotubes, biomaterials, etc., the initial positions of the constituent atoms can be identified from their known structure. The second condition required to start the MD simulation is the initial velocity. The initial velocity of each and every atom in the system should be defined. The velocity values can be randomly assigned to individual atoms using statistical mechanical distribution functions, like Maxwell Boltzmann statistics. In addition to the initial conditions, we should also provide the boundary conditions for each direction of the system. One of the most commonly used boundary conditions is the periodic boundary condition, which assumes that if an atom leaves the system from one side of the simulation box, it is made to re-enter the box from the opposite end. This boundary condition is suitable for mimicking the behavior of systems that are part of a macroscopic system and helps to remove the surface effects in such systems. The boundary condition should be selected based on its closeness to the actual physical system that we intend to study.

2. Select the Suitable Interatomic Potential
   One of the important steps in the MD simulation is identifying the interatomic potential that is most suitable to the system under study. The interatomic potential is a mathematical model representing the interactions between the atoms in the MD simulation. For example, if we consider the calculation of the thermal conductivity of copper, we need to have a mathematical model that represents the interaction between the copper atoms, how one copper atom influences the behavior of the other copper atoms, or what the force exerted by the copper atoms on each other. One of the most commonly used and simplest interatomic potentials is the Lennard-Jones 12-6 potential. It is a pair-wise interaction potential and is suitable for modeling inert gases and noble metals. While investigating the physical behavior of systems with more than one atom type, we must identify the interatomic potential that models the interactions between the different atom types. For example, in the case of water, we have one oxygen and two hydrogen atoms in a water molecule. To perform the MD simulation, we require single or multiple interatomic potentials for modeling oxygen-oxygen, hydrogen-hydrogen, and oxygen-hydrogen interactions. The simple point charge (SPC) model is one of the interatomic potentials suitable for modeling water. SPC has a Lennard-Jones term to model the pair-wise interaction of the constituent atoms and also a Coulombic term to model the polar nature of the water molecules. In addition, while investigating the behavior of molecular systems, the structural information about the system, like bonds, angles, dihedrals, etc, should also be considered. For example, in the case of water molecules, there are two OH bonds and one H-O-H angle that needs to be considered. The SPC model only considers the non-bonded interactions of the constituent atoms and does

not consider the structure of the individual water molecule. In such cases, we have to define additional potential functions to model the structural aspects of the molecules. There are also potential models that consist of terms to take into consideration the non-bonded interactions and the molecular structure of the constituents.

3. Calculate the Interatomic Forces
   For calculating the trajectory of the atoms, MD solves Newton's second law (Equation 9.11) for each and every atom considered in the simulation. In order to solve Equation 9.11, we have to calculate the force acting on the individual atoms. The force can be caused by the interaction with other atoms present in the simulation domain or from any external fields. The force due to interaction with neighboring atoms can be calculated from the interaction potential selected in the previous step; it is defined as the negative derivative of the interaction potential with respect to the relative position of the atoms. In this manner, the total force acting on each individual atom can be calculated.
4. Calculate the Acceleration of Atoms
   From the force calculated in step 3 and using Equation 9.11, the acceleration of each and every atom can be obtained. This acceleration helps us in evaluating the movement of the individual atoms.
5. Move the Atoms—Integration
   As mentioned earlier, Newton's second law (Equation 9.11) is solved to evaluate the trajectory of the atoms. This translates to integrating the acceleration term and obtaining the position and velocities of the constituents at a later instant. In MD simulation, this step is performed using integration algorithms based on Taylor series expansions. First, using the initial conditions and the interatomic potential, the force acting on each atom is calculated, and then, using Newton's law, the acceleration at this instant is obtained. Using the acceleration, velocity, and position values at this instant of time, as well as Taylor series expansion, these values are calculated after a small timestep. There are many integration algorithms commonly used in an MD simulation, such as Verlet, velocity Verlet, Leapfrog, predictor-corrector, etc. The timestep is chosen based on the system characteristics and the available computational power.
6. Repeat From Step 3 Till Equilibration Is Achieved
   From step 5, we get the new positions and velocities of each and every atom present in the simulation after a selected time step. These calculations work as a loop until the system reaches an equilibrium or minimum energy state. This is defined as the equilibration period. After this, the loop is continued to calculate the desired properties. From the positions and velocities of the atoms obtained in each timestep in an MD simulation, we can calculate various properties like thermal conductivity, coefficient of viscosity, elastic modulus, etc., and also understand the dynamic behavior of the atoms/molecules.

In certain situations, we may also need to control parameters like pressure, temperature, etc, while performing the MD calculations. For example, suppose we need to understand the variation of thermal conductivity of carbon nanotubes with temperature. In that case, we have to perform the MD simulation with an additional constraint to maintain the temperature at the desired level. In such cases, we have to use algorithms like thermostats, barostats, etc., to maintain the parameters at the desired values.

There are many open-source and commercial software packages, like LAMMPS, NAMD, gromacs, Material Studio, etc., that are available for performing MD calculations.

EXAMPLE 9.1: STRUCTURE OPTIMIZATION OF GRAPHENE SHEET

This section explains the implementation of an example system using quantum ESPRESSO (Giannozzi et al., 2009). Here, we are performing a structure optimization for a graphene sheet, as shown in Figure 9.1, using DFT.

There are two types of structural optimization calculations that can be performed using quantum ESPRESSO: "relax" and "vc-relax." In "relax," only the atomic positions are allowed to vary, while in "vc-relax," both atomic positions and lattice constants vary. In this particular example, we will be performing the "relax" calculation on the graphene sheet. The input data for quantum ESPRESSO calculations are mentioned under different name lists like control, system, electrons, etc. Each name list consists of different flags (parameters) relevant to the calculations. The name lists are followed by sections known as input cards consisting of fields like atomic positions, k points, etc. A typical quantum ESPRESSO code is shown below, where the code is written in italics and the explanation in normal text in brackets.

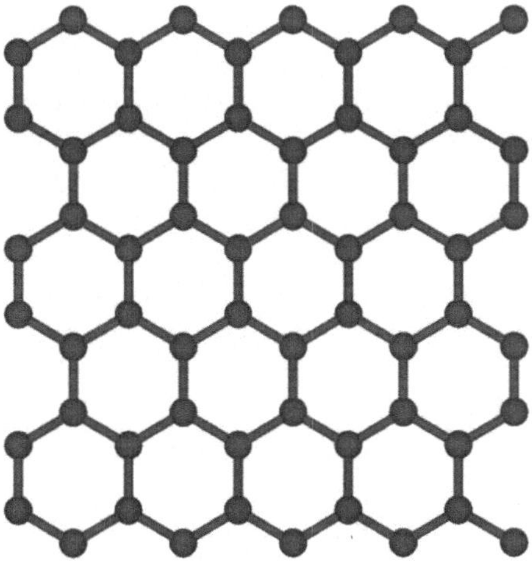

FIGURE 9.1 Graphene sheet.

Code With Explanation

&CONTROL (The general variables that control the run are listed as flags under this name list)
calculation = *"relax"* (The type of calculation)
max_seconds = *1.72800e+05* (The maximum CPU time for the calculation in seconds)
nstep = *100* (Number of structural optimization calculations that will be performed in this run)
pseudo_dir = *"/home/user/.burai/.pseudopot"* (Location of the pseudopotential files)
/
&SYSTEM (The input variable for the system like the lattice type, lattice constants etc are listed here)
a = *2.10520e+01*
b = *2.13440e+01*
c = *1.00000e+01*
degauss = *1.00000e-02*
ecutrho = *2.25000e+02*
ecutwfc = *2.50000e+01*
ibrav = *8* (Index for the Bravais lattice)
nat = *60* (Number of atoms)
ntyp = *1* (Number of atom types)
occupations = *"smearing"*
smearing = *"gaussian"*
/
&ELECTRONS (The parameters used for electronic calculations are listed here)
conv_thr = *1.00000e-06* (Threshold for convergence for self consistent field calculations)
electron_maxstep = *200* (Maximum number of iterations for the self consistent field calculations)
mixing_beta = *7.00000e-01* (Mixing factor for the self consistent field calculation)
/
&IONS (The parameters used for ionic calculations are listed here)
ion_dynamics = *"bfgs"* (The type of algorithm to be used for structural relaxation)
/
K_POINTS {automatic} (Input card to specify the parameters for the K-grid)
2 2 2 0 0 0
ATOMIC_SPECIES (Input card with the details of the atom types, their mass, and name of the pseudopotential file)
C 12.01070 C.pbe-rrkjus.UPF

ATOMIC_POSITIONS {angstrom} (Input card with the initial positions of all the atoms in the simulation domain)

C 6.228000 5.000000 5.000000
C 5.000000 5.709000 5.000000
C 5.000000 7.127000 5.000000
C 6.228000 7.836000 5.000000
C 8.684000 5.000000 5.000000
C 7.456000 5.709000 5.000000 ....

EXAMPLE 9.2: THERMAL CONDUCTIVITY OF GRAPHENE SHEET

This section explains the implementation of an example system using LAMMPS (Thompson et al., 2022). Here, we are performing an MD calculation to obtain the thermal conductivity of the graphene sheet shown in Figure 9.1. The Green–Kubo formula (Kang and Wang, 2017), given in Equation 9.12, is employed to compute the thermal conductivity of graphene.

$$k = \frac{V}{k_B T^2} \int_0^{\infty} \langle J_x(0) J_x(t) \rangle dt = \frac{V}{3k_B T^2} \int_0^{\infty} \langle J(0).J(t) \rangle dt \tag{9.12}$$

Where J is defined as

$$J = \frac{1}{V}\left[\sum_i e_i v_i - \sum_i S_i v_i \right] \tag{9.13}$$

Here, e_i denotes the per-atom energy, encompassing both potential and kinetic components, while S_i represents the stress tensor. $\frac{1}{V}$ serves as the scaling factor for J. This method, which involves calculating the temporal correlation function of the heat flux, has been widely used in MD studies. The integral of this temporal correlation function is directly related to thermal conductivity. MD simulations are used to track the positions and velocities of particles over time, which aids in the integration process.

Code With Explanation

(Variable section: these definitions help in using the same parameters in multiple commands in the file.)

variable damp equal 0.5
variable seed equal 456789 (Random number)
variable T equal 300 (Temperature value)

variable V equal vol (Volume of the simulation box)
variable P equal 1.01325 (Pressure in bar units)
variable dt equal 0.001 (Timestep)
variable p equal 1000
variable s equal 1
variable d equal $p*$s (p, s, d values are used for calculating the correlation functions in the Green–Kubo formula)
variable kB equal 1.3806504e-23 (Boltzmann constant)
variable eV2J equal 1.60218e-19 (Conversion factor – eV to Joule)
variable ps2s equal 1.0e-12 (Conversion factor – ps to sec)
variable A2m equal 1.0e-10 (Conversion factor – Angstroms to m)
variable convert equal ${eV2J}*${eV2J}/${ps2s}/${A2m} (Combined conversion factor for converting from LAMMPS "metal" units to SI units)
(Initialization variables)
dimension 3 (Its a 3 dimensional simulation)
boundary p p p (Boundary conditions)
units metal (LAMMPS has various options of unit systems to specify the parameters, 'metal' is one of them)
atom_style full (The attributes of atoms are selected here)
read_data graphene.data (The positions of each and every carbon atom are read from this file. This method helps us to create molecular systems using different softwares or codes and read their structural information into the LAMMPS simulation.)
mass 1 12.0107 (Mass of the atom type used in the present simulation)
pair_style airebo 2 (The interatomic potential model used in the present simulation)
*pair_coeff * * CH.airebo C* (The parameters for the interatomic potential function)
timestep ${dt} (Timestep for the calculation)
velocity all create 300 789543 mom yes rot yes dist gaussian (Defining the initial velocities of the atoms)
fix 1 all npt temp 300.0 300.0 0.2 iso 0.0 0.0 0.1 (Conduct the MD simulation with temperature and pressure constraints)

run 2500 (Equilibration run – The number of steps have to be decided iteratively)
unfix 1
fix 1 all nve (Remove the temperature and pressure constraints while performing the integrations)
reset_timestep 0
dump myDump1 all atom 100 dump.all (Write the positions of all the atoms to a file every 100 steps)

(Green–Kubo calculation)
thermo $d

```
compute     KE all ke/atom
compute     PE all pe/atom
compute     stress all stress/atom NULL virial
compute     heatflux all heat/flux KE PE stress
variable    Jx equal c_heatflux[1]/$V
variable    Jy equal c_heatflux[2]/$V
variable    Jz equal c_heatflux[3]/$V
fix         Jeff all ave/correlate $s $p $d c_heatflux[1] c_heatflux[2] c_
heatflux[3] type auto file Jeff.dat ave running
variable    scale equal ${convert}/${kB}/$T/$T/$V*$s*${dt}
variable    kxx equal trap(f_Jeff[3])*${scale}
variable    kyy equal trap(f_Jeff[4])*${scale}
variable    kzz equal trap(f_Jeff[5])*${scale}
thermo_style    custom step temp v_kxx v_kyy v_kzz
run    1000000
variable    k equal (v_kxx+v_kyy+v_kzz)/3
print       "Effective thermal conductivity: $k[W/mK]"
```

REFERENCES

Frenkel, D., Smit, B., 2001. Understanding Molecular Simulation: From Algorithms to Applications, 2nd ed. Academic Press, San Diego.

Giannozzi, P., Baroni, S., Bonini, N., Calandra, M., Car, R., Cavazzoni, C., Ceresoli, D., Chiarotti, G.L., Cococcioni, M., Dabo, I., Corso, A.D., Gironcoli, S. de, Fabris, S., Fratesi, G., Gebauer, R., Gerstmann, U., Gougoussis, C., Kokalj, A., Lazzeri, M., Martin-Samos, L., Marzari, N., Mauri, F., Mazzarello, R., Paolini, S., Pasquarello, A., Paulatto, L., Sbraccia, C., Scandolo, S., Sclauzero, G., Seitsonen, A.P., Smogunov, A., Umari, P., Wentzcovitch, R.M., 2009. QUANTUM ESPRESSO: a modular and open-source software project for quantum simulations of materials. J. Phys. Condens. Matter 21, 395502. https://doi.org/10.1088/0953-8984/21/39/395502

Kang, J., Wang, L.-W., 2017. First-principles Green-Kubo method for thermal conductivity calculations. Phys. Rev. B 96, 020302. https://doi.org/10.1103/PhysRevB.96.020302

Marx, D., Hutter, J., 2009. Ab Initio Molecular Dynamics: Basic Theory and Advanced Methods. Cambridge University Press, Cambridge. https://doi.org/10.1017/CBO9780511609633

Merkle, R.C., 1991. Computational nanotechnology. Nanotechnology 2, 134. https://doi.org/10.1088/0957-4484/2/3/005

Ohno, K., Esfarjani, K., Kawazoe, Y., 2018. Computational Materials Science: From Ab Initio to Monte Carlo Methods. Springer-Verlag GmbH, Berlin, Germany.

Rapaport, D.C., 2004. The Art of Molecular Dynamics Simulation, 2nd ed. Cambridge University Press, Cambridge. https://doi.org/10.1017/CBO9780511816581

Sholl, D., Steckel, J.A., 2009. Density Functional Theory: A Practical Introduction, 1st ed. Wiley-Interscience, Hoboken, NJ.

Srivastava, D., Atluri, S.N., 2002. Computational nanotechnology: a current perspective. Comput. Model. Eng. Sci. 5, 531–538.

Thompson, A.P., Aktulga, H.M., Berger, R., Bolintineanu, D.S., Brown, W.M., Crozier, P.S., in 't Veld, P.J., Kohlmeyer, A., Moore, S.G., Nguyen, T.D., Shan, R., Stevens, M.J., Tranchida, J., Trott, C., Plimpton, S.J., 2022. LAMMPS – a flexible simulation tool for particle-based materials modeling at the atomic, meso, and continuum scales. Comput. Phys. Commun. 271, 108171. https://doi.org/10.1016/j.cpc.2021.108171

10 Nanomaterials in Transportation

10.1 NANOTECHNOLOGY IN AUTOMOBILES

The automobile industry rapidly adapts to technological advancements to enhance vehicles' performance, safety, and sustainability. Nanotechnology offers unlimited opportunities in air, water, and land transportation. Incorporating nanotechnology in transportation will provide passengers with better safety, speed, and comfort. Nanotechnology plays a significant role in transport systems' design, manufacturing processes, and performance. Nanomaterials find applications in the fabrication of body parts, filters, windows, engine components, coatings, sensors, etc. (Figure 10.1) in automobiles, aircraft, and marine applications. This chapter covers the applications of nanotechnology in the various means of transportation.

10.1.1 Engine Components and Structures

Advancements in materials with nanoarchitecture can improve automobiles' efficiency without compromising vehicles' safety and performance. Reduction in the overall weight of automobiles improves fuel efficiency remarkably. As the vehicles become lighter, the comfort of driving the vehicle increases. At the same time, better crash protection can be expected. As a result, the service life of the vehicles is increased. If we reduce the weight of the vehicle by 10%, the fuel economy can be improved by 6–8%. Modernizing the materials used by the automobile industry can reduce the overall weight of the vehicles by 50%. The cast iron and steel components can be replaced by lightweight composites of magnesium (Mg) alloys and aluminum (Al) alloys, carbon fiber-reinforced composites, and polymer nanocomposites, thus increasing fuel efficiency.

Nanostructured metals and alloys are used in engine components such as bearings, valves, springs, etc. The nanostructuring of metals and alloys means controlling the microstructure to obtain nanosized grains to improve the strength through grain size refinement. These materials with nanosized grains can significantly enhance their strength. Moreover, incorporating nanosized ceramic materials into the metal matrix strengthens the material by a dispersion-strengthening mechanism. For instance, the nanocomposites of aluminum and magnesium reinforced with nanostructured SiC, TiC, TiB_2, MWCNTs, etc., exhibit mechanical properties such as tensile strength,

DOI: 10.1201/9781003317272-10

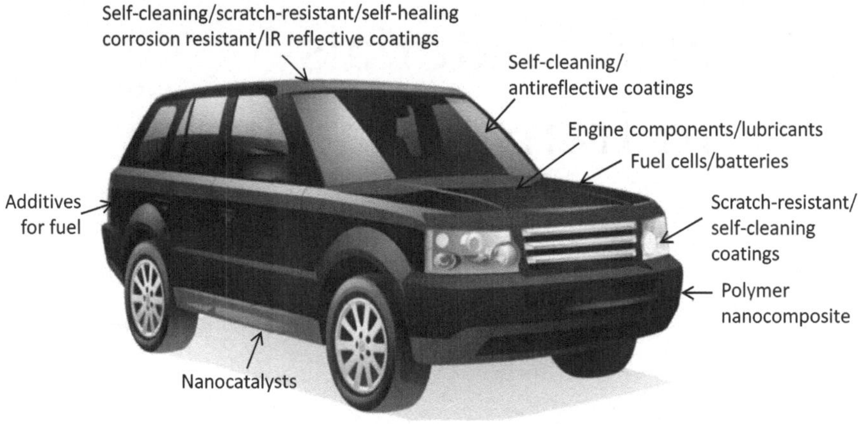

FIGURE 10.1 Application of nanotechnology to the various parts of an automobile. (Image courtesy: www.freepik.com.)

yield strength, and ductility superior to aluminum or magnesium alone (Ceschini et al., 2017). Such composites can replace conventional steel or cast iron-based components in aerospace, automobile, and marine applications. Additionally, these composites exhibit high wear resistance, thermal stability, and low coefficient of friction as compared to the matrix metal. Ti, Al, and Mg alloys can provide lightweight components for automobiles and aerospace applications. Ti-6Al-4V and Ti-Al-Fe nanoalloys fabricate engine valves, connecting rods, etc. (Lou and Alpas, 2019).

10.1.2 Body Panels and Interiors

Novel lightweight materials with high strength and durability are used for automobile and aerospace body panels, interiors, etc. Polymer nanocomposites are the most suitable candidates for such applications. A nanocomposite is a material with a continuous matrix reinforced with nanosized materials with different shapes and morphology. By adding nanoparticles, one can improve the matrix's chemical, mechanical, and thermal properties without altering the matrix materials' unique characteristics. Moreover, the quantity of nanosized reinforcement is minimal enough to reach the required properties; therefore, the matrix's aesthetics, such as flexibility and transparency, are not severely affected. Many natural and synthetic nanostructured materials are reinforced in various polymer and metallic matrices. Carbon nanotubes (CNTs), graphene, clay, cellulose nanocrystals, layered hydroxides, metal oxides, carbides, nitrides, etc., are nanostructured reinforcements in polymer nanocomposites. Nanosized reinforcements will improve the mechanical and barrier properties of polymers. Besides the properties above, polymer nanocomposites exhibit enhanced electrical conductivity, degradation stability, antimicrobial activities, sensing, flame retardancy, etc. HIPS, PVC, PS, PP, HDPE, PET, PE, PA, EVOH, MET, BOPP, PEN, PMMA, LDPE, VMPET, (HIPS, high-impact polystyrene; PVC, poly(vinyl chloride); PS, polystyrene; HDPE, high-density polyethylene; PET, polyethylene terephthalate;

PE, polyethylene; PA, polyamide; AL, aluminum; PP, polypropylene; EVOH, ethylene vinyl alcohol copolymer; MET, methyl–ethylene terephthalate; LDPE, low-density polyethylene; PMMA, poly(methyl methacrylate); BOPP, biaxially oriented polypropylene; VMPET, vacuum metalized PET.), etc. are some thermoplastic polymers used as the matrix in polymer nanocomposites.

The interiors of modern transportation systems are extensively made of polymers. Different polymers have different applications, as in Table 10.1. The nanocomposites of polymers listed in Table 10.1 are successfully prepared using different nanofillers/reinforcements; irrespective of the fillers used, all the nanocomposites exhibited superior properties than the virgin polymers. However, the multifunctionality of the composite depends on the reinforcing nanoparticles. For instance, if the interiors

TABLE 10.1
Polymers and Their Potential Applications in Automobiles

Polymer	Application
Polypropylene (PP)	Bumpers, various tanks, insulation, door panels, console, trunk liners
Polyurethane (PU)	Foam seats, rigid foam insulation panels, headliners, suspension insulators, bumpers
HDPE	Batteries, oil tank, brake fluid tank, interior linings, bumpers, storage compartments, fender liners, windshield washer bottles
Polyvinyl chloride (PVC)	Body-side moldings, interior lining, under-the-hood wiring, car abrasion protection coatings, carpet matting, sealants, and adhesives
Acrylonitrile Butadiene Styrene (ABS)	Exterior body panels, wheel covers, dashboard components, and interior trims
Acrylonitrile Styrene Acrylate (ASA)	Body panels, spoilers, grilles, etc.
polyethylene terephthalate (PET)	Laminating glass windows, foam for interior, door panels, instrument panels, and headliners
LDPE	Fuel tanks, underbody shields, and interior trim
Polyamide (PA)	Air intake manifolds, engine covers, rocker valve covers, airbag canisters, plastic gears
Polystyrene (PS)	Body-side moldings, bumpers, grilles, and other trim pieces
Polyoxymethylene (POM)	Fuel system, Interior and exterior, and engine components
Polycarbonates (PC)	Headlight lenses, window and windshield glazing, dashboard and interior trim, and exterior body panels.
Polymethyl methacrylate (PMMA)	Headlamp lenses, taillight lenses, and instrument panels
Polybutylene terephthalate (PBT)	Connectors, switches, relays, sensors, engine covers, oil filter housings, radiator end tanks, grilles, mirror housings, wheel, door handles, seat belt components, and airbag housings

of automobiles are made of antimicrobial materials, we can avoid the formation of molds, foul smells, and, most importantly, the transmittance of diseases. Many nanostructured materials, such as silver, gold, titanium oxide, zinc oxide, copper oxide, and copper nanoparticles, exhibit antimicrobial characteristics. Reinforcing these nanoparticles in the polymer matrices used for interior fabrication can prevent colonization of the microorganism in the vehicle's interior. It is essential if the vehicle is used for public transportation. Incorporating these nanomaterials into the polymer matrices induces antimicrobial properties and enhances the mechanical properties of the polymer used.

The accumulation of dust and dirt on plastic components is due to the static charges. The accumulated static charges can lead to a spark and become a source of fire in vehicles. Due to static electricity, one may experience shock from vehicles as well. So static electricity must be removed as it is generated. Polymers can generate a large amount of static electricity. To remove static charges, nanomaterials with anti-static can be incorporated into polymers used in automobiles. Silver nanoparticles can drain the static charges from the surface. CNTs are also effective antistatic agents for polymers.

Many nanomaterials enhance the thermal degradation temperature and flame retardancy of the polymers. Nanoparticles of Al_2O_3, layered double hydroxides, montmorillonite clay, etc., can act as flame retardant agents against fire, thus improving passenger safety. Additionally, the barrier properties of these nanomaterials help arrest the propagation of reactive species from the surface under fire to the unburnt regions, thus preventing the spread of fire. Fabrics made of these materials are fire resistant; thus, the seats are fabricated using them.

The nanoparticles are reinforced in thermoset plastics such as polyester, silicone, melamine, polyurethane, epoxy, urea formaldehyde, etc. These polymers are often reinforced with glass, Kevlar, and carbon fibers to enhance the mechanical properties far superior to the polymer alone. These composites are called fiber-reinforced plastics (FRPs). The addition of nanomaterials to FRPs can improve the mechanical properties of the composites further, and they come under the category of hybrid nanocomposites. Whether it is glass fiber-reinforced (GFRPs) or carbon fiber-reinforced (CFRPs) composites, there will be an enhancement in the mechanical property. The fibers can surely increase the strength, but the poor adhesion of the fibers in the matrix limits the composite from achieving the theoretical maximum strength of the composite. The nanomaterials can promote the interfacial interaction between the polymer and the fiber (Xu et al., 2023). As a result, one can observe superior properties in these hybrid composites. The steps in preparing nanoparticle-reinforced CFRP hybrid composite are shown in Figure 10.2.

For instance, the GFRPs reinforced with nanodiamonds enhance the strength by 17%. Reinforcing 0.1 wt% of nanosized iron oxide in GFRP increases the tensile strength by 84.6% and impact strength by 127%. Adding 1.0 wt% of MWCNT improves the toughness by 78.06% and the flexural strength by 69%. 0.5 wt% nano-silica reinforced GFRPs exhibit an increment in tensile strength of 35% and the flexural modulus by 85% when compared with GFRP without nano-silica. Similar improvements can be observed in the case of CFRP in the presence of nanomaterial reinforcements (Rajhi, 2022). Figure 10.3 shows the improvement in the flexural

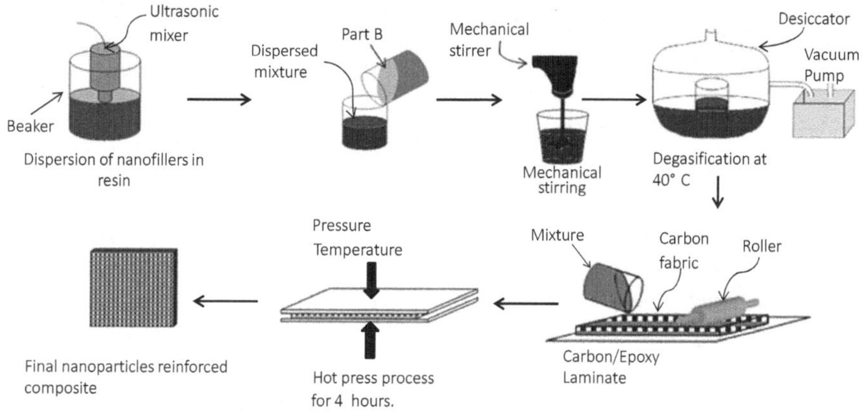

FIGURE 10.2 Stages in the fabrication of nanofiller-reinforced GFRPs. (Reprinted with permission from Hosur et al., 2018. Copyright © 2018 Springer.)

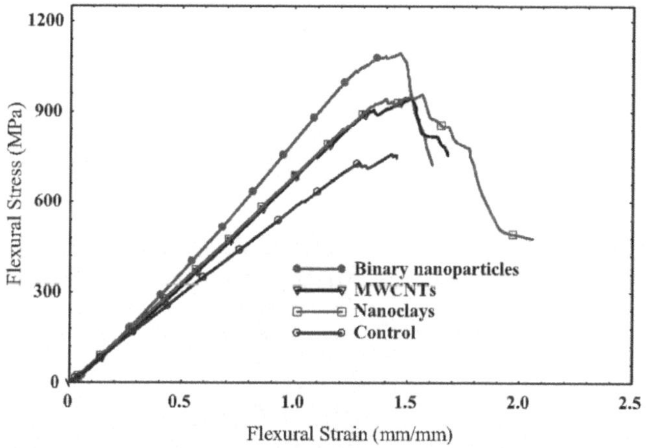

FIGURE 10.3 Flexural stress versus strain response of control and nanoparticle-reinforced carbon fiber/epoxy composite samples. (Reprinted with permission from Hosur et al., 2018. Copyright © 2018 Springer.)

strength of the CFRP-epoxy composite with the addition of MWCNTs and nano clay. Both nano clay and MWCNTs enhanced the mechanical properties of CFRPs. However, the best improvement in the properties was obtained when they were used together (Hosur et al., 2018).

The mechanical properties in the nanoparticles-reinforced FRP composites are due to an improvement in the thermoset polymer resin matrix adhesion and the fibers in the presence of nanomaterials. Moreover, the nanoparticles can hinder the dislocation movements and the crack propagation. These composites are suitable for automotive, marine, and aerospace applications, and one can customize these hybrid composites from a large variety of organic and inorganic fibers and nanomaterials with different morphologies and sizes.

10.1.3 Instrument Panel

Instrument panel also known as dashboards in automobiles is an integral part of a car's structure. It displays gauges such as an odometer, tachometer, temperature, fuel level, etc. (Figure 10.4) to represent the status of an automobile and warning signals to alert the driver about various issues. Customers worldwide seek an attractive look for the instrumental panel. The materials used to produce dashboards are polypropylene, polyurethane, ABS, PVC, etc. Every material used in dashboards has some drawbacks, such as the high cost of polyurethane, sharp corners of ABS, etc. Therefore, nanoparticles can be incorporated into the best suitable polymer to achieve the required properties suitable for automobile dashboards, primary flight displays, airplane control panels, and various panels in the cockpit of boats and ships. Moreover, bionanomaterials such as cellulose nanomaterials or industrial by-products such as fly ash can improve the properties and reduce the carbon footprint of producing these composite structures.

10.1.4 Fuel Tanks

Fuel tanks are essential components of a fossil-fuel-powered vehicle to store and protect the fuel from evaporating. Tanks are airtight and have adequate ventilation. The automotive fuel tank is conventionally made of steel tank. Steel tanks can easily

FIGURE 10.4 A representative image of the instrument panel of an automobile. (Image courtesy: www.freepik.com.)

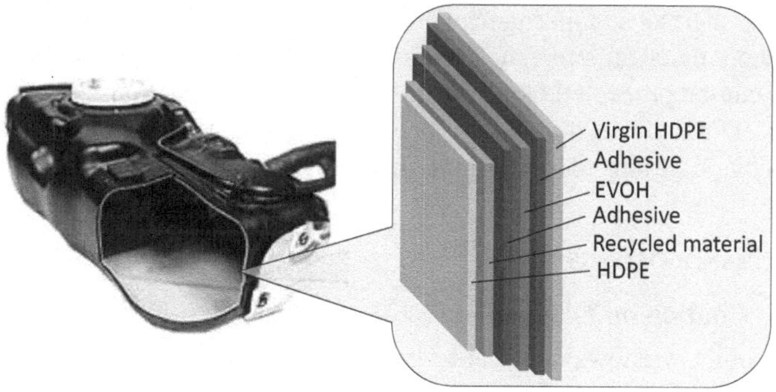

FIGURE 10.5 Construction of multi-layered fuel tank.

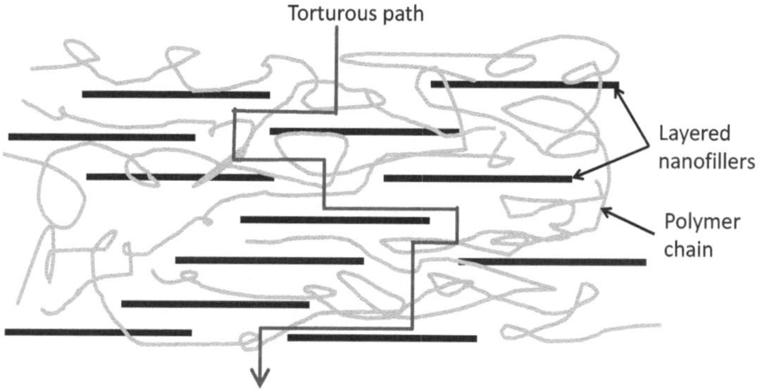

FIGURE 10.6 The torturous path offered by layered nanofillers to the permeating gases.

be subjected to corrosion, expensive, and inflexible, but they have excellent impermeability to fuel. Nowadays, fuel tanks are replaced by lightweight polymers as they address many issues associated with steel tanks. Multi-layered fuel tanks based on high-density polyethylene (HDPE) were introduced first; the production of the tank used an expensive thermoforming process, resulting in substantial waste. The multi-layer tanks provided only a temporary solution, but additional problems were raised. The multi-layer technology uses an ethylene vinyl alcohol (EVOH) layer in the middle to make the tank impervious to fuel. EVOH layer is brittle and can easily crack if the tank is distorted; refer to Figure 10.5.

The drawbacks of multi-layered HDPE fuel tanks can be overcome by reinforcing HDPE with nanofiller high surface area and 2D layered structure. HDPE nanocomposite reinforced with inorganic clay composed of layered silicate is suitable for one-step fabrication of single-layer fuel tanks. The highly exfoliated silicate layers in the composite create a torturous path to permeating molecules, thus making HDPE impermeable to fuels, as shown in Figure 10.6.

The amount of reinforcement and the degree of exfoliation determines the impermeability of the composites. The addition of nanoparticles to the HDPE matrix enhances the thermal and mechanical properties of the HDPE in addition to impermeability without much affecting the processability of HDPE. The nanocomposite-based fuel tanks can be processed using single-step blow molding at a low cost. Graphene is another 2D nanolayer that also increases HDPE's stiffness and other mechanical properties by providing adequate impermeability.

10.1.5 Nanocoatings

10.1.5.1 Coating on Glasses

The automobile windows can scatter light when water droplets, dirt, or oil adhere to their surface. The scattering of light can lead to glare and unexpected accidents. The same effect is observed with headlights and mirrors. Therefore, automobile headlights, windows, and mirrors can be designed with antifogging, anti-reflecting, self-cleaning, UV shielding, transparent, scratch-resistant, impact strength, and water-repellant characteristics. Epoxy reinforced with TiO_2 nanoparticles transparent coatings can improve glass surfaces' antifogging and super hydrophilic properties. The super hydrophilic property of the composite layers is based on the capillary action with a contact angle of less than 5°. As a result, there is an increased tendency for the water droplets to get absorbed immediately and spread uniformly to the surface, reducing the scattering of light significantly. Moreover, such coating can fill flaws or cracks on the glass surfaces, preventing the propagation of cracks. SiO_2-TiO_2 nanofilm coatings deposited on glass substrates through the sol-gel technique provide self-cleaning glasses with photocatalytic properties (Shokuhfar et al., 2012). Magnetron sputtered TiO_2 and SiO_2 thin films exhibit anti-reflective properties. Nano SiO_2/PDMS composite coatings are self-cleaning in nature with high transparency. Polysiloxane/nano SiO_2 coatings exhibit good mechanical properties while being superhydrophobic and anti-reflective (Ramezanzadeh et al., 2011); this coating is helpful for rearview mirrors. Many such transparent coatings are commercially available.

Infrared (IR) reflective nanocoating is another rapidly growing area, and some products are already available in the market. Such coatings are aimed to reflect/absorb the IR region of the solar spectrum, thus reducing the effect of radiation on automobile heating. Nanoparticles of LiB_6 and Cs-doped WO_3 are excellent candidates for NIR blocking. Interestingly, WO_3 is a popular electrochromic material. Applying a small electric field can control the transmittance/ absorbance of solar radiation through electrochromic materials. This is also useful in rearview mirrors; when the light falls on the mirror, a voltage is applied to the electrochromic layer, making it absorb most of the incoming light and less intense. A sensor controls the voltage applied to the electrochromic layer by measuring the intensity of the incident light. A photograph in Figure 10.7 shows the effect of the electrochromic layer on dimming the intense light.

The light filtering properties of WO_3 can be modified by doping different transition metals into WO_3. Incorporating these materials as a coating in automobiles can prevent UV and IR radiation from entering the passenger cabin. Additionally, applying these electrochromic nanophotonic crystals to rear view mirrors can control the reflection of light from other vehicles and provide a clear view irrespective of the

FIGURE 10.7 Rearview mirrors with an electrochromic layer for auto-dimming. (Image courtesy: www.freepik.com.)

TABLE 10.2
Various Nanoparticles and Their Role in Coating Numerous Applications

Nanomaterials	Functional Properties
SiO_2, Al_2O_3, ZrO_2	Scratch resistance
Nanoclay, graphene	Gas barrier
CuO, TiO_2, ZnO	Antimicrobial
Nanoclay	Corrosion and fire retardant
TiO_2, ZnO, $BaSO_4$, CeO_2, graphene	Ultraviolet stability
SiO_2, $CaSiO_3$, CNTs, TiO_2	Impact resistance
CNTs, ZrO_2, nanoclay	Heat resistance
SnO_2, CNTs, graphene	Electrical conductivity

light available on the street. The nanoparticles are reinforced with a suitable polymer matrix as an IR coating, which can reduce the heat developed inside the automobile and thus reduce the fuel usage for air conditioning. Such coatings are ideal for buildings as well. ZnO nanoparticles also exhibit properties suitable for IR reflectance (Benitha et al., 2022).

Additionally, polysiloxane or acrylate paints reinforced with nanoparticles can be applied to improve the scratch resistance of polycarbonate glasses in headlights. Adding nanoparticles such as SiO_2, Al_2O_3, etc., is suitable reinforcement in scratch-resistant transparent coatings to retain the transparency of the headlights. Table 10.2 lists the reinforcing nanoparticles in nanocoating and their functional properties.

10.1.5.2 Coating for Surfaces

Nanotechnology-based coatings enhance the protection against corrosion, scratches, and environmental damage. Nanocoatings are composed of nanoparticles dispersed in a polymer matrix. This thin layer coating exhibits superior barrier properties and aesthetic appeal. Some nanocomposite coatings exhibit superhydrophobic properties, repelling water, dirt, and other contaminants. Thus, the need for

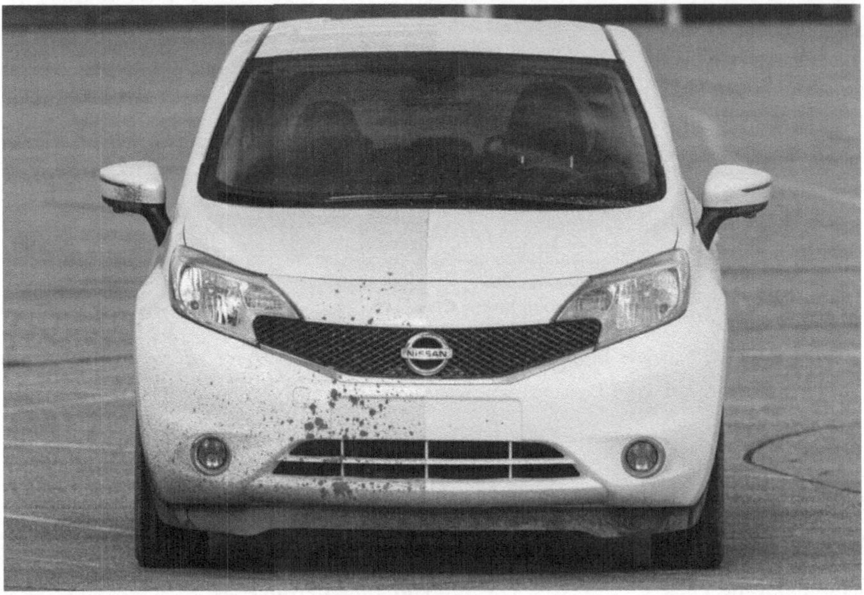

FIGURE 10.8 Superhydrophobic coating developed by Nissan, the right-hand side is coated with a superhydrophobic coating free from dirt. (Image courtesy: www.usa.nissannews.com.)

frequent washing can be avoided. The automobile manufacturer Nissan introduced superhydrophobic coating in their cars. A comparison of the accumulation of dirt with and without superhydrophobic coating is shown in Figure 10.8. Figure 10.9 shows the mechanism of nanostructured superhydrophobic coating. In general, adding nanoparticles to the coatings' appearance and scratch and wear resistance thus increases the durability of paints and underlying coatings. These coatings are effective under extreme weather conditions. The presence of inorganic nanoparticles in the coating increases the hardness of the coatings, thus preventing scratches, wear, abrasion, and cracking. Moreover, the nanoparticles avoid the propagation of cracks, hence the failure of the coating. Adding nanoparticles such as SiO_2, SiC, ZrO_2, ZnO, Al_2O_3, and TiO_2 are promising candidates for scratch and abrasion-resistant coatings.

The recent addition to the coatings is self-healing coatings, which can repair minor scratches and blemishes on the surface. Such coatings require an external stimulus, such as heat or light, to trigger the chemical process behind the self-healing action. The self-healing coating can increase the life span of the vehicle paint and aesthetics. Moreover, the self-healing coatings can protect the underlying substrate from corrosion for an extended period. Core/shell nanoparticles loaded with corrosion inhibitors can also be used as a reinforcement in active coatings, as the corrosion inhibitors can be released gradually when exposed to corrosive electrolytes. These coatings have nanosized containers filled with healing agents. The healing agent is

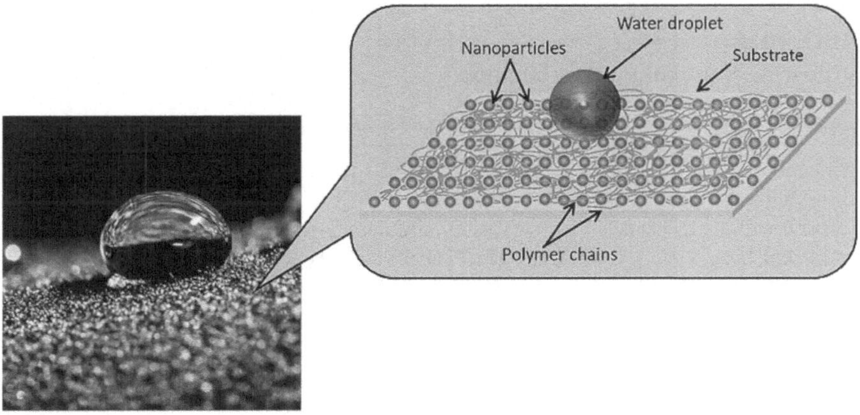

FIGURE 10.9 Photograph of a water droplet on a superhydrophobic coating and the principle of nanotechnology-based superhydrophobic coating.

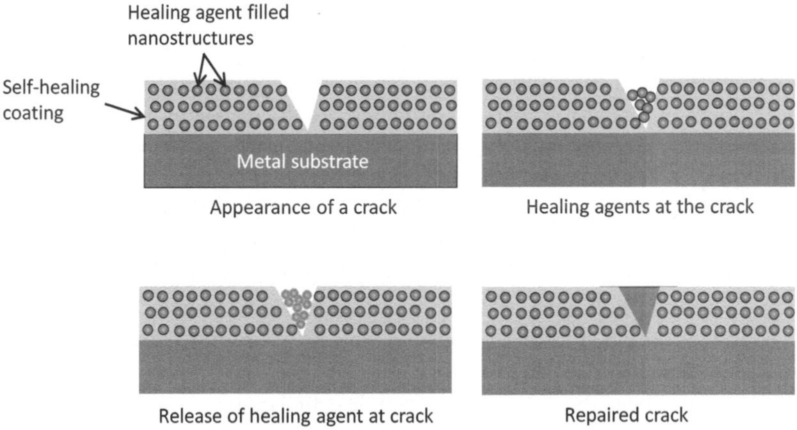

FIGURE 10.10 Mechanism of nanotechnology-based self-healing coating.

released when a crack forms on the coating and fills the cracks to protect the surface underneath, as shown in Figure. 10.10.

Additionally, nanocoatings can offer multifunctionality, antifogging, anti-icing, anticorrosive, and antimicrobial properties. Nanoparticles dispersed within coatings can alter surface characteristics to prevent fog formation on windows, reduce ice buildup on mirrors and sensors, and inhibit the growth of bacteria and fungi on interior surfaces. These functionalities enhance driving visibility, safety, and cabin hygiene, providing a more comfortable and enjoyable driving experience. Some nanocoatings can offer protection to paints against harsh environments and UV rays to enhance the appearance and durability of paints.

TABLE 10.3
Purpose of Nanocoatings in Engines

Property	Coating Composition
Lubrication	Ni-Mo–MoS_2, Ni–BN, Graphite–Ni.
Corrosion resistance	Mo/Cr
Wear and scuffing resistance	Al_2O_3, SiC and Fe-oxides
Surface finishing	Fe/FeO and stainless steel/BN
Thermal barrier coating	Yttria stabilized zirconia (YSZ) on nickel-based super-alloys

10.1.5.3 Coating on Piston

As lightweight aluminum alloys replaced the cast-iron engine blocks, poor wear resistance and low seizure loads were the primary concerns when using them for cylinder bores. The car manufacturers used cast-iron liners along the cylinder bore to overcome the difficulty because they have good wear resistance. Due to the liners' large wall thickness, the engine's weight is increased. The friction between the cylinder bore and piston, as well as the cylinder bore/piston ring, primarily contributed to engine friction losses. Various nanoscale ceramic coatings are proposed to replace the cylinder liners in internal combustion engines. Replacing liners with nanocoatings not only reduces the engine weight but also reduces its dimensions (Dahotre and Nayak, 2005).

One can expect a high coefficient of thermal expansion (CTE), lower thermal diffusivity and hysteresis, and greater hardness and toughness compared to traditional (microscale) coatings. The decreased thermal diffusivity results from increased phonon scattering grain boundary imperfections at the nanoscale. Furthermore, the nanocoating is expected to exhibit plasticity through grain sliding (Hahn et al., 1997).

Al_2O_3 nano ceramic coating on the engine cylinder can act as a thermal barrier coating; thus, one can attain increased power, less fuel consumption, and less exhaust gas emissions and smoke density. The deposition of copper and nickel nanocoatings on piston tops can act as a catalyst for reducing hydrocarbon emissions, smoke levels, and NO_X emissions, besides enhancing brake thermal efficiency and reducing specific fuel consumption. The nanocoatings of zirconia (ZrO_2), aluminum titanate, Al_2O_3, yttria stabilized zirconia, TiO_2, etc., on the piston heads can drastically enhance the engine performance and reduce emissions. Various nanocoatings on engine components and their applications are listed in Table 10.3.

10.1.6 Filters

Filters are a part of human life with different perspectives. Filters are used in water purifiers, protective masks, air conditioners, automobile air filters, etc. Filters are also critical in many engineering and chemical industries for absorption, separation, wastewater treatment, etc. For all of these applications, the performance of the filters can be improved tremendously by increasing the fineness of the fiber structure used in

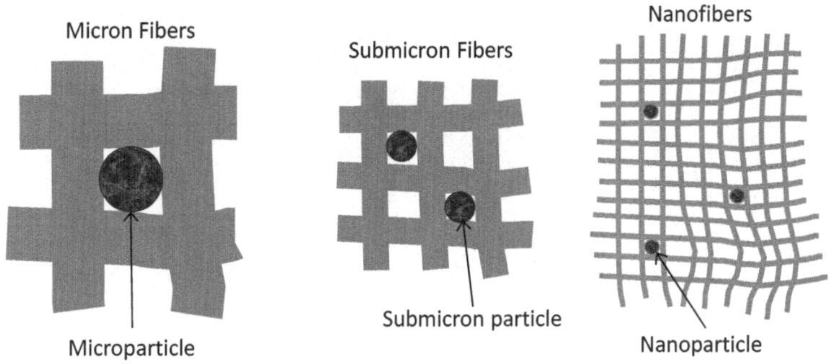

FIGURE 10.11 The effect of fiber size composing the filters on the particle size filtered.

the filter. Figure 10.11 depicts the significant improvement in the effective area available for filtration as the size of the fibers constituting the filter reduces. The same filter efficiency is achieved by a small area as the fineness of the fiber increases.

A filter is efficient when it can remove unfavorable particles with sizes in all ranges, including sub-micron ranges. The filter can remove sub-micron particles only with that range's structure and channel size. Electrospun nanofibers can be used to develop effective filters. The electrospun nanofibrous structured filters have a high surface-to-volume ratio, thereby a higher surface cohesion, aiding the tiny particles in the sub-micron range to entrap in the filter. The strength of the nanofiber web alone is often too low to use as a filter, so spun bonded non-woven or melt-blown non-woven fabrics are always used as sublayers to support the nanofibrous web. Electrostatically charged electrospun fibers enable the electrostatic attraction of the particles and improve the filtration efficiency. Selective filtration is also possible by incorporating specificity agents during the production of the nanofibers using electrospinning. These filters can be used as molecular filters or countermeasures against chemical and biological weapons and their detection.

10.1.7 Catalytic Converters

The catalytic converter is an integral part of automobiles to meet the stringent norms for pollution control. The exhaust gas from the engine passes through a three-way catalytic converter, where the hydrocarbons (HC), carbon monoxide (CO), and nitrogen oxides (NO_x) are catalytically converted to less harmful gases. The catalytic converter is used in the exhaust manifold combined with the exhaust system, so that the catalytic reaction occurs as the exhaust gas flows over the catalyst surface. A three-way catalytic converter (Figure 10.12) can reduce the emission of CO, HC, and NO_x in the exhaust gases with a conversion efficiency of up to 90%. The catalytic converts are efficient when sufficient oxygen is present in the exhaust gases. Therefore, an oxygen sensor is placed in the exhaust manifold to regulate the air/fuel ratio at the inlet.

Noble metals are ideal for catalytic conversion, and a wash coat of porous Al_2O_3 supports the catalyst on a monolith or honeycomb structure. The catalytic converters

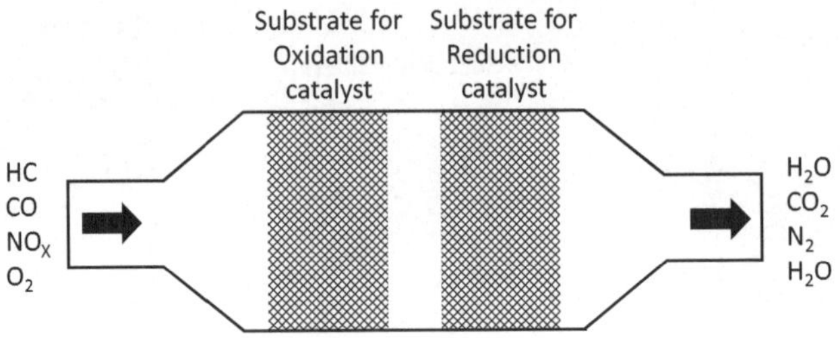

FIGURE 10.12 Construction of a catalytic converter.

based on noble metals such as platinum (Pt), palladium (Pd), and rhodium (Rh) can face problems such as converter meltdown, carbon deposit, catalyst fracture, and poisoning. Recently, nanostructured oxides of transition and rare-earth metals have been identified as potential candidates for oxidation and reduction catalysts to replace noble metals. MnO_2, TiO_2, CeO_2, NiO, CuO, etc., are some of the nanostructured transition oxides for catalytic conversion.

10.1.8 Brake System

A brake is an essential component of an automobile. In a brake system, a brake pad is used to convert the kinetic energy of the automobile to heat energy. Brake pads are made of a metal backing plate sandwich structure, friction material, and an intermediate layer. The friction material comprises fibrous reinforcements, abrasives, fillers, and solid lubricants, thus forming a composite. Metal matrix composites (MMCs) are commonly used as friction materials in brake pads owing to their mechanical properties and wear resistance. Therefore, brake pads have metallic matrix, asbestos, semimetallic compounds, non-asbestos organic materials, and low-steel, low-carbon materials (Akıncıoğlu et al., 2018). Natural fibers and existing fillers can replace the carcinogenic asbestos fibers; some are barites, mica, Allium sativum, and cashew dust. However, these new materials must meet the criteria for the brake pads, such as consistent friction, high wear resistance, reduced noise, and damping across a wide range of operating conditions.

Incorporating nanomaterials into the brake pads can improve the various properties of brake pads with desired characteristics better than asbestos-reinforced brake pads. Carbon-based nanomaterials, single and multi-walled carbon nanotubes (MWCNTs), graphene, etc., are effective friction modifiers besides enhancing strength and wear resistance. The brake pads formed using surface-modified carbon nanofibers and multi-walled carbon nanotubes (MWCNTs) improve damping characteristics and reduce the noise. Hybrid nanocomposite materials, such as those incorporating carbon nanospheres, offer improved thermal stability and frictional performance. The diffusion of oxygen into breakpads can weaken them because it will be very reactive as the temperature at the breakpads increases during breaking. Iron sulfide can reduce oxygen diffusion, thus enhancing the stability of brake pads and the friction coefficients at high temperatures. Incorporating nano-silica particles improves the hardness and roughness of copper-based friction materials.

Additionally, kaolin-TiO_2 decreases the wear rates, thus improving the durability. Aluminum-based MMC of Al/(B_4C+SiC) exhibits wear resistance suitable for breakpads (Shorowordi et al., 2006).

The incorporation of nanomaterials into brake pads can improve several properties of the brake pads. Carbon nanomaterials enhance the coefficient of friction, resulting in better braking performance and wear resistance. Heat dissipation is crucial as it is generated during braking; if the heat is not dissipated properly, the brake can weaken or fail. Incorporation of nanomaterials with high thermal conductivity can improve heat dissipation and thereby improve the overall performance and safety of the brake pads. Integrating nanomaterials reduces the weight of lightweight brake components without compromising strength and durability. This can contribute to reducing the vehicle's overall weight and improving fuel efficiency and handling. Nanocoatings can offer corrosion and oxidation resistance, extending their lifespan and maintaining performance, especially in harsh environments. Nanomaterials can dampen the vibrations and reduce noise generated during braking, providing a quieter and more comfortable driving experience. Adding self-healing nanomaterials in brake pads can repair minor damage or wear automatically and increase the life span of the components.

10.1.9 Tires

Tires are the essential component of an automobile since they are the only components in contact with the ground. Generally, tires are composed of rubber, steel threads, nylon threads, reinforced fillers, etc. The correct formulation of rubber is critical to attain the desired properties. The rubber used in tires contains reinforcements and fillers 30% by weight. The reinforcements like carbon black harden the rubber, wear and abrasion resistance, grip, reduce friction, resistance to tear, etc. An ideal tire should have less rolling resistance without compromising on the grip. Carbon black, soot, and silica are the most common additives in tires to obtain the best properties and increased lifetime. The tire properties significantly improve when these particles are added to the rubber.

The tire industry uses the scope of nanotechnology extensively to improve traction, tire performance, durability, wear resistance, and sustainability. Incorporating nanoparticles, such as silica and carbon black, into the rubber compound used for tire tread to make it soft, improve traction, wear resistance, and grip on wet surfaces. Combining carbon black with silica makes the tread softer, decreasing rolling resistance and lowering fuel consumption. It also allows you to drive in temperatures beyond freezing, as the tread remains pliable throughout the performance. Adding nanoscale soot improves tires' fuel efficiency and durability because of less rolling resistance and inner grip. The high surface area of soot particles leads to a strong interaction of soot with rubber. Nanoparticles and ultimately improve the overall properties of tires. Adding MWCNTs to rubber improves its tensile strength. Adding montmorillonite clay to rubbers ensures the isotropic behavior of rubber, which can enhance the comfort and handling of vehicles. Thus, incorporating nanoparticles into the rubber matrix reduces rolling resistance, increases durability and self-healing properties, and improves wet traction, noise reduction, and weight reduction. As a result, one can achieve low fuel consumption, more comfort, ease of handling the vehicles, longer life span, less maintenance cost, etc.

10.1.10 Electronic Components

Nanotechnology enables the miniaturization of electronic components with superior functionality. Nanoelectronics applications include vehicle electronic control units, propulsion control, safety, power management, and connectivity. Sensors work on the principle of nanotechnology and allow continuous monitoring of vehicle performance, exhaust gas, temperature, pressure, environmental conditions, etc., to create a feedback system that optimizes the operating conditions of the vehicles and improves specific fuel consumption. Moreover, such a system can alert the driver about the vehicle's safety, and preventive maintenance can be done if needed. Moreover, energy-efficient lighting and displays based on light-emitting diodes (LEDs) and organic light-emitting diodes (OLEDs) can reduce energy consumption significantly.

10.1.11 Energy Storage and Conversion

Recently, vehicle manufacturers have been focusing on producing vehicles that run on environmentally benign techniques, such as electricity, fuel cells, etc. Attention is paid to electric vehicles (EVs) daily, and customers are looking for high-performance batteries. The high performance of an EV depends on its battery performance. Batteries with increased energy density, faster charging times, and longer lifespan can be designed using nanostructured materials, as discussed in Chapter 5. Redesigning Li-ion batteries using nanomaterials can overcome the limitations of Li-ion batteries and increase the acceptance of EVs in the global market.

Fuel cells are another sustainable and eco-friendly means of powering transportation systems with low emissions. Nanostructured catalysts play a remarkable role in the hydrogen oxidation and oxygen reduction reactions at the electrode surfaces, as we discussed in Chapter 5. The fuel cell life span, cost, and efficiency can be reduced by applying nanotechnology to the fuel cells.

10.1.12 Additives for Fuels

In internal combustion engines, the chemical energy is converted to mechanical energy. Fuels supply the chemical energy to the internal combustion engines. Many additives are used in automotive, aviation, and marine fuels. The additives and their role are listed in the Table. 10.4.

Nanoparticles are added to the fuels to improve the combustion characteristics. These additives are added to diesel, gasoline, biodiesel, alcoholic and blended fuels. In this study, standard base fuels are diesel, gasoline, biodiesel, alcoholic, and blended. The various additives for diesel, biodiesel, and their blend fuels are more extensive than gasoline fuels. Nanostructured Al_2O_3, CeO_2, MnO, and CuO particles resulted in a remarkable reduction in the ignition delay in combustion ignition engines. The presence of nanoparticles in diesel/biodiesel can lead to finer atomization of fuel

TABLE 10.4
List of Additives in Fuels and Their Functions

Additive	Function
Fuel Stabilizers	Prevent the oxidation and degradation of fuel during storage, making the fuel usable for a long time.
Cetane Boosters	Increase the cetane number, thus improving the ignition quality of the fuel. Ensure better engine starting, smoother running, and reduced emissions.
Octane boosters	Improve octane rating, thus preventing premature ignition of air/fuel mixture during the combustion.
Biocide	Kill microbes (bacteria and fungi) due to the presence of water/ moisture in the fuel, or else they can clog the fuel system and lead to corrosion.
Static dissipators	Minimize the hazardous effects of static charges that build up during the movement of jet fuels.
Combustion additives	Improve combustion efficiency, thus reducing harmful emissions.
Metal deactivator	Mitigate the effects of dissolved trace metals, especially copper, which can impair the thermal stability of jet fuels.
Detergents	Added to gasoline and diesel fuels to clean and prevent deposits in the fuel injectors and on intake valves. They remove deposits and avoid clogging.
Corrosion inhibitors	Corrosion inhibitors protect the various components of the fuel system from corrosion due to moisture and other corrosive elements of fuel.
Reodorants	Make the smell of diesel fuel more pleasant and less intense.
Demulsifiers	help to separate water from fuel to prevent microbial growth, corrosion, and fuel system damage.
Friction modifiers	To reduce the mechanical friction between the various components of the fuel system.
Dispersants	To break up and disperse particles and deposits that can form in the fuel system to prevent clogging.
Cold flow improvers or Anti-Gel Additives	prevent wax gel/crystal formation in cold temperatures.

droplets with uniform size. Moreover, the oxygen in the nanoparticle and water can lead to a micro-sized explosion due to rapid evaporation of water and enhanced combustion properties of fine fuel droplets. Nanostructured graphene, CNTs, Al_2O_3 and CeO_2, and TiO_2 are used. Nano additives in diesel and biodiesel fuels increase the thermal efficiency of the fuel, thereby enhancing the engine power. Moreover, due to the improved ignition characteristic of adding nanoparticles, harmful emissions such as unburned hydrocarbons (UBHC), carbon monoxide (CO), and particulate matter (PM) are minimized in the exhaust gas. In general, adding nanoparticles increased the density, viscosity, cetane number, and heating value of the fuel compared to fuel without nano additives (Hatami et al., 2020a). Adding TiO_2 and Fe_2O_3 nanoparticles to gasoline decreases the viscosity by 16.7% and the octane index by 10.9%.

10.1.13 Coolant Additives

Coolants are essential for internal combustion engines as they help in heat transfer to optimize engine conditions and get the best performance. The engine and its components harness most of the heat generated in the internal combustion engine, and coolants prevent them from melting down or overheating, thus the failure of the engines. The addition of alcoholic additives and ethylene glycol improves the performance of coolants.

The efficiency of engine cooling systems can be increased significantly by adding nanostructured materials to conventional coolants forming nanofluids. The nanofluids can be used in radiators directly. Adding CuO or Al_2O_3 nanoparticles to water enhances the cooling efficiency of water, exhibiting uniform temperature distribution under all operating conditions. The same results can be observed using ethylene glycol with increased heat transfer rate as base fluid. Adding carbon nanotubes to water also improves the Nusselt number compared to pure water, indicating an improvement in heat transfer rate. CNT nano coolants are outperforming commercial coolants on the market.

Though many nanomaterials can be used in nanocoolants, MWCNTs, graphene, SiO_2, TiO_2, Al_2O_3, GNP, CNT, Cu/ CuO, and hybrid nanoparticle-added coolants are identified as the most suitable candidates as listed in Table 10.5, irrespective of the base fluids (Hatami et al., 2020b). The significant improvement in thermal conductivity and heat transfer efficiency, in addition to corrosion prevention, clogging, etc., reveals the potential of nano coolants in engine cooling systems.

Nanocoolants can play a critical role in the thermal management of batteries in electric vehicles (EVs) by improving battery performance, lifespan, and safety. The nano coolants can help maintain optimal operating temperatures within the battery pack, power electronics, and electric drivetrain components. The nano coolant can effectively mitigate temperature variations and hot spots within the battery pack, thus reducing thermal stress on individual cells and ultimately increasing their lifespan and overall battery health. Optimal thermal management can be effective during the charging processes. The nanocoolants can prevent overheating and thermal runaway. As a result, the overall safety of EVs can be improved.

10.1.14 Additives for Lubricants

Nanoparticles can be added to lubricants to improve their effectiveness for internal combustion engines, such as engine oils, compressor oils, journal-bearing lubricants,

TABLE 10.5
List of Nanofluids for Vehicle System Cooling

Nano Fluids	Improvement
Al_2O_3–EG	Improvement in thermal conductivity and heat transfer.
Al_2O_3–water	Better heat transfer coefficient, heat transfer rate, and Nusselt number compared to pure water.
CuO–water,	Enhanced the heat transfer coefficient by up to 9%.
TiO_2–water	Heat transfer coefficient and Nusselts number
SiO_2–water	Heat transfer enhancement

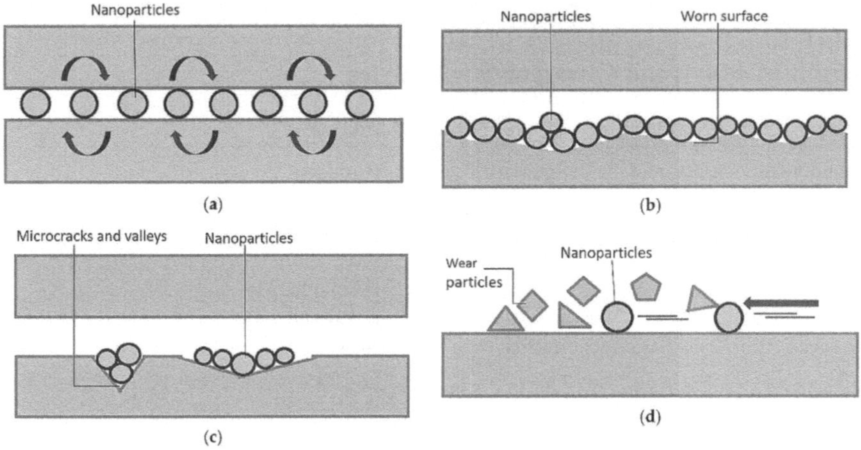

FIGURE 10.13 Mechanism of nanolubricants; (a) Rolling/ball-bearing mechanism, (b) protective film/tribo-film formation, (c) mending effect, and (d) polishing/smoothing effect. (Morshed et al., 2021)

grease, etc. Nanolubricants have the potential to reduce friction and wear, enhance energy efficiency, etc., and thus extend the lifespan of mechanical systems significantly. The mechanisms of lubrication by nanoparticles are the ball-bearing effect, protective film, mending effect, and polishing effect, as shown in Figure 10.13. The ball-bearing effect is exhibited by nanoparticles with spherical morphology. The spherical nanoparticles convert the sliding motion to the rolling motion of the mating surfaces, creating nanosized ball bearings. The nanoparticle-containing lubricating films are stable and act like a protective film. Nanoparticles can fill the irregularities on the mating surfaces to create a smoothing or mending effect. The nanoparticles can act as abrasives to remove the surface irregularities on the mating surfaces, creating a polishing effect (Hao et al., 2020). Nanomaterials such as TiO_2, ZnO, CuO, SiO_2, ZnO, ZrO_2, graphene, carbon nanotubes, cellulose nanoparticles, etc., in various base lubricants like SAE 20, SAE30, and SAE 40 reduce the coefficient of friction and wear rate (Hatami et al., 2020c).

Adding carbon nanomaterials such as multi-walled carbon nanotubes (MWCNT) and graphene nanoplatelets (GNP) can increase the heat transfer properties of the lubricants. Nanolayered MoS_2 in mineral oils enhances anti-wear, anti-friction, and thermal conductivity enhancements. The nanoparticulate TiO_2 and Al_2O_3 reduce the coefficient of friction significantly from 50 to 80% depending on the weight percentage of nanoparticles. The layered nanostructures, such as graphene, MoS_2, etc., improve the wear and -friction characteristics.

10.1.15 Sensors

A sensor is a device that detects or measures a physical quantity and converts it into a measurable form. Sensors are used in various applications, such as pressure,

TABLE 10.6
Sensing Materials and Corresponding Analytes

Composition of Materials	Type of Sensing	Analyte
Lithium Chloride doped TiO_2	Resistive	Humidity
Sb-doped SnO_2	Resistive	H_2, CO, CH_4, $C_3H_8$and C_2H_5OH.
MoS_2	Resistive	VOCs such as alcohol, ketone, aldehyde
WO_3	Resistive	NH_3
Hollow SnO_2	Resistive	Formaldehyde
Pd Loaded SnO_2	Resistive	Hydrogen
Mxenes	Resistive	Ammonia
CeO_2	Resistive	O_2 and CO
WO_3	Resistive	Ammonia
Cobalt doped ZnO	Photoelectric	Oxygen
ZnO	Photoelectric	Ethanol vapor
Indium tin oxide (ITO)	Photoelectric	NO_2
Cadmium titanate	Capacitive	Humidity
Pt doped CeO_2	Amperometric	CO and C_3H_8 detection

temperature, humidity, hazardous gases, drinking water quality, weather, explosives, medical diagnostics, food inspection, etc. Several parameters govern the quality of a sensor. The high specific surface area of the nanostructured materials makes it the most conceivable candidate for sensor applications. Nanomaterials with high porosity, interconnectivity, high axial strength, and flexibility make the suitable materials for the sensors. Sensors can be classified based on detection mechanisms followed in the sensors. Nanomaterials are used in sensors like acoustic, resistive, photoelectric, optical, amperometric, etc. A more detailed discussion on sensors can be found in Chapter 13. In an automobile, the sensors find application in sensing oxygen, humidity, temperature, carbon monoxide, nitrous oxide, volatile organic compounds (VOCs), etc. Some nanomaterials for sensing various analytes in the transporting industry are listed in Table 10.6.

10.2 APPLICATIONS IN THE AEROSPACE INDUSTRY

Aerospace transportation is the fastest mode of transportation of people and goods worldwide, making it an indispensable mode of transportation. However, the risk involved in aerospace transportation using aircraft is higher than other modes of transportation. Stringent safety requirements and fail-safe components are critical for manufacturing aircraft components. Materials with high yield strength, tensile strength, corrosion resistance, and low density are used in aircraft construction. Currently, the focus is on designing lightweight materials with excellent strength and efficient and benign systems for the environment. With the help of nanotechnology, the aircraft's weight can be significantly decreased, and ultimately, it can achieve higher speed, better maneuverability, and reduced fuel consumption. The

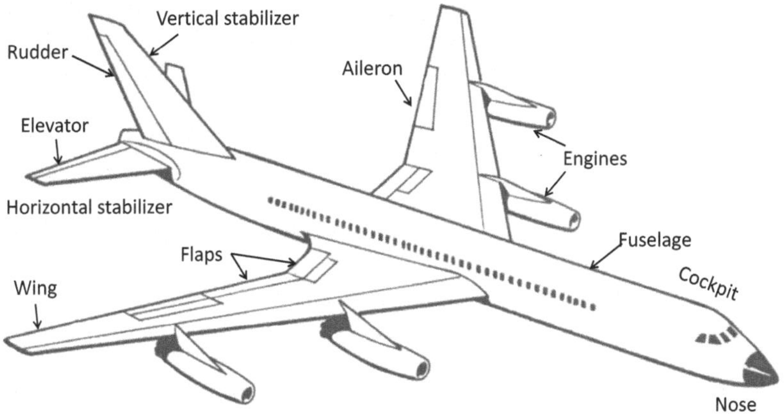

FIGURE 10.14 Aircraft components potentially made of nanohybrid FRP composites.

application of nanotechnology on different parts of an airplane is schematically shown in Figure 10.14.

10.2.1 Lightweight Composites

The hybrid nanocomposites of FRPs reinforced with nanomaterials exhibit excellent properties suitable for high-strength, light-weight applications. GFRPs and CFRP are appropriate candidates for the fabrication of hybrid composites. These composites are replacing conventional materials for aerospace-related production as lightweight CFRPs exhibit mechanical properties superior to GFRP and traditional materials. FRP hybrid composites are used to fabricate fuselage structures in modern aircraft, thus not only reducing the overall weight but also having good impact resistance and fatigue life. When used in aircraft wings, high strength and stiffness can be achieved. FRP wings can be designed for advanced aerodynamics and, thus, improved performance. CFRPs can be used to fabricate horizontal and vertical stabilizers (tail components) of the aircraft as they can withstand various stresses exerted on these components during the flight. CFRPs can fabricate control surfaces such as ailerons, elevators, and rudders. Due to the lightweight of CFRPs, these components can be controlled quickly and accurately. The control surface determines the maneuverability of an aircraft. FRPs can be used for fabricating fairings (covers for various parts of the aircraft) and engine nacelles (enclosures around the aircraft's engines). Landing gear components such as doors and supports can be made from CFRP due to their durability and resistance to fatigue and corrosion. CFRP is also used for engine components, such as fan blades and casings. CFRP fuel tanks can offer weight savings and resistance to fuel degradation, making them an excellent choice for specific aircraft models. FRPs are used in specialized parts such as brackets, fasteners, and fittings where strength, lightweight, and resistance to fatigue and corrosion are required. Specific components in the aircraft's interior are made of FRPs, such as seats, cabin panels, overhead storage compartments, etc. The material's light weight contributes to overall weight savings, while its strength and durability ensure passenger safety.

The FRP composites are helpful in helicopters and drones. The CFRPs are suitable for the fabrication of rotor blades. CFRPs have high stiffness and strength and are lightweight, essential for better flight control. The CFRP is used to construct the helicopter fuselage, tail boom, landing gear components, interior structures, and components such as seats, panels, partitions, tail rotor, engine components, fairings and cowlings, etc. FRPs are used to construct drone frames and structural components such as arms and landing gear. The lightweight and strong nature of FRPs allows drones to carry payloads efficiently and maintain stability during flight. A drone's propellers can be made using FRPs due to their strength, flexibility, and weight. The body panels and fairings of the drones can be made using FRPS with streamlined aerodynamics. FRPs are used for battery and payload housings in drones. It fabricates various frames and mounts for hardware, motor, etc. FRPs can be used to construct wings and control surfaces such as ailerons and rudders. The addition of nanomaterials in FRPs improves the properties of the composites significantly without increasing the weight, as we discussed earlier in this chapter.

10.2.2 Protective Coatings

Aircraft can be easily subjected to icing and lightning strikes, and the above issues must be addressed correctly for a safe flight. Aircraft icing is the frozen water droplet on the aircraft's surface during the flight at low temperatures, which can lead to stability and control issues. The icing can increase the drag, so a significant reduction in the aircraft's performance can be expected. Anti-icing properties can be achieved using superhydrophobic coatings embedded with nanomaterials, as discussed in the previous section. Such coatings offer resistance to scratch and abrasion. CNTs, TiO_2, Al_2O_3, SiO_2, TiN, Si_3N_4, SiC, etc., are the potential nanostructured reinforcements for superhydrophobic and scratch-resistant coatings. A study has shown that composites reinforced with CNTs can mitigate icing effects on aircraft.

Lightning strikes substantially threaten aircraft safety, as an untreated strike can generate electrical currents as high as 200,000 amperes, potentially igniting fuel vapors and causing hazardous conditions. Using CNTs and carbon nanofillers with non-conductive polymers is a promising approach for protecting aircraft from lightning strikes. Buckypaper, a form of entangled CNTs, can improve current carrying capacity and safeguard aircraft electrical systems during a strike. CNT-reinforced polymers enhance the electrical conductivity of composite panels, allowing current to flow throughout the entire aircraft structure and providing protection against electrical discharge. The nanocomposite coating with metal nanoparticles exhibits electrostatic discharge and electromagnetic interference (EMI) shielding properties, making them resistant to lightning strikes. Additionally, recent research has applied synthesized carbon nanotube paper with a self-healing composite base, enhancing electrical conductivity and deicing capabilities. Additionally, coating loaded with CNTs can act as electromagnetic shielding for sensitive aircraft components. Additionally, nanomaterials such as CNTs, SiO_2, TiO_2 nanoparticles, and graphene in polymeric coatings can minimize surface degradation and extend the lifespan of aircraft components.

10.3 MARINE TRANSPORT

Marine transportation is the primary means of transporting goods and people safely and comfortably. Corrosion of ships is a significant concern as the ship is continuously exposed to harsh saline environments. The major maintenance costs of ships are associated with the corrosion and the removal of barnacles attached to the ship hulls, which otherwise create drag that reduces fuel efficiency. Though stainless steel is corrosion resistant under a standard atmosphere, it will undergo corrosion easily in seawater. Moreover, the bottom of the ship can be subjected to erosion and fouling. All these factors can affect the performance of vessels over time. The corrosion-resistant coatings can prevent corrosion and biofouling, thus retaining the structural integrity of the ships.

Reinforcing the primer coating with TiO_2 nanoparticles lowers the corrosion rate, revealing the significance of nanoparticles in corrosion-resistant coatings. Nanoparticles, such as CNTs, MgO, ZnO, SiO_2, and Al_2O_3, can also increase the corrosion resistance of the coatings remarkably. Reinforcing carbon nanomaterials to coatings can reduce biofouling effectively. The barrier properties of such coatings can be enhanced by the addition of 2D nanomaterials, such as clay, layered double hydroxides, MoS_2, and graphene; thus, the permeability of seawater through the coating can be reduced, thus increasing the life of the underlying metallic structures from corrosion. See Figure 10.15.

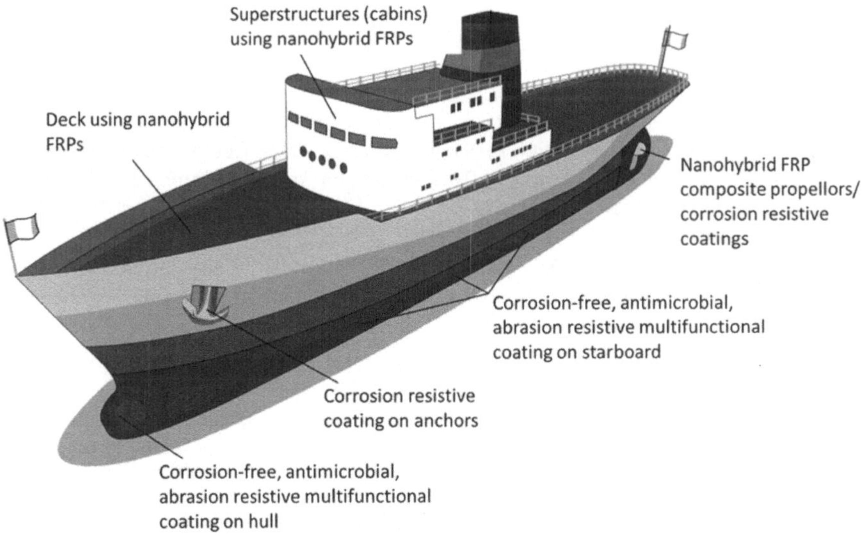

FIGURE 10.15 Application of nanotechnology in ships.

10.3.1 Applications of Nanotechnology in Ships

In cruise ships, the antimicrobial/antiviral coating is essential to reduce the contamination of various surfaces onboard. The cruise ship industry was most affected during the Covid-19 outbreak. Adding nanoparticles such as copper, silver, ZnO, TiO_2, etc., in the coatings can prevent bacterial and viral infections from transferring through the surfaces, e.g., SARS-CoV-2 viruses. These nanomaterials are harmful to viruses but not to human beings. Hybrid FRP composites reinforced with nanomaterials are used for shipbuilding. GFRPs and CFRPs are used to construct hulls for yachts and high-performance boats. In ships, they are used for building decks, cabins, wall panels, furniture, propellers, and pipes for water, fuel, etc., to reduce the overall weight of the ship. FRP composites, such as surfboards, canoes, and kayaks, are also used for recreational purposes. The performance of the FRPs in this application can be improved by hybridizing them with nanomaterials. Various applications of nanotechnology in ships are schematically shown in Figure 10.14.

Many applications discussed earlier under nanotechnology in automobiles are relevant in aerospace and marine transportation.

10.4 ENVIRONMENTAL HEALTH AND SAFETY CONCERNS

Nowadays, nanotechnology is used in various scientific and engineering fields. Though nanomaterials have superior properties, the extensive use of nanomaterials and their release into the atmosphere can harm humans and other biological organisms. Recent studies show that nanomaterials can have a toxic impact on human health. Therefore, sufficient care must be taken while handling the nanomaterials. The effect of only a few nanomaterials is explored now; many others need to be studied thoroughly before they are applied to any applications. Strong regulations must be implemented to control the use of nanomaterials for the pertaining applications if necessary.

REFERENCES

Akıncıoğlu, G., Öktem, H., Uygur, I., Akıncıoğlu, S., 2018. Determination of friction-wear performance and properties of eco-friendly brake pads reinforced with hazelnut shell and boron dusts. Arab. J. Sci. Eng. 43, 4727–4737. https://doi.org/10.1007/s13369-018-3067-8

Benitha, V.S., Jeyasubramanian, K., Prabhin, V.S., 2022. Enhanced near-infrared reflectance and functional characteristics of nano metal oxide embedded alkyd coatings. Mater. Res. Express 9, 056404. https://doi.org/10.1088/2053-1591/ac6fbd

Ceschini, L., Dahle, A., Gupta, M., Jarfors, A.E.W., Jayalakshmi, S., Morri, A., Rotundo, F., Toschi, S., Singh, R.A., 2017. Mechanical Behavior of Al and Mg Based Nanocomposites, in Ceschini, L., Dahle, A., Gupta, M., Jarfors, A.E.W., Jayalakshmi, S., Morris, A., Rotundo, F., Toschi, S., Singh, R.A. (Eds.), Aluminum and Magnesium Metal Matrix Nanocomposites. Springer, Singapore, pp. 95–137. https://doi.org/10.1007/978-981-10-2681-2_4

Dahotre, N.B., Nayak, S., 2005. Nanocoatings for engine application. Surf. Coat. Tech. 194, 58–67. https://doi.org/10.1016/j.surfcoat.2004.05.006

Hahn, H., Mondal, P., Padmanabhan, K.A., 1997. Plastic deformation of nanocrystalline materials. Nanost. Mater. 9, 603–606. https://doi.org/10.1016/S0965-9773(97)00135-9

Hao, L., Wang, Z. [Zheng], Zhang, G., Zhao, Y., Duan, Q., Wang, Z. [Zhenni], Chen, Y., Li, T., 2020. Tribological evaluation and lubrication mechanisms of nanoparticles enhanced lubricants in cold rolling. Mech. Ind. 21, 108. https://doi.org/10.1051/meca/2019085

Hatami, M., Hasanpour, M., Jing, D., 2020a. Recent developments of nanoparticles additives to the consumables liquids in internal combustion engines: Part I: Nano-fuels. J. Mol. Liq. 318, 114250. https://doi.org/10.1016/j.molliq.2020.114250

Hatami, M., Hasanpour, M., Jing, D., 2020b. Recent developments of nanoparticles additives to the consumables liquids in internal combustion engines: Part III: Nano-coolants. J. Mol. Liq. 319, 114131. https://doi.org/10.1016/j.molliq.2020.114131

Hatami, M., Hasanpour, M., Jing, D., 2020c. Recent developments of nanoparticles additives to the consumables liquids in internal combustion engines: Part II: Nano-lubricants. J. Mol. Liq. 319, 114156. https://doi.org/10.1016/j.molliq.2020.114156

Hosur, M., Mahdi, T., Jeelani, S., 2018. Studies on the performance of multi-phased carbon/epoxy composites with nanoclay and multi-walled carbon nanotubes. Multiscale and Multidiscip. Model. Exp. and Des. 1, 255–268. https://doi.org/10.1007/s41939-018-0017-9

Lou, M., Alpas, A.T., 2019. High temperature wear mechanisms in thermally oxidized titanium alloys for engine valve applications. Wear, 22nd International Conference on Wear of Materials 426–427, 443–453. https://doi.org/10.1016/j.wear.2018.12.041

Morshed, A., Wu, H., Jiang, Z., 2021. A Comprehensive Review of Water-Based Nanolubricants. Lubricants 9, 89. https://doi.org/10.3390/lubricants9090089

Rajhi, A.A., 2022. Mechanical Characterization of Hybrid Nano-Filled Glass/Epoxy Composites. Polymers (Basel) 14, 4852. https://doi.org/10.3390/polym14224852

Ramezanzadeh, B., Moradian, S., Tahmasebi, N., Khosravi, A., 2011. Studying the role of polysiloxane additives and nano-SiO2 on the mechanical properties of a typical acrylic/melamine clearcoat. Prog. Org. Coat. 72, 621–631. https://doi.org/10.1016/j.porgcoat.2011.07.003

Shokuhfar, A., Zamani, M., Eghdam, E., Sahnesarayi, M., Mastali, S., 2012. SiO2-TiO2 Nanostructure Films on Windshields Prepared by Sol-Gel Dip-Coating Technique for Self-Cleaning and Photocatalytic Applications. NanoSci. & NanoTech. 2, 16–21. https://doi.org/10.5923/j.nn.20120201.04

Shorowordi, K.M., Haseeb, A.S.M.A., Celis, J.P., 2006. Tribo-surface characteristics of Al–B4C and Al–SiC composites worn under different contact pressures. Wear 261, 634–641. https://doi.org/10.1016/j.wear.2006.01.023

Xu, H., Hu, J., Liu, X., Wu, H., Jiang, Y., Xu, Z., Chen, S., Li, T., Zhang, J., Zhang, D., 2023. Interface strengthening and high-value recycling of epoxy resin/carbon fiber fabric composites. Chem. Eng. J. 465, 142998. https://doi.org/10.1016/j.cej.2023.142998

11 Military Applications

11.1 INTRODUCTION

Millions of years ago, humans led a nomadic life, relying on hunting, gathering, and natural resources for survival. Living in the open or in caves, they used grass or animal skins for protection. Over time, they learned to create fire, practice agriculture, and domesticate animals. Innovations like the wheel, stone tools, and weapons like bows and arrows emerged. Historical records reveal a continuous pursuit of understanding materials for tools, weapons, and protection. Eras like the Stone Age, Bronze Age, and Iron Age were characterized by significant progress in material development. The Iron Age, lasting till the 20th century, brought significant changes in living standards and warfare technology.

Advancements continued with the development of explosives, leading to small firearms and artillery guns. Metallic shields were introduced for protection, and the 20th century witnessed tanks and chemical weapons in warfare. After World War II, the uncovering of polymers brought about a materials revolution, leading to their widespread use across multiple industries. The Polymer Age, marked by multifunctional polymers, composites, and nanotechnology, extended into the Nano Age. Anthropological studies show the correlation between weapon technology, civilization, and contemporary materials.

Figure 11.1 illustrates the progression of defense technology, showcasing the transition from traditional bows and arrows to contemporary missiles and directed-energy weaponry. The historical perspective explores the interconnected development of materials and weapon technologies. Determining the exact beginning of humanity or the Stone Age presents difficulties, but it is generally believed to have extended from around 2.5 million years ago to 6000 BC.

11.2 EVOLUTION OF MATERIALS: A HISTORICAL OVERVIEW

Substances, in the forms of gas, liquid, or solid, are composed of atoms or molecules and encompass both living and non-living matter on Earth. The diverse array includes animal and plant species, water, air, clay, minerals, and rocks, each with their own unique characteristics. Materials are classified based on their known properties and usefulness into categories such as inorganic, organic, or biological, depending on

DOI: 10.1201/9781003317272-11

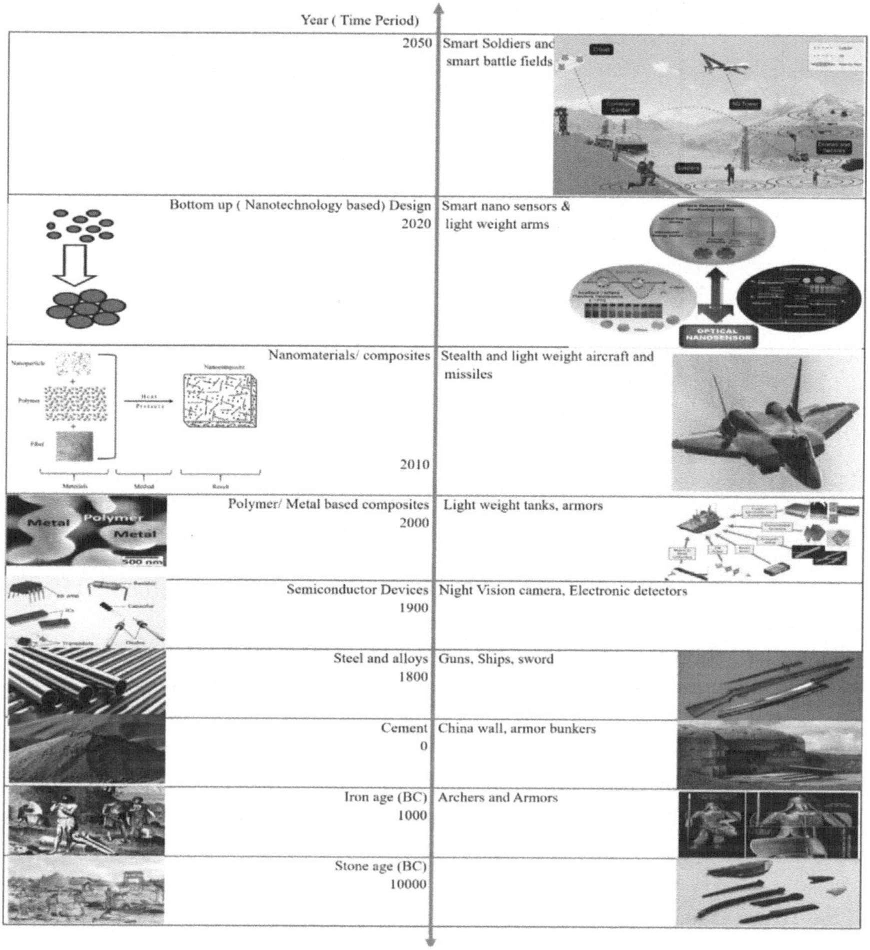

FIGURE 11.1 Shows the progression of defense technology, showcasing the transition from traditional bows and arrows to contemporary missiles and directed-energy weaponry.

their composition and origin. Throughout human history, materials have played a vital role in survival and in enhancing daily life.

Materials find applications when their physical properties align with specific purposes. The utilization of stone marked the early Stone Age, characterized by rough and irregular shapes. Subsequent ages saw the evolution of materials, notably metals like bronze and iron, leading to the Bronze and Iron Ages.

The developmental timeline, depicted in Figure 11.2, illustrates the gradual progression of metals and their respective strengths, with significant advancements occurring after 1900. Over the past decade, there has been a significant advancement in material development, including high-temperature refractory materials, electronic components like silicon and semiconductors that have propelled the information

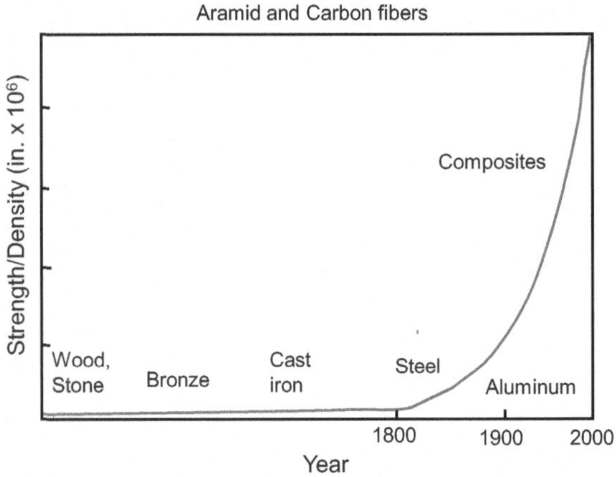

FIGURE 11.2 Progression of metals and their respective strengths. (Reprinted with permission from National Research Council, 1989, Copyright © 1989, The National Academies Press.)

revolution, and synthetic materials such as polymers, plastics, and carbon-reinforced fibers, igniting another industrial revolution.

This exponential growth in material innovation after 1900 allowed for the processing of materials with desired properties, addressing limitations of existing materials and enabling the creation of new systems previously deemed impossible. These advancements have influenced a wide range of sectors, spanning civil and non-civil areas, and extending to strategic military applications. The current era underscores the importance of designing, developing, and utilizing a diverse array of materials customized for particular applications.

11.3 EVOLUTION OF DEFENSE TECHNOLOGY

It's reasonable to argue that the progression of civilization, material science, and defense technology has been intricately linked since the early stages. Beginning with basic weapons crafted from natural elements like stones, wood, or bones in prehistoric times, modern defense technology has evolved into a complex and multifaceted field. This complexity arises from integrating offensive and defensive components, operating across various war theaters (land, sea, aerospace, and space), facing diverse threats, including lethal weapons, and adapting to evolving threat perceptions enabled by precision detection technology.

The ongoing journey in defense technology development and evolution can be divided into five distinct phases based on concurrent material ages:

1. Prehistoric weapons (the Stone Age), crafted from rough stones, later refined with polished stones, wood, and bones.
2. Weapons of the Bronze Age, attributed to civilizations like the Germans, British, Scandinavians, Gauls, and Kelts, characterized by advancements in ancient manufacturing.

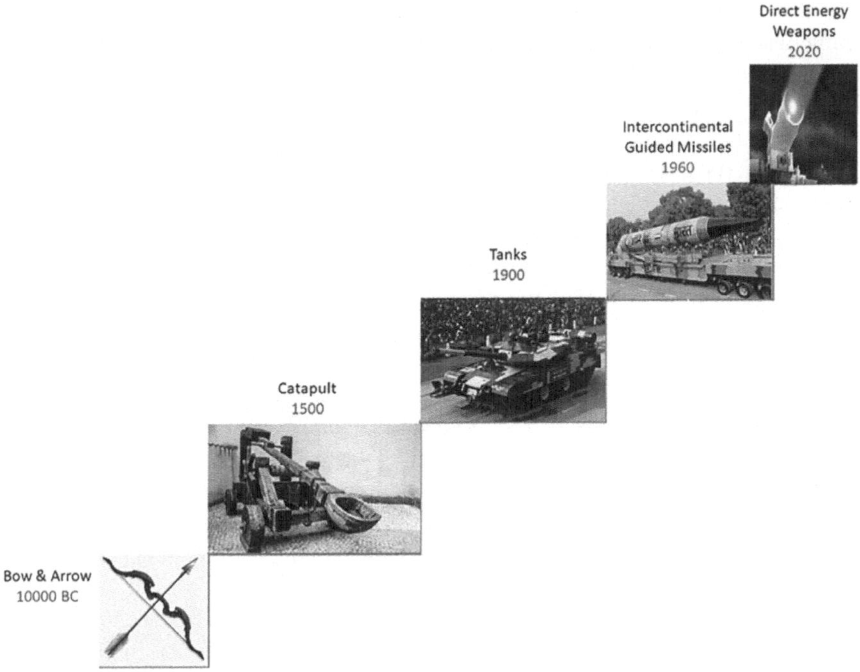

FIGURE 11.3 Different types of projectile weapons.

3. Weapons of the Iron Age, marking the transition from ancient to medieval times.
4. Medieval weapons during the seventeenth and eighteenth centuries.
5. Pre- and post-World Wars weapons eras, as illustrated in Figure 11.3.

This vast journey encompasses the evolution from bows and arrows, effective at relatively short distances of a few hundred meters, to modern guided intercontinental ballistic missiles capable of reaching targets thousands of kilometers away, and directed-energy weapons (DEW) with the ability to eliminate targets even in space. This advancement in weapon and defense technologies is intrinsically linked to the progress of materials over millennia. In the following sections, we explore the historical development of weapons and associated defense technology, beginning from the prehistoric era.

11.4 NANOTECHNOLOGY IN DEFENSE APPLICATIONS

Nanotechnology entails the manipulation of materials at the nano scale, employing substances within the nanometer range in different configurations to yield favorable results across various scientific and technical domains (Drexler, 1986). Utilizing newly devised tools, nanotechnology facilitates the creation of synthetic materials and goods with heightened attributes, adaptable to a wide array of fields, ranging from agriculture to space exploration, encompassing natural sciences, biology, and engineering disciplines. The precise control of atoms at the nanometer level holds promise for substantial breakthroughs in healthcare, agriculture, energy, water, quantum, and information technology, potentially surpassing existing technologies or providing enhanced solutions to present challenges.

Nanomaterials, recognized as the foundation of nanoscience and nanotechnology, are characterized as materials where at least one dimension falls within the range of 1 nm to 100 nm. The interactions between atoms within nanomaterials exhibit distinct differences from their larger-scale counterparts, enabling the engineering of nanomaterials with distinctive physical and chemical attributes. The field of nanomaterials encompasses a wide range of geometries, including one-dimensional systems like nanotubes, nanorods, and nanowires; two-dimensional systems like monolayers, nanoribbons, and nanosheets; and three-dimensional systems such as nanoparticles, clusters, and diverse molecules. Nanostructured materials demonstrate unique properties and behaviors compared to structures and sizes at larger length scales, exhibiting traits like lightweight properties and exceptional mechanical strength. Moreover, nanostructured systems present distinctive electronic, optical, magnetic, thermal, and chemical reactivity characteristics (Bayda et al., 2020; Khan et al., 2022).

Acknowledged as a prominent technology of the twenty-first century, nanotechnology harnesses nanoparticles derived from various sources, including natural emissions like volcanic eruptions, synthetic methods tailored for specific nanomaterial characteristics, or unintentional generation from sources such as diesel exhaust. Over recent decades, scientists and researchers have been captivated by nanoscience and nanotechnology, striving to replicate natural processes through bottom-up or top-down approaches, meticulously assembling nanomaterials atom by atom or molecule by molecule. A plethora of engineering techniques enable the creation and advancement of functional nanostructured materials and devices at nano/microscopic scales, with applications spanning a wide array of fields (Malik et al., 2023).

11.5 NANOMATERIAL CLASSIFICATION

Nanomaterials, distinguished by their dimensions on the nanometer scale, are categorized according to their geometry, morphology, and composition. The alteration of dimensions impacts electron behavior, resulting in unique quantum mechanical phenomena like the quantum confinement effect. This classification is illustrated in Figure 1.15 (Chapter 1), depicting 0D, 1D, 2D, and 3D nanostructures (Sajanlal et al., 2011). Each dimensionality unveils specific properties and potential applications, which will be further examined in the subsequent sections.

11.5.1 Zero-Dimensional (0D) Nanomaterials

Zero-dimensional (0D) nanomaterials confine all three dimensions in space, usually exhibiting diameters of less than 10 nm and nearly spherical shapes. Instances encompass nanospheres, clusters, quantum dots, and nano-onions, comprised of metallic, semiconducting, or insulating materials in crystalline, polycrystalline, or amorphous structures.

11.5.2 One-Dimensional (1D) Nanomaterials

One-dimensional (1D) nanomaterials confine two geometric directions within nanometer-length scales. While preserving long-range periodicity in one direction, they break it in the other two. Examples include nanowires, nanotubes, nanofibers, and nanorods, which can exist in crystalline, polycrystalline, or amorphous structures.

11.5.3 Two-Dimensional (2D) Nanomaterials

Geometrical confinement in one direction leads to 2D nanomaterials, exhibiting planar geometries. Examples include nanometer-thick films, monolayer or multilayer structures, nanoribbons, and other planar structures, often experimentally realized on substrates.

11.5.4 Three-Dimensional (3D) Nanomaterials

Three-dimensional (3D) nanomaterials, larger in scale (up to 100 nm), can exhibit anisotropy along any of the three dimensions and typically display crystalline characteristics. Examples include low-dimensional components dispersed in larger media, as well as larger nanoparticles such as nanocubes, hollow nanocages, nanoporous materials, and flower-like structures. The distinctive attributes of 3D nanomaterials, such as their large specific surface area and quantum mechanical properties, render them valuable for a wide range of applications.

Amorphous nanoparticles typically assume spherical shapes, whereas crystalline nanoparticles can take on various morphologies like rods, fibers, planar structures, and 3D clusters. Anisotropy in nanomaterials contributes to the emergence of new functional properties, often not present in or difficult to achieve with spherical nanoparticles. Asymmetry is characterized by the "aspect ratio" in one-dimensional nanostructures, defined as the ratio of diameter to length. Nano systems with spherical, helical, cubic, or oval shapes have low aspect ratios, whereas nanorods exhibit larger ratios. In 3D nanoparticles, the "surface-to-volume ratio" is crucial, featuring significantly higher ratios compared to bulk counterparts. The geometric size and shape profoundly influence the physical and chemical properties of nanomaterials, making control of morphologies an essential engineering parameter to achieve desired properties.

Understanding the diverse dimensions and structures of nanomaterials is essential for harnessing their unique properties across various applications. This section will explore the application of nanomaterials in military contexts.

11.6 SMART COATINGS

Nanotechnology has revolutionized various industries, and its impact on military applications is profound. Nano coatings, which involve the application of nanoscale materials onto surfaces, play a crucial role in enhancing the performance and capabilities of military equipment. These coatings are designed to provide functionalities such as corrosion resistance, camouflage, self-healing, ballistic protection, and more.

11.6.1 Types of Nano Coatings

11.6.1.1 Corrosion-Resistant Coatings

Nano coatings are extensively used to protect military equipment from corrosion caused by harsh environmental conditions. These coatings create a barrier at the nanoscale, preventing the penetration of corrosive agents and enhancing the lifespan of military assets. Because of its notable advantages such as superior mechanical strength, decent weldability, and relatively cheap carbon, steel is extensively utilized as a raw material in marine engineering applications, petrochemical industries, military applications and transmission pipelines (Abdeen et al., 2019; Farooq et al., 2022). However, the corrosion rate of unprotected carbon steels in naval and offshore environments presents a significant threat, resulting in structural deterioration, mechanical failures, substantial economic losses, environmental harm, or even sudden safety issues.

In order to reduce the likelihood of corrosion and prolong the lifespan of carbon steels, a range of techniques are frequently utilized either separately or in conjunction. These methods encompass the selection of corrosion-resistant materials], surface treatments (Shri Prakash and Balaraju, 2024), cathodic protection (Mobin and Zehra, 2023), oxidation inhibitors (Chigondo and Chigondo, 2016), protective coatings (Eduok et al., 2020; Wei et al., 2018), and more. Of these strategies, protective coatings are particularly popular because of their cost-effectiveness, simplicity of application, and versatility in combating corrosion.

Inspired by the lotus effect, renowned for its superhydrophobic properties characterized by water contact angles (WCA) surpassing 150 degrees and sliding angles (SA) below 10 degrees, as first observed by German scientists W. Barthlott (Barthlott and Neinhuis, 1997) and C. Neinhuis (Neinhuis and Barthlott, 1997), researchers have been captivated by the interfacial non-wetting characteristics and distinctive multiphase contact behaviors exhibited by such coatings. A variety of methodologies have been utilized to fabricate superhydrophobic coatings, including chemical etching, laser treatment, electrospinning dip/spray coating, electrodeposition, among others, leveraging the synergistic effects of micro-/nano-/micro-nano structural roughness and the attainment of low surface energy (Caldona et al., 2017; Cheng et al., 2018; Chi et al., 2019; Mo et al., 2015; Pan et al., 2016). Figure 11.4 (a) illustrates lotus leaves demonstrating repellent behavior, with (b) Scanning Electron Microscope (SEM) images of the leaf surface, revealing micrometer-sized epidermal cells nestled between two grooves, and (c) exhibiting wax tubules of nanometer scale on the upper leaf side (Ensikat et al., 2011).

For example, Wu and colleagues (2020) engineered a versatile superhydrophobic coating by blending Al_2O_3 nanoparticles with fluorinated epoxy resin via an inverse infiltration technique. Zhao et al. (2021) produced a transparent superhydrophobic coating using spin-coating silica-oligomer hybrid films, followed by oxygen plasma treatment and vapor deposition of 1H,1H,2H,2H-perfluorodecyltriethoxysilan

FIGURE 11.4 (a) Illustrates lotus leaves demonstrating repellent behavior, with (b) Scanning Electron Microscope (SEM) images of the leaf surface, revealing micrometer-sized epidermal cells nestled between two grooves, and (c) exhibiting wax tubules of nanometer scale on the upper leaf side. (Ensikat et al., 2011)

molecules. Pagliero et al. (2020) devised a superhydrophobic polyvinylidene fluoride (PVDF) coating through the non-solvent induced phase separation method. Mahadik and Mahadik (2021) formulated a versatile superhydrophobic coating via a sol-gel synthesis approach utilizing organically modified silica.

Advancements in both theory and technique within this realm of research have illustrated the potential of superhydrophobic materials across various applications, including self-cleaning, oil–water separation, anti-icing, water harvesting, radiation cooling, microdroplet manipulation, drag reduction, advancements in modern medicine, photocatalysis, and notably, in the realms of anticorrosion and anti-biofouling (Parvate et al., 2020).

The Principles of Hydrophobic Surfaces

Superhydrophobic surfaces are fundamentally rooted in the behavior of water droplets on a given surface, a phenomenon heavily influenced by surface energy and roughness. There are numerous avenues for modifying substrate surface structures to enhance repellency. Detailed studies of natural surfaces by Zisman and collaborators have shown that surface energy is predominantly dictated by surface chemical composition, with an observed decrease in surface energy corresponding to higher fluorine content. Moreover, all surfaces containing fluorine exhibit reduced surface

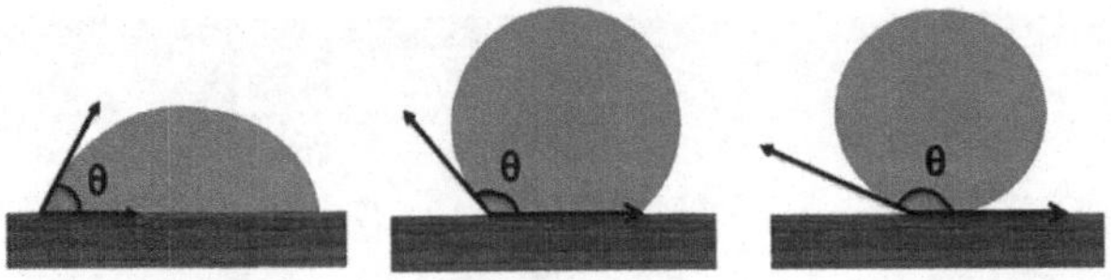

FIGURE 11.5 Shows the water droplet at different types of surfaces. (Ahmad and Kan, 2016)

free energy (Hare et al., 1954), a trait that can be further optimized by sequentially substituting fluorine, hydrogen, and carbon. The alignment of CF_3—groups in a close hexagonal packing configuration yields surfaces with the lowest energy, thereby showcasing outstanding hydrophobic characteristics that facilitate bonding between the substrate and fluorine-based hydrophobic coatings via weak Van der Waals forces (Sharma et al., 2020).

Numerous research groups have documented the incorporation of fluorinated atoms, with a particular focus on maximizing CF_3 group functionalization of surfaces. However, despite these efforts, the highest contact angle (CA) achieved cannot surpass 120° (Mahadevan, 2001; Nishino et al., 1999). Figure 11.5 shows the water droplet at different types of surfaces.

Different Types of Corrosion-Resistant Coatings

Synthetic coatings represent a versatile technique employed across diverse industrial applications, including anticorrosion, anti-dust, anti-ice, anti-mud deposition, antifouling, and drag reduction coatings. Drawing inspiration from natural sources such as lotus plant leaves and water striders, researchers worldwide seek to emulate the exceptional wetting properties observed in these natural entities by replicating their surface features on various substrates. Extensive laboratory-scale studies have been undertaken to drive forward the advancement of superhydrophobic coatings (SHCs) tailored for different substrates, with promising potential for practical deployment across various industries.

For instance, Mo and colleagues (2015) pioneered an inventive approach to produce superhydrophobic nanocomposite coatings on steel substrates. This method involved blending modified nano titanium oxide particles with fluoromethylhydrosilicone oil, synthesized from a dehydrogenation coupling reaction of amidmethylhysilicone oil and dodecafluoroheptylpropyl-trimethoxysilane. Subsequently, the nanocomposite superhydrophobic coatings were formed by applying the resultant solution onto a steel substrate using the dip-coating technique, followed by a heating process as illustrated in Figure 11.6 a). The resultant composite coating exhibited a water contact angle (CA) of 151.5° (Figure 11.6 b) and demonstrated effective corrosion inhibition properties.

Another significant category of hydrophobic corrosion-resistant coatings comprises silica-based coatings. Among these, nano-silica stands out as the preferred nanomaterial for fabricating superhydrophobic coatings (SHCs), owing to its cost-effectiveness, wide availability, high transparency, and ease of functionalization with various types of agents (Caldona et al., 2017; Fihri et al., 2017; Gao et al., 2014;

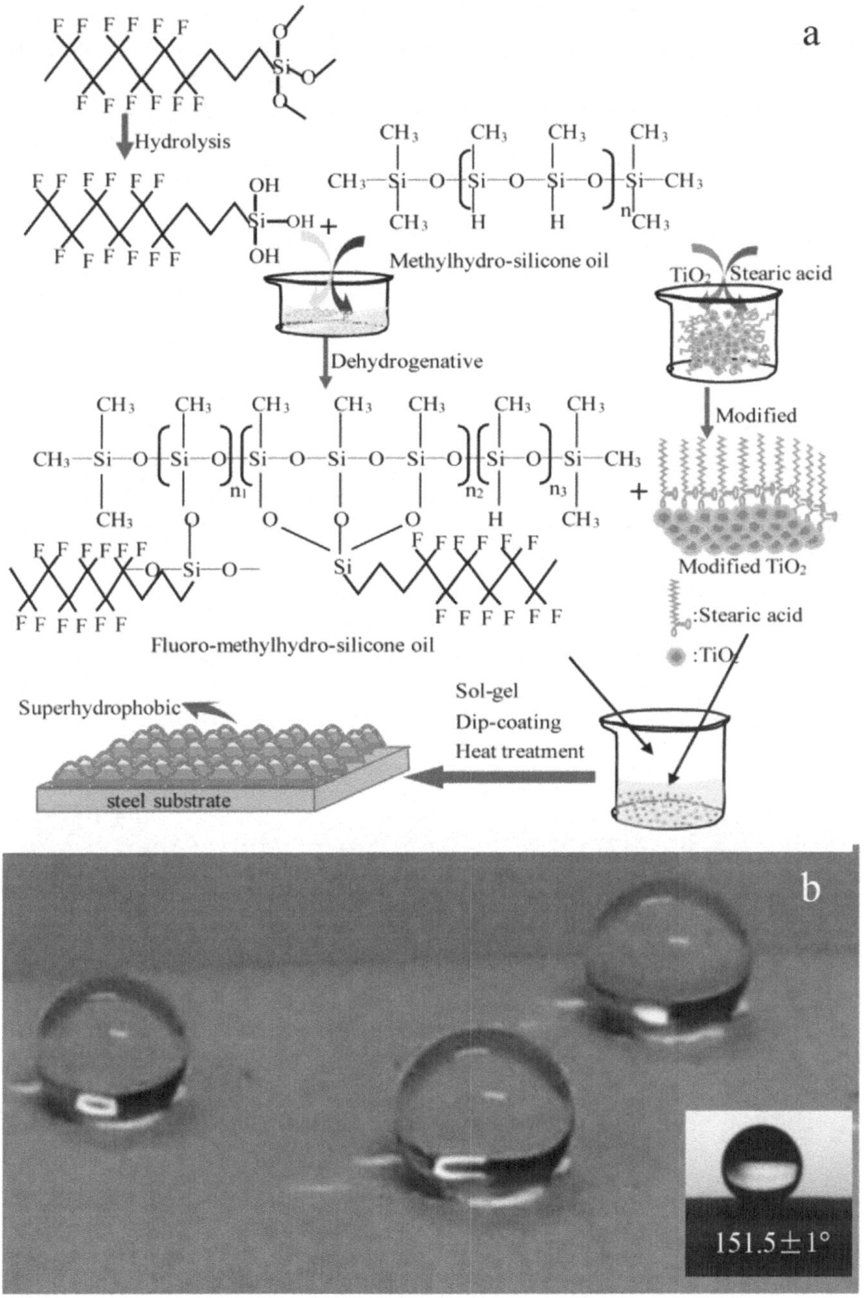

FIGURE 11.6 (a) Steps in Sol-gel dip-coating, (b) hydrophobic and demonstrated effective corrosion inhibition properties of coatings. (Mo et al., 2015)

Li et al., 2012; Polizos et al., 2018). Zhang and collaborators (Zhang et al., 2016) introduced a mechanically robust SHC employing a dual-layer approach tailored for self-cleaning and anticorrosion applications. Initially, silicon dioxide nanoparticles (10–20 nm) were functionalized with trichloro (1H,1H,2H,2H-perfluorooctyl) silane. These functionalized particles were then incorporated into a poly-dimethylsiloxane (PDMS) resin with the addition of a curing agent. The resulting material was applied onto glass and copper substrates via spray coating, followed by an additional layer of sprayed fluorinated silica deposition to achieve the desired functionality. Subsequently, the coating underwent drying at 120 °C for 1 hour to ensure mechanical durability. Figure 11.7 (a) depicts the dual-layer method for SHC production, while Figure 11.7 (b) showcases tightly coated glass and metal substrates, facilitated by the superior adhesion provided by PDMS. The altered wettability of the coated glass and metal post-spray coating is illustrated in Figure 11.7 (c).

In a separate investigation, Liu and co-authors (Liu et al., 2015) devised a direct and innovative technique for crafting a superhydrophobic coating (SHC) using raspberry-like silica nanoparticles, particularly tailored for self-cleaning applications. Silica nanoparticles underwent functionalization with fluoroalkoxysilane via condensation in ethanol at ambient conditions. The sol-gel approach was employed to embed these functionalized nanoparticles within a silica matrix, which was subsequently applied onto a glass substrate. The resulting coating manifested a water contact angle (CA) of 152° alongside a sliding angle (SA) of under 10°. Self-cleaning efficacy was evaluated by dust application onto the coating, followed by observation

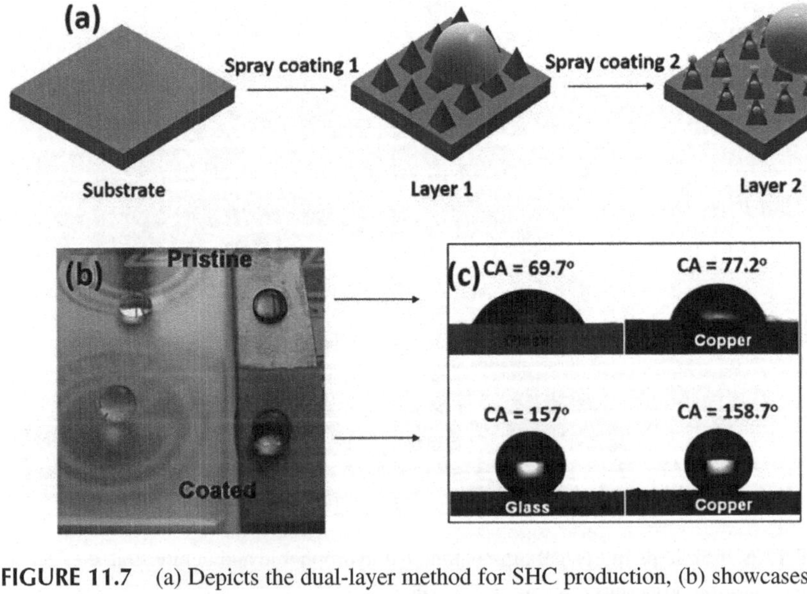

FIGURE 11.7 (a) Depicts the dual-layer method for SHC production, (b) showcases tightly coated glass and metal substrates, (c) the altered wettability of the coated glass and metal post-spray coating. (Reprinted with permission from Zhang et al., 2016, Copyright © 2016, Elsevier.)

of its subsequent cleaning process. The low SA facilitated the effective removal of deposited dirt particles from the SHC surface. Coating stability was assessed via scratch resistance testing and water stream or water jet impact tests, demonstrating the coating's resilience against scratch testing under an applied force of 150 mN. Such coatings are particularly efficacious, especially when applied to etched metal substrates.

Numerous researchers have advocated for the utilization of TiO_2 nanoparticles in SHC development due to their manageable structure, requisite stability, photocatalytic activity, and biocompatibility (Bai et al., 2014; Frank and Bard, 1977; Ohko et al., 2001). In a study by Qing and team (2015), a straightforward, cost-effective, and environmentally friendly technique was employed to fabricate a hybrid superhydrophobic fluorinated polymethylhydrosiloxane/TiO_2 nanocomposite coating via spray coating onto steel substrates. Hydrophobic groups such as CH_3 and CH_2 were introduced onto the surface of TiO_2 nanoparticles to confer hydrophobicity. Optimal weight ratio conditions (7:3) between surface-modified TiO_2 nanoparticles and fluorinated polymethylhydrosiloxane yielded a static water contact angle (CA) of 161° accompanied by a sliding angle (SA) of 5°, attributed to the coating's low surface free energy and micro/nano binary structure.

The resilience of these coatings was tested against varying pH levels, while their corrosion-resistant performance was evaluated on steel substrates through immersion in a 3.5 wt% NaCl aqueous solution. Impressively, even after exposure to air under ambient conditions or immersion in a corrosive salt solution, the superhydrophobic coatings (SHCs) maintained consistent wetting properties, underscoring the exceptional durability and superhydrophobicity of the developed coating on steel substrates. These coatings demonstrate significant promise across a spectrum of research and industrial domains. Furthermore, Huang and collaborators (2015) detailed the development of superhydrophobic TiO_2-nanotube coatings aimed at reducing platelet adhesion and bolstering the corrosion resistance of 316 L stainless steel.

Vertically aligned superhydrophobic TiO_2 nanotube (TNT) arrays were synthesized on steel substrates via electrochemical anodization of deposited Ti films, followed by calcination and fluorosilanization. These coatings exhibited a maximum water contact angle (CA) of 152.2° ± 0.8° and displayed enhanced corrosion resistance. Additionally, they positively impacted platelet bonding and fibrin formation upon contact with platelet-rich rabbit plasma, suggesting improved blood compatibility. These findings hint at the potential utilization of such coatings in biomedical devices requiring blood-contacting corrosion-resistant materials.

Zinc oxide (ZnO) nanoparticles have garnered considerable research attention due to their extensive study, attributed to their low cost, abundance, tunable morphology, wide bandgap semiconductor properties, chemical, mechanical, and thermal stability, as well as their eco-friendly characteristics (Pogrebnjak et al., 2019; Srivastava et al., 2009). Researchers have actively tailored the hydrophilic nature of ZnO nanoparticles to achieve superhydrophobicity (Feng et al., 2004; Pesika et al., 2002).

In a recent study, Rezayi and Entezari (2016) detailed the development of a durable superhydrophobic aluminum surface via etching and the application of ZnO

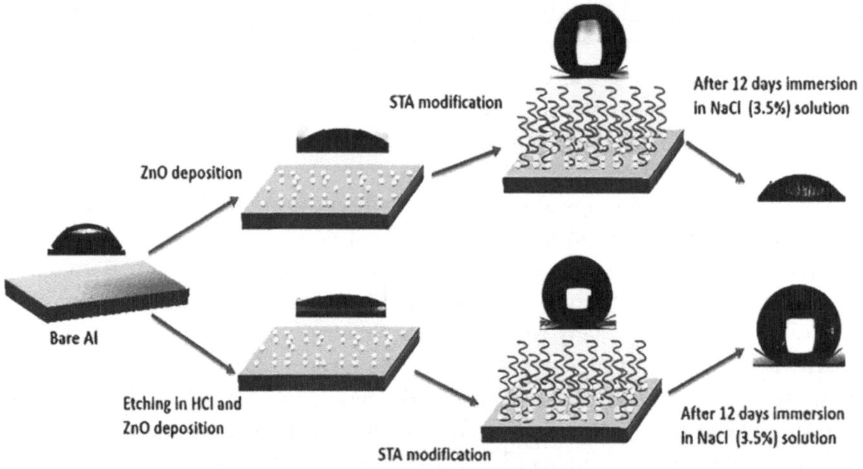

FIGURE 11.8 A schematic diagram illustrating the sequential process of fabricating corrosion-resistant superhydrophobic coatings (SHCs) on aluminum surfaces, with or without etching. (Reprinted with permission from Rezayi and Entezari, 2016, Copyright © 2016, Elsevier.)

nanoparticles. The deposition of ZnO nanoparticles onto the aluminum surface was achieved through techniques such as immersion and ultrasound. To reduce the surface energy of the aluminum substrate, a stearic acid (STA) ethanol solution was employed. It was observed that the ultrasound method (STA–ZnO–Al–U) outperformed the immersion method (STA–ZnO–Al–I) in producing stable superhydrophobic aluminum surfaces. A schematic diagram illustrating the sequential process of fabricating corrosion-resistant superhydrophobic coatings (SHCs) on aluminum surfaces, with or without etching, is depicted in Figure 11.8.

In a separate investigation, Radwan and colleagues (2016) unveiled the creation of a corrosion-resistant superhydrophobic coating (SHC) utilizing polyvinylidene fluoride (PVDF) and ZnO nanoparticles via a one-step electrospinning process designed for aluminum (Al) substrates. They achieved a water contact angle (CA) of $155 \pm 2°$ with a CA hysteresis of $4.5 \pm 2°$. Notably, despite the electrospun coating containing only 1/6th the concentration of ZnO nanoparticles compared to the spray-coated counterpart, it still attained superhydrophobicity. The electrospun coating showcased a more uniform dispersion of pristine nanoparticles without the need for any functionalizing agent. Corrosion tests indicated a significant enhancement in the corrosion resistance of the PVDF-ZnO composite coatings on Al substrates compared to PVDF alone, attributed to the introduction of ZnO nanoparticles. The pivotal role of ZnO nanoparticles in enhancing hydrophobicity was attributed to the reduction in pore size of the coating and the augmentation of air trapping within the surface structure. The developed nanocomposite coating on Al substrates demonstrated a corrosion resistance 35 times higher than that of uncoated aluminum substrates, as evidenced by the substantial disparity in water CAs between coated and uncoated substrates, as depicted in Figure 11.9.

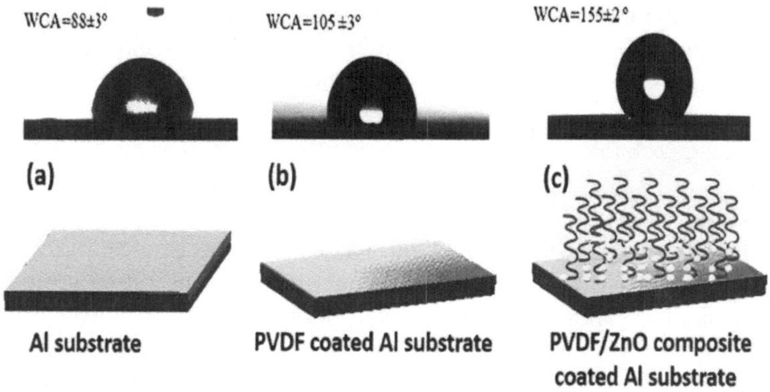

FIGURE 11.9 The developed nanocomposite coating on Al substrates and their corrosion resistance. (Reprinted with permission from Radwan et al., 2016, Copyright © 2016, Elsevier.)

11.6.1.2 Camouflage Coatings

Camouflage nano coatings employ advanced materials capable of adapting to diverse environments. By manipulating light at the nanoscale, these coatings facilitate effective concealment for military personnel, vehicles, and equipment, seamlessly blending them with their surroundings. Specifically designed for military use, textiles require resilience against ballistic threats and durability in various environmental conditions, such as ultraviolet (UV) light, moisture, fire, heat, and wind. Additionally, they must prioritize comfort and lightweight properties (Xiao et al., 2015). These textiles must offer effective camouflage and concealment across different environmental scenarios, covering a broad range of wavelengths on the electromagnetic spectrum, including the visible region (400–800 nm), near-infrared region (NIR) (750–1,200 nm), and thermal or far-infrared region (FIR) (3–5 and 8–14 μm) (Degenstein et al., 2021).

Adherence to military textile standard specifications and requirements, such as colorfastness to light, washing, and perspiration, is essential to ensure the visibility or infrared concealment properties remain uncompromized during use and are unaffected by cleaning processes (Md. Abdul Hannan and Rahman, 2020). Camouflage properties in military textiles are manifested through various forms, including clothing, lightweight flexible nets, garnishing, and covers (Degenstein et al., 2021). Textiles, being pliable three-dimensional materials, are fashioned from fibers spun into yarns and woven (interlaced) or knitted (interlooped) into fabrics. Alternatively, non-woven textiles can be produced by bonding fibrous webs through mechanical entanglement, resin application, or thermal and chemical fusion of fibers. Embroidery serves as another method for crafting technical textiles, involving the stitching of yarns or threads onto the surface of a base material. Enhancing functional and aesthetic properties in textiles is achievable through surface finishes, coatings, or lamination techniques (Bilisik et al., 2016).

Traditionally, military textiles were crafted from woven cotton fabrics, gradually transitioning to blends with synthetic fibers like nylon for enhanced lightweight attributes and quicker drying. As advancements in fiber manufacturing technology unfolded, the incorporation of high-performance fibers, such as para-aramids and ultra-high molecular weight polyethylene (UHMWPE), became integral in crafting uniforms for ballistic protection. Continuous improvement of protective fabrics remains an ongoing process, adapting to emerging threats and incorporating evolving technologies into textile systems (Degenstein et al., 2021; Faruk et al., 2023).

Integration of smart technologies into fabrics serves to safeguard against various threats, encompassing biological, mechanical, or chemical hazards, while also enhancing fabric functionality, such as altering appearance or improving thermal regulation. These technologies can be applied through diverse methods and mediums like nanoparticles, microencapsulation, lamination, being woven or knitted into textiles, or introduced at the polymer stage before fiber extrusion. Smart textiles are classified into three generations based on the stage in the manufacturing process at which they are added. First-generation smart textiles are added to fully assembled garments, second-generation textiles incorporate the active component during textile manufacturing (e.g., weaving or knitting), and third-generation smart textiles integrate active components within the fiber or yarns (Hossain et al., 2024; Ruckdashel et al., 2021).

Despite the vast potential for smart textile applications, the functionality of these textiles must align with their real-world usage and maintenance. Hence, it is crucial to consider how smart solutions may impact the performance of textiles in practical applications.

Camouflage in Thermal Infrared

Designing thermal infrared camouflage presents significant challenges due to several inherent factors:

1. All macroscopic objects, including living organisms, emit thermal infrared radiation above 0 K.
2. The thermal signature of a target depends on various factors, such as object geometry, dimensions, emissivity of target materials, and the nature and state of the background. Even a slight temperature difference of a fraction of a degree Celsius can render an object easily detectable with current thermal sensor technology.
3. The dynamic nature of the infrared signature is influenced by factors such as solar radiation during the day and skyline radiation at night.
4. Camouflaging active systems like missiles, aircraft, tanks, or vehicles in motion is particularly challenging due to the release of waste heat into the environment.

Addressing these challenges necessitates tailored solutions, as there is no universal method for camouflaging different types of targets in the thermal infrared region. Modern thermal imaging systems can detect temperature differences as small as 1 °C. Vehicles made of different materials, whether composite or metal, exhibit similar detectability in thermal imaging. Hence, specialized paints remain necessary to reduce contrast in the thermal infrared spectrum (3–5 μm and 8–14 μm).

Real military vehicles, even when stationary, continue to emit heat from their engines and batteries. To counteract this, low-emissivity coatings are applied to minimize thermal contrast. These coatings, designed to match emissivity with background elements like vegetation and soil, are critical for effective thermal camouflage. Metallic flakes, particularly aluminum, are preferred as pigments in thermal camouflage coatings due to their low emissivity and high infrared reflectance.

For active systems, employing active cooling methods such as circulating cold air or water droplets (mist) around hot spots like engines and exhausts are viable solutions to reduce thermal signatures. Shape tailoring and surface treatment with thermally stable low-emissivity coatings for both passive and active targets offer another avenue to diminish the contrast in thermal emission between the target and its background. Shape tailoring involves strategic design of exhaust and engine positions, directing emissions upward for aircraft and downward for ships, for instance (Zhu et al., 2020).

11.6.1.3 Self-healing Coatings

Utilizing nanoparticles with autonomous repair capabilities facilitates the development of self-healing coatings. In military contexts, such coatings can autonomously mend minor damages or scratches, thereby enhancing the longevity and resilience of equipment in demanding operational settings. Self-healing involves the detection and repair of damage on-site, representing an innovative strategy for prolonging the lifespan and robustness of materials across diverse applications. The fundamental mechanism of self-healing relies on the utilization of external healing agents, as shown in Figure 11.10. However, memory polymers present a unique approach for addressing

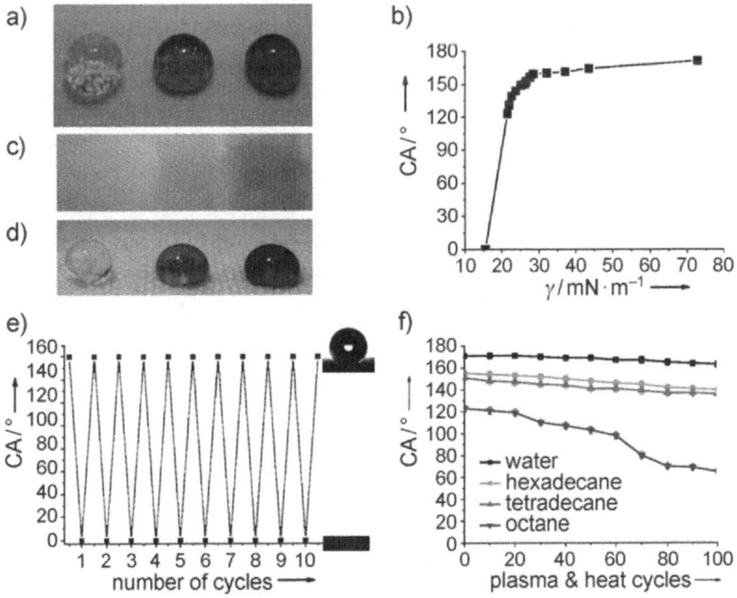

FIGURE 11.10 Self-healing nanoparticle coatings and their efficiency after different cycles. (Reprinted with permission from Wang et al., 2011, Copyright © 2011, John Wiley & Sons.)

damage at the structural level, offering potential for repeatability and controllability through preprogramming (Kausar et al., 2023).

Memory polymers exhibit the capability to repair external scratches, surface imperfections, and damage in thin materials, as well as internal cracks and flaws within objects. For instance, a composite system comprising a cross-linked PCL network alongside linear PCL interpenetrating the network has demonstrated both memory effects and self-healing abilities. nanolayered, graphene-filled, epoxy-based memory composites have shown significant enhancements in scratch resistance and thermal healing capacity, surpassing unfilled polymers (Pilate et al., 2016).

Segmented block shape memory polyurethane (PU) coatings have been employed for corrosion protection, demonstrating physical healing of defects upon damage. In this process, soft segments relax under higher temperatures, filling in damaged areas, while hard segments maintain the film's mechanical properties. Furthermore, memory polymers have exhibited the ability to self-repair large structural-scale damage, albeit requiring two steps: crack closure followed by repair. External confinement enhances the efficiency of this repair process (González-García et al., 2011).

Research has delved into exploring the self-healing attribute of memory polymers in engineered textiles, particularly in the advancement of smart coating textiles. Fabric coatings incorporating hydrolyzed fluoroalkyl silane (FAS) with fluorinated-decyl polyhedral oligomeric silsequioxane (FD-POSS) have demonstrated superamphiphobicity and notable self-healing capabilities against chemical damage. Furthermore, the addition of hydrophobic nanoparticles alongside FAS/FD-POSS has yielded multiple self-healing abilities against physical damage, such as blade scratching and abrasion, as well as chemical damage (Heyse et al., 2016).

The extension of coated textiles' lifespan relies on their ability to heal at an early stage of damage formation, such as the emergence of microcracks. The capability for self-healing within the coating not only rectifies functionality deterioration but also prolongs the durability of coated products. This self-repair process can either occur autonomously or be triggered by an external stimulus, such as heat or radiation (Nik Md Noordin Kahar et al., 2021).

In recent years, various concepts have emerged and are continually being refined. Notably, considerable attention has been focused on employing self-healing concrete, composites, and coatings for corrosion prevention. However, the exploration of self-healing coatings for textiles remains relatively underexplored (Obayi and Nnamchi, 2020).

Two primary systems have been developed: extrinsic and intrinsic. Extrinsic approaches entail the use of a container filled with a healing agent. Upon damage, the container ruptures, releasing the healing agent into the crack where it reacts with a catalyst or curing agent. These containers can assume various forms, including spherical microcapsules, hollow structures, compartmented fibers, or even vascular systems, as shown in Figure 11.11.

In textile coatings, the incorporation of microcapsules stands as the most feasible extrinsic approach. Microcapsules containing healing agents can be easily integrated into existing coating pastes. However, this method suffers from a limitation in its recurring healing capability, as damage in the same area can only be repaired once.

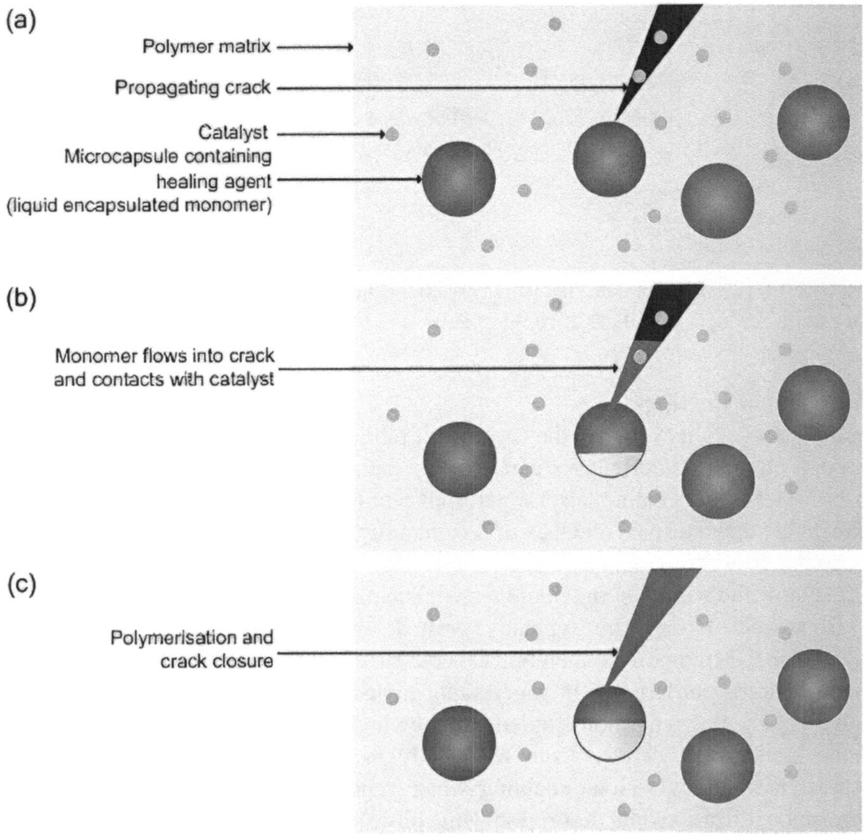

FIGURE 11.11 Mechanism of self-healing polymer matrix composites. (Reprinted with permission from Heyse et al., 2016, Copyright © 2016, Elsevier.)

Additionally, the amount of available healing agent is restricted by the size of the capsules.

For scenarios requiring multiple repair actions, intrinsic systems based on reversible covalent chemistry, physical interactions, or supramolecular interactions present a viable solution. In such cases, the damaged coating can autonomously repair itself through increased mobility, particularly noticeable at elevated temperatures. The improved flow of the coating facilitates the refilling of scratches, leading to restored strength attributed to the reestablishment of bonds, whether chemical or physical.

In real-world applications like laundering, fabric coatings undergo degradation due to mechanical forces and chemical agents. However, robust coatings have been developed to endure repeated washing cycles and abrasion without compromising superamphiphobicity. Moreover, damaged coatings have exhibited repeatable self-healing capabilities through thermally activated processes, ensuring the restoration of liquid repellency even after prolonged wear and tear (Lee et al., 2021).

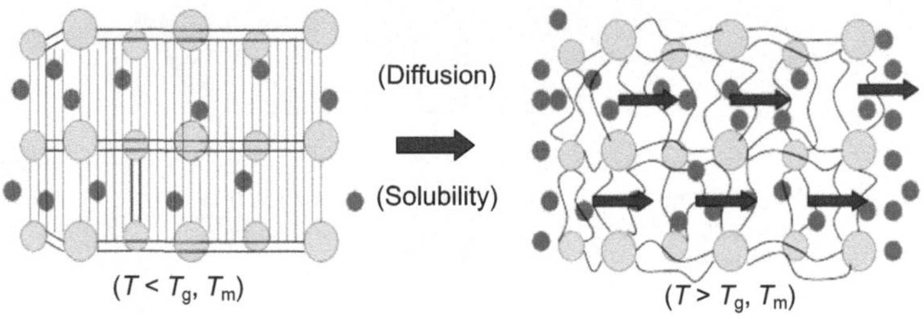

FIGURE 11.12 Diffusion mechanism in breathable fabrics. (Reprinted with permission from Hu and Lu, 2016, Copyright © 2016, Elsevier.)

Water Proof Coatings

Textile breathability refers to the capacity of fabrics to permit moisture vapor to pass through them. This attribute is particularly crucial in functional garments claiming to be waterproof. Traditionally, waterproof textiles were often deemed uncomfortable due to their stiffness and lack of breathability, hindering perspiration vapor from escaping. However, breathable fabrics alleviate this issue by allowing water vapor diffusion while still repelling liquid water penetration (Kim, 2021).

Breathable fabrics are typically crafted using microporous or hydrophilic membranes. Microporous membranes comprise interconnected channels smaller than raindrops but larger than water vapor molecules, commonly constructed from polymers like polytetrafluoroethylene or polyvinylidene fluoride. Conversely, hydrophilic membranes, derived from chemically modified polyester or polyurethane, possess amorphous regions enabling water vapor molecules to pass through via absorption, diffusion, and desorption while obstructing liquid water due to their solid nature, as shown in Figure 11.12.

Memory polymers introduce an innovative approach to achieving adaptable water vapor permeability (WVP) in breathable textiles. Research indicates that memory polymers can enhance comfort in both cold and warm climates. For instance, a Swiss company introduced a membrane sensitive to heat, which opens to release excess heat and closes to retain heat close to the body at low temperatures. Similarly, Japanese company Mitsubishi Heavy Industries developed breathable fabrics utilizing memory membranes, delivering excellent waterproofing and breathability across various environments (Sikdar et al., 2022).

Moreover, companies such as Toray Industries and Marmot Mountain Works have been at the forefront of developing PU film breathable clothing systems. They achieve this by integrating shape memory polymers with microporous coatings and garment lamination technology. Shape memory PU coatings have emerged as viable options for providing breathability to nonporous fabrics due to their sensitivity to temperature and humidity, facilitating water vapor permeability (WVP).

Research endeavors have also concentrated on fine-tuning the WVP of shape memory PU coatings by considering factors such as the incorporation of hydrophilic groups and crystalline soft segments. Fabrics coated with adjustable WVP films offer adaptable solutions for breathable clothing, suitable for diverse applications like sportswear, footwear, gloves, and socks. These advancements underscore the

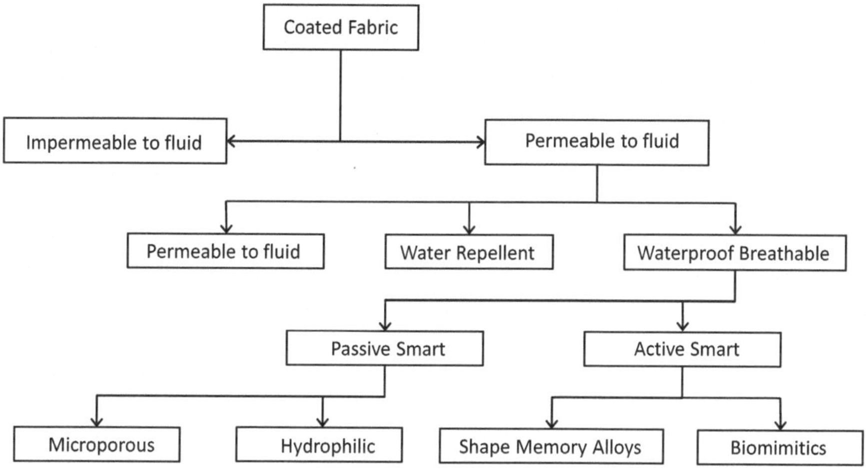

FIGURE 11.13 Classification of different types of breathable fabrics.

potential of memory polymers in augmenting the comfort and utility of breathable textiles (Jahid et al., 2018). Figure 11.13 illustrates the various types of breathable fabrics.

11.6.1.4 Bulletproof Coatings

Integrating nanomaterials such as graphene or carbon nanotubes enhances the effectiveness of armor, bolstering the protection of military personnel against projectiles and thus improving safety in combat situations. A state-of-the-art coating technology has emerged, capable of adjusting its characteristics in response to changes in the environment. These adjustments could be activated by variations in temperature, humidity, pH levels, light exposure, mechanical impacts, or contact with different chemicals. Implementing such adaptive coatings onto textiles offers a promising avenue for the advancement of smart textiles, delivering augmented protection and comfort within textile systems and protective attire. Recent progress in nanotechnology has spurred the development of numerous innovative smart coating materials and application techniques. This section delves into the latest breakthroughs in smart coatings applied to textiles, with a specific emphasis on their potential use in protective clothing. Such clothing encompasses a variety of gear, including body armor, hazardous material (HAZMAT) suits, healthcare attire, and firefighter turnout coats (Benzait and Trabzon, 2018; Wu et al., 2023).

Bulletproof Body Armor: Essential Protection for Military and Law Enforcement Personnel

Body armor plays a pivotal role in safeguarding military and law enforcement personnel, its efficacy contingent upon the selection of ballistic materials employed. It encompasses both flexible soft armor, composed of lightweight textiles, and rigid hard armor, which incorporates ballistic textile materials with inflexible insert plates crafted from ceramics, metals, or fiber-reinforced composites.

Flexible ballistic textile materials comprise aramid fibers such as Kevlar® (DuPont), Twaron® (Teijin), and Technora® (Teijin); highly oriented polyethylene fibers like Spectra® (Honeywell) and Dyneema® (DSM); and polybenzobisoxazole (PBO) fibers such as Zylon® (Toyobo) (Phoenix and Porwal, 2003). Soft body armor typically consists of around 20 to 50 layers of ballistic fabrics, tailored to meet diverse protection needs against varying ballistic threats (Lee et al., 2003). While soft armor effectively shields against shrapnel, it may lack adequate defense against high-velocity bullets, which can penetrate the fabric despite multiple layers. In contrast, hard armor, fortified with rigid ceramic plates placed at the front and back, offers significant protection against bullets (Gadow and Von Niessen, 2006; Kosashvili et al., 2005).

The introduction and adoption of body armor have notably contributed to saving the lives of law enforcement officers and military personnel. Since its inception in 1972, soft body armor alone has been credited with saving over 2,900 lives, with officers wearing such armor estimated to be 14 times more likely to survive a gunshot to the torso (Wilhelm and Bir, 2008). However, utilizing body armor also presents challenges concerning wearer comfort, mobility, and overall performance. Heavy body armor often induces excessive thermal stress, with numerous layers of materials, like 32 layers of Kevlar, significantly reducing breathability (Starr et al., 2015).

Enhancing the protective capabilities of ballistic materials holds promise for reducing the number of layers required to meet specific protection standards, thus improving the comfort and mobility of armor users. Shear thickening fluid (STF) exhibits non-Newtonian flow behavior, where its viscosity substantially increases under critical shear rates or stress levels, transforming it into a solid-like (Lee et al., 2003). This reversible shear thickening behavior, attributed to mechanisms like the hydrocluster mechanism, positions STF as an ideal candidate for smart coating applications in body armor. Upon bullet impact, the shear stress prompts STF to transition from a flexible, fluid-like state to a rigid, solid-like state, thereby enhancing protection. Furthermore, the fluid's thickening transition is reversible, allowing it to maintain stability in liquid form without compromising material flexibility under normal conditions (Sun et al., 2013).

Pioneering work led by Norman Wagner's research group at the University of Delaware introduced STF technology to ballistic materials, resulting in the development of liquid body armor. Lee et al. (2003) demonstrated the efficacy of impregnating Kevlar® woven fabric with colloidal silica-ethylene glycol STF, showing that four layers of STF-Kevlar® provided comparable weight and ballistic protection performance to ten layers of neat Kevlar®. Subsequent research by Srivastava, Majumdar, and Butola (2011) explored the coating of Kevlar® fabric with silica-polyethylene glycol STF, investigating parameters such as padding pressure and STF concentration on impact energy absorption. Their findings underscored the importance of uniform distribution and penetration of STF within the yarn structure, with higher padding pressures contributing to improved impact energy absorption.

Body Armor with Stab Resistance

Soft body armor remains a staple choice for law enforcement personnel due to its lightweight design and effective defense against handgun threats. According to the

National Institute of Justice (2008), soft armor types II, IIA, and IIIA are suitable for continuous wear during duty shifts, while harder armor types III and IV are typically reserved for tactical situations or specific threats necessitating such protection. However, ballistic fibers offer limited protection against non-ballistic threats like knives or sharp objects (Gadow and Von Niessen, 2006). Introducing stab-resistant properties into soft ballistic materials could offer comprehensive protection against various threats.

Applying hard coatings onto soft ballistic fabrics enhances their resistance to stabs and cuts while maintaining fabric flexibility. Gadow and von Niessen (2006) utilized a thermal spray process to coat cermet (WC Co 83/17) and oxide ceramics (aluminum oxide [Al_2O_3] and titanium oxide [TiO_2]) onto aramid Twaron® fabric, with AlSi 12 serving as an additional bond coat to improve bonding strength. Among the investigated coating systems, Al_2O_3-AlSi 12 and WC Co 83/17 demonstrated high bonding strength and microhardness. Comparing untreated Twaron® fabrics to those coated with multilayer Al_2O_3-AlSi 12 and single-layer WC Co 83/17, the weight gain was minimal (less than 150%), while stab penetration work increased significantly (by a factor of 4 for Al_2O_3-AlSi 12 and 5 for WC Co 83/17).

Atanasov et al. (2014) utilized atomic layer deposition (ALD) to coat conformal dielectric Al_2O_3 and TiO_2 thin films onto Kevlar® fibers. While a single layer of Al_2O_3 reduced fabric cut resistance, a single layer of TiO_2 moderately increased it. However, applying a conformal 50-50 Å bilayer of TiO_2-Al_2O_3 thin film significantly increased cut resistance by 30%, with minimal weight gain of 0.87%. The Kevlar fabric with ALD coatings retained durability and flexibility (Atanasov et al., 2014).

Rosen et al. (2007) investigated the stab resistance of Kevlar® fabrics treated with kaolin clay-glycerol (K-STF) and spherical silica (S-STF) shear thickening fluids (STF) against various threats. They found that K-STF and S-STF-coated Kevlar® fabrics significantly enhanced stab resistance against knives, spikes, and hypodermic needles, with minimal weight gain (16.7% for K-STF and 20% for S-STF).

Additionally, Hassan et al. (2010) coated Kevlar® and nylon fabrics with nanoparticle silica-polyethylene glycol (STF) and conducted dynamic and quasistatic stab resistance tests. STF-coated fabrics exhibited improved stab resistance against knives and spikes in both dynamic and quasistatic tests, despite slight changes in fabric thickness and flexibility (Hassan et al., 2010).

In 2013, Sun and colleagues applied a coating of silica-ethylene glycol (EG) shear thickening fluid (STF) onto ultra-high-molecular-weight polyethylene (UHMWPE) and assessed stab resistance against knife and spike threats. Their study revealed that STF-coated UHMWPE fabric offered comparable protection against spike threats with significantly lower weight and thickness and better flexibility compared to uncoated fabric (Sun et al., 2013).

Consistent with previous research, STF-coated fabrics demonstrated greater improvement in stab resistance against spike threats compared to knife threats due to rapid aggregation of STF particles upon impact, enhancing fabric compactness and impeding spike penetration (Sun et al., 2013)

11.6.1.5 Radar-Absorbing Coatings

Coatings developed to absorb radar waves play a crucial role in diminishing the radar signature of military vehicles and aircraft. This technology significantly improves stealth capabilities, enabling covert operations and reducing the risk of enemy detection. Radar Absorbent Material (RAM) coatings and structures constitute a highly researched area for nanomaterial applications, primarily owing to their lightweight properties, high strength, and exceptional impact resistance, all of which are essential for creating effective radar-absorbing coatings and structures. Nanomaterials integrated into these structures possess remarkable attributes, including superior absorption capabilities across a broader range of frequencies compared to their bulk counterparts. For clarity, nanomaterials are typically classified into four main types: (1) nanocrystal RAMs, (2) core–shell nanocomposite RAMs, (3) nanocomposites comprising nanostructured carbon combined with magnetic and polymer (conducting or insulating) RAMs, and (4) thin metallic films (Kumar and Singh, 2018).

Crystalline Nano RAMs

The increased surface area characteristic of nanomaterials is attributed to the greater availability of atoms with dangling bonds on the surface (Narendra Kumar and Sunita Kumbhat, 2018). These unsaturated surface atoms enhance interface polarization and multiple scattering, leading to the effective absorption of microwave radiation. Both dielectric and magnetic nanomaterials contribute to absorption through the dielectric and magnetic loss mechanisms discussed in previous sections. For instance, nanostructured $La_{0.8}\ Ba_{0.2}\ MnO_3$ particles, with dimensions of approximately 80 nm, exhibit extensive microwave-absorbing properties across a wide frequency range (2–18 GHz). In this context, the geometric morphology and grain dimensions are critical parameters influencing the microwave reflection loss (RL) (Sharma et al., 2008; ZHOU et al., 2007).

Core–Shell Nanocomposite RAMs

Improving the functionality of nanomaterials involves incorporating a distinct layer of materials, commonly known as core–shell nanostructured materials. The size and configuration of both the core and shell serve as design parameters for customizing properties such as reflectivity within the desired electromagnetic spectra range (Ghosh Chaudhuri and Paria, 2012). In these particles, the shell material can enhance the chemical and thermal stability of the core or serve as a protective layer with various functionalities. Core–shell nanomaterials find diverse applications, including radar-absorbing materials, solar photovoltaics, drug delivery in biomedical applications, and nanoelectronics, due to the in-situ protection of the core material and the added value provided by the shell materials.

The combination of a magnetic nanoparticle core and a dielectric shell, or vice versa, in core–shell nanocomposites offers an advantage by incorporating contributions from two types of radar-absorbing materials simultaneously. Core–shell nanocomposites with a magnetic core and dielectric shell may demonstrate more efficient microwave-absorbing characteristics owing to their complex permittivity and permeability values (Sharma et al., 2008). Recent research and development efforts have focused on synthesizing and investigating the microwave absorption

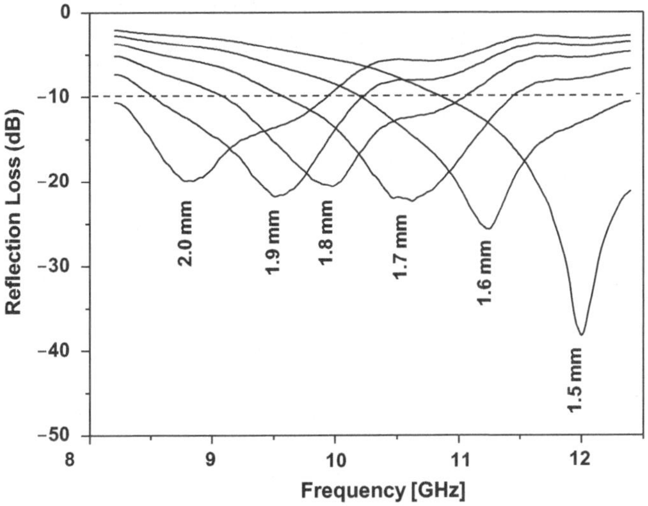

FIGURE 11.14 Microwave absorption properties. (Reprinted with permission from Kumar and Dixit, 2019, Copyright © 2019, Springer.)

characteristics of core–shell nanomaterials. Noteworthy examples include hexagonal nanoferrite with a Ni-P-coated core, $MnFe_2O_4$ spinal nano coated with nano TiO_2, carbon fibers covered with cupric oxide, and an FeCo alloy with a carbon-coated shell, all displaying intriguing microwave-absorbing properties (Sharma et al., 2009; Xiao et al., 2006). For instance, FeCo core particles with a diameter of approximately 25 nm and a dielectric graphitic shell with a thickness of around 10–15 nm were synthesized via the pyrolysis of a nanocomposite of Fe-Co oxides in an inert environment. When dispersed in an NBR matrix, this material exhibited excellent microwave absorption properties in the X-band (8–12 GHz) (Figure 11.14), leveraging contributions from both components. Due to their unique electromagnetic properties and lightweight nature, core–shell FeCo (C) nanoparticles hold potential applications as radar absorbers for stealth purposes.

Nanocomposite Coatings of Carbon Nanomaterials

Over recent decades, carbon-based nanomaterial coatings, particularly carbon nanotubes (CNTs) and graphene coatings, have undergone extensive exploration. Their composites with insulating metal oxides, polymer matrices, or conducting metal and polymer matrices offer a diverse range of physical properties, including electrical conductivity, high mechanical strength coupled with thermal stability, and tunable microwave absorption characteristics across a broad frequency range (Sun et al., 2011). Both single-walled and multi-walled carbon nanotubes, whether unfilled or filled with metals or coated with magnetic metals or oxides (ferrites), have been thoroughly investigated regarding their microwave absorption characteristics (Silva et al., 2013). Studies indicate that microwave absorption increases with CNT concentration (> 4 wt%), surpassing the phase transition of the composite from an insulator to a conducting phase. Additionally, radar-absorbing materials (RAM) containing

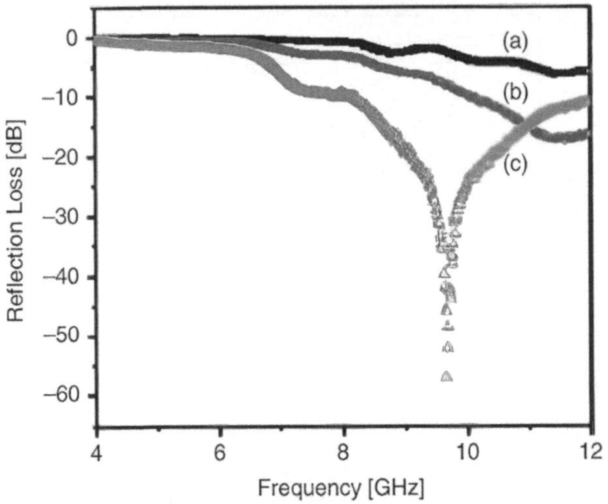

FIGURE 11.15 Microwave absorption characteristics of reduced graphene oxide in an NBR matrix. (Kumar and Dixit, 2019, Copyright © 2019, Springer.)

CNTs offer corrosion resistance, while metal-filled CNTs serve a dual purpose by providing electromagnetic interference (EMI) shielding and microwave absorption.

Graphene, another prominent member of nanocarbon species, exhibits comparable or superior microwave absorption characteristics compared to CNTs. Its monolayer structure of closely packed carbon atoms in a honeycomb two-dimensional (2D) planar arrangement makes graphene unique, showcasing superior chemical and physical properties and demonstrating microwave absorption properties in conjunction with magnetic materials. Following the initial report on the synthesis and investigation of microwave absorption characteristics, as illustrated in Figure 11.15, of a composite comprising reduced graphene oxide in an NBR matrix (Singh et al., 2012), numerous composites of graphene with metal oxides have been reported, displaying microwave absorption over a broad frequency range (2–18 GHz) (Varghese et al., 2013; Wang et al., 2014). Simulation studies by Kumar et al. indicated that a 10 wt.% concentration of graphene oxide in the NBR matrix exhibits high reflection loss values (> 10 dB) across a wide frequency range of 7.5–12 GHz, with a maximum loss of 57 dB at 9.6 GHz for a 3 mm composite absorber sheet.

Metallic Thin Film Coatings

Metallic films, when deposited in nanometer thicknesses on suitable polymeric substrates, exhibit distinct physical characteristics compared to their bulk counterparts. While metals are recognized as efficient microwave reflectors, at nanometer thicknesses, they can transition to behave as microwave absorbers. For instance, thin Kantal films with thicknesses ranging from 10 to 200 nm, integrated into the internal walls of waveguides, demonstrate effective radar-absorbing material (RAM) properties (Soethe et al., 2011). The fundamental mechanism involves the generation of electric dipoles in metallic films with very low thickness, interacting with electromagnetic waves, leading to their absorption. This absorption results from the polarization of the film by the electric field of the microwave, causing the formation of eddy currents responsible for electrical component absorption. Consequently, heat is generated due to the Joule effect, as defects in structures introduce electrical resistance to the current (Yoshikawa, 2010).

Moreover, it is crucial for the thickness of the metal layer to be smaller or equal to the skin depth (δ) of the metal. The skin depth is contingent on the electrical conductivity of metals and the frequency of incident waves. Therefore, if the metal layer's thickness is sufficient, the generated electrical currents will be confined within the film, leading to microwave losses. The skin depth (δ) can be estimated using Eq. (Equation 11.1) as provided below:

$$\delta = \frac{1}{\sqrt{\pi f \mu \sigma}} \tag{11.1}$$

where f is the frequency; μ is the magnetic permeability of the vacuum ($\mu = 4\pi \cdot 10^{-7}$H/m), and σ is the electric conductivity.

11.6.1.6 Antibacterial Coatings

Maintaining hygiene in field conditions is paramount for military personnel. Nano coatings infused with antibacterial properties offer a solution by inhibiting the growth of harmful microorganisms on gear and equipment, thereby preventing the spread of infections.

Textiles, due to their large surface area, rough texture, and ability to absorb moisture, are highly susceptible to rapid colonization by microorganisms. These organisms not only degrade the fibers and cause discoloration but also emit unpleasant odors, ultimately impacting the textile's longevity. Consequently, there has been a significant push to develop antimicrobial treatments for textiles used in garments, bedding, and sportswear. This drive is primarily motivated by the necessity to curb odor propagation and halt the spread of infectious diseases, including nosocomial bacteria in healthcare settings.

Numerous methods have been devised to impede bacterial growth (bacteriostatic) or eradicate bacteria (bactericidal) on textiles. These treatments range from the release of toxic compounds from the fabric to mechanisms that target microorganisms upon contact with the fiber surface. While some treatments provide broad-spectrum protection, others are specifically effective against bacteria or fungi. Antimicrobial compounds include metals, organometallic derivatives, phenolic compounds, nitrogen and halogen compounds, and nanoparticles, with bio-based products gaining increased attention(Dunne and Crijns, 2017; Singh et al., 2023).

Metallic Coatings for Antimicrobial Activity

Metal-based antimicrobials, including silver, titanium, zinc, and copper, are commonly incorporated into finishes, coatings, and during fiber extrusion processes. While silver demonstrates superior efficacy against bacteria and copper against fungi, both rely on the release of metal ions from the core to exert their antimicrobial effects.

These ions form complexes with essential microbial components, disrupting critical cellular functions. Notably, the multifaceted attack of metal ions makes the development of resistance unlikely, although isolated resistant strains have been reported in specific environments.

The active metal ions can be released either directly from the metal or through ion exchange mechanisms. Although leaching tests may indicate limited or no leaching from silver-treated fabrics, metal-based antimicrobials are often considered non-leaching due to the tendency of silver and copper ions to form complexes with organic substances in their vicinity. Consequently, metal-based antimicrobials are distinguished from toxic chemical compounds, providing an effective and environmentally friendly solution to microbial control on textiles (Birkett et al., 2022; Vaidya et al., 2017).

Halogen Compounds

Halogen compounds, renowned for their potent disinfectant properties, are commonly found in cleaning and disinfectant products such as bleach, chlorhexidine, and Betadine, and are also utilized in textiles. However, their persistent presence in the environment, aggressive nature, and association with carcinogenic disinfectant by-products, particularly chlorine, limit their usage in textiles. Pentachlorophenol, dichlorodiphenyl methane, iodine, and iodophors are among the commonly used halogen-based products.

Nitrogen Compounds

Nitrogen compounds, including urea, amine, and guanidine, disrupt proteins, thereby displaying antimicrobial properties. Quaternary ammonium salts, or quats, represent a special category of nitrogen compounds utilized for antimicrobial treatment. These positively charged molecules feature a lengthy organic chain that enables them to penetrate microbial cell membranes, resulting in microbial destruction. Quats are odorless and capable of forming chemical bonds with textile fibers, reducing leaching and enhancing durability compared to smaller nitrogen compounds (Heyse et al., 2016).

Phenolic Compounds

Phenolic compounds, historically employed as disinfectants, have been refined into less toxic derivatives to mitigate human health risks. Triclosan serves as a well-known example, widely appreciated for its effective disinfection properties. Nonetheless, its extensive use has led to the emergence of bacterial strains resistant to its effects (Rahman et al., 2022).

Aldehyde Compounds

Aldehyde compounds, such as formaldehyde and glutaraldehyde, are commonly utilized for cross-linking in chemistry. They also cross-link proteins and enzymes, rendering them ineffective for microorganisms. However, due to their carcinogenic nature, stringent regulations govern their use in textiles (Sewell et al., 1984).

Bio-Based Products

Bio-based products, such as chitosan derived from chitin, offer a renewable source of antimicrobial agents. These polysaccharide chains possess an array of amine functionalities similar to nitrogen compounds, providing antimicrobial activity. The development of new antimicrobial products from plant and animal extracts, including menthol, carvacrol, eucalyptol, and sericin, demonstrates the growing interest in bio-based alternatives. However, challenges such as durability, applicability, availability, and potential discoloration or odor limit their industrial application in textiles.

Additionally, short-chain amino acids produced through improved biotechnology tools have shown promise as bio-based antimicrobial finishes for textiles. Despite the potential of bio-based alternatives, further research is needed to address these challenges and facilitate their widespread industrial adoption in textile applications (Yan et al., 2021).

11.6.1.7 Chemical and Biological Agent Protection

Specially engineered nano coatings play a crucial role in shielding against chemical and biological agents, providing essential protection for military personnel in diverse operational scenarios. To counter these threats effectively, military, law enforcement, and first responders such as firefighters and HAZMAT workers rely on HAZMAT protective clothing in situations like urban combat, terrorism incidents, and chemical spills. Typically, HAZMAT protective clothing comprises multilayer laminated fabric, acting as a barrier to absorb or block contaminants, including toxic chemicals or microorganisms (Schreuder-Gibson et al., 2003). However, absorbed toxic chemicals may pose a secondary contamination risk during use or decontamination (Chen et al., 2009), and biological warfare agents can adhere to HAZMAT clothing surfaces, potentially endangering nearby personnel (Schreuder-Gibson et al., 2003).

To enhance the protective capabilities of HAZMAT clothing, significant research has been devoted to developing smart coatings capable of detoxifying chemicals and neutralizing biological agents. These coatings necessitate a high surface area with numerous reactive sites, leading to extensive exploration into nanotechnology. Sundarrajan et al. (2010) identified electrospinning as a promising nanotechnology for producing continuous nanometer-sized fibers with a high surface area (Sundarrajan et al., 2010). Lee et al. (2009) anticipated that coating nanoparticles onto nanoscale electrospun fibers would significantly increase the number of reactive sites (Lee et al., 2009).

A prevalent method for achieving nanoscale composition and structure in thin films is layer-by-layer (LbL) assembly, involving the alternating adsorption of positively and negatively charged species from an aqueous solution. Titanium dioxide (TiO_2) exhibits photocatalytic properties effective for decomposing chemicals, making it a viable coating material for enhancing HAZMAT protection (Fujihira et al., 1981; Fujishima and Honda, 1972). Grandcolas et al. (2009) successfully coated cotton fibers with WO_3-titanate nanotubes (WO_3-TiNT) using the LbL deposition method, demonstrating rapid photocatalytic degradation of harmful chemicals and warfare agents under simulated solar light (Grandcolas et al., 2009).

Similarly, Lee et al. (2009) coated TiO_2 nanoparticles onto electrospun poly(dimethylsiloxane-b-etherimide) (PSEI) fibers via LbL deposition, resulting in flexible, photocatalytic fabric capable of degrading organic compounds under UV light (Lee et al., 2009). Krogman et al. (2009) employed a vacuum-assisted spray LbL process to coat nylon mats with bilayers of TiO_2 nanoparticles and polyelectrolytes, yielding a water vapor-permeable yet protective coating against chemical warfare agent simulants (Krogman et al., 2009).

Chen et al. (2010) utilized LbL assembly to create multifunctional protective fabric capable of detoxifying both chemical and biological agents. By alternating the deposition of bactericidal and chemically reactive polyelectrolyte coatings onto electrospun polyacrylonitrile (PAN) fibers, they achieved a fabric with enhanced hydrolysis rates of chemical warfare agent simulants and bactericidal properties against various pathogens, while maintaining breathability (Chen et al., 2010).

Overall, these advancements underscore the potential of smart coatings and nanotechnology in enhancing the protective capabilities of HAZMAT clothing against chemical and biological threats..

11.7 INFRARED CAMERA-THERMAL IMAGING

During the late 1960s, real-time thermal imaging technology began to gain prominence, particularly in defense systems, where significant advances in infrared technology have been made. One notable advancement is the development of infrared thermal imaging cameras, capable of capturing images of invisible infrared energy, which is part of the electromagnetic spectrum and perceived as heat but invisible to the naked eye. Infrared technology can detect thermal energy emitted by all living organisms, objects, and materials, representing it as thermograms or images. These thermal imaging cameras excel in penetrating environments such as smoke, light fog, snow, rain, and extreme darkness, making them valuable tools for various applications, including use aboard aircraft and satellite devices.

Forward Looking Infrared (FLIR) technology, originally developed as a sensor system for fighter aircraft, serves as a forward sensor primarily fixed to capture what lies directly ahead of the aircraft. The image captured by FLIR is displayed to the pilot through the aircraft's heads-up display (HUD). The predecessor FLIR technology, initially developed by the United States Navy for identifying and targeting enemy forces, was characterized by its large, expensive, and heavy systems. However, with the advancement of technology, more rugged, efficient FLIR cameras with reduced size and cost have become available (Faust et al., 2014).

11.7.1 Basics of Thermal Imaging

The principle of thermal imaging revolves around the detection and representation of infrared radiation emitted by objects at temperatures above absolute zero. This infrared radiation falls within the electromagnetic spectrum, typically ranging from 0.9 to 14 micrometers. An infrared (IR) camera captures this radiation and translates it into a thermogram, which is a grayscale or pseudo-color thermal image depicting thermal variations across objects or scenes. The operation of IR cameras is governed by two fundamental laws: the total radiation law and Stefan-Boltzmann's law. The total radiation law dictates that the amount of incident energy equals the sum of absorbed, reflected, and transmitted energy. Stefan-Boltzmann's law, expressed as

$$E = \varepsilon \sigma T^4 \tag{11.2}$$

It states that the total radiant energy (E) emitted by a body is proportional to its emissivity (ε) and the fourth power of its temperature (T). Thermal imagers resemble digital cameras, but instead of charge-coupled devices (CCD) or CMOS detectors, they employ focal plane arrays (FPA) consisting of micrometer-sized pixels made of various materials sensitive to IR radiation.

Advancements in IR cameras primarily stem from improvements in detector technology, transitioning from single detectors to linear and array detectors. FPAs come in two main types: thermal detectors and quantum detectors. Quantum detectors, composed of materials such as InSb, InGaAs, PtSi, HgCdTe (MCT), and layered GaAs/AlGaAs for QWIP detectors, offer enhanced sensitivity and speed but require cooling to cryogenic temperatures. On the other hand, thermal detectors, like uncooled microbolometers made of metal or semiconductor materials sensitive to temperature, are cost-effective and possess a broader IR spectral response. Thermal imaging systems typically operate within two wavelength windows: the mid-wave infrared (MWIR) band (3 to 5 microns) and the long-wave infrared (LWIR) band (8 to 12 microns). MWIR is prone to atmospheric attenuation, primarily due to water vapor, making it unsuitable for imaging. In contrast, LWIR experiences minimal atmospheric attenuation and is thus widely used in challenging environments (Rogalski, 2019).

11.7.2 Applications

Thermal imaging finds extensive applications in military operations due to its ability to provide valuable situational awareness and enhance mission effectiveness. Some key applications of thermal imaging in the military include:

Surveillance and Reconnaissance: Thermal imaging allows military personnel to conduct surveillance and reconnaissance missions more effectively, especially during low-light conditions or adverse weather. It enables them to detect and monitor enemy movements, positions, and activities from a safe distance, enhancing overall battlefield awareness.

Target Acquisition: Thermal imaging helps military forces identify and acquire targets with greater precision, even in challenging environments such as dense foliage or urban areas. By detecting the heat signatures emitted by potential targets, thermal cameras assist in distinguishing between friendly and hostile forces, reducing the risk of friendly fire incidents.

Search and Rescue Operations: Thermal imaging aids in search and rescue efforts, particularly in locating missing or injured personnel in remote or hazardous environments. The ability to detect body heat signatures enables rescue teams to quickly identify individuals, even in darkness or obscured conditions, improving the chances of successful rescue missions.

Border Surveillance: Thermal imaging is valuable for monitoring and securing borders, coastlines, and other sensitive areas. By detecting human and vehicle movement across borders, thermal cameras help border patrol agents detect unauthorized crossings, smuggling activities, and other security threats, enhancing border security and control.

Intelligence Gathering: Thermal imaging provides valuable intelligence to military commanders by revealing enemy positions, fortifications, and activities. By integrating thermal imagery with other intelligence sources, military analysts can gain a comprehensive understanding of the battlefield environment, enabling informed decision-making and strategic planning.

Weapon Systems: Thermal imaging technology is integrated into various weapon systems, including rifles, machine guns, and missile guidance systems. By providing enhanced targeting capabilities, thermal sights enable soldiers to engage targets accurately, even in challenging visibility conditions, increasing the effectiveness of military firepower.

Battlefield Medicine: Thermal imaging assists military medical personnel in assessing and treating battlefield injuries more effectively. By detecting differences in body temperature, thermal cameras help identify injured soldiers and prioritize medical treatment, improving outcomes on the battlefield.

Overall, thermal imaging plays a critical role in enhancing military operations across various domains, from reconnaissance and surveillance to target acquisition and battlefield medicine. Its ability to provide clear, real-time imagery in diverse environmental conditions makes it an indispensable tool for modern military forces.

11.8 THERMOCHROMIC AND ELECTROCHROMIC MATERIALS

11.8.1 Electrochromic Materials

In recent decades, electrochromic (EC) materials have garnered significant attention from researchers due to their reversible and multi-color changing abilities, high energy capacity, and controllable modulation of solar irradiation. These versatile materials hold immense potential for applications such as electronic labeling, electronic displays, "smart" windows, multifunctional capacitors, and more. Additionally, advancements in electrochromism with novel materials have led to active exploration in the realms of intelligent stealth and camouflage, particularly within military contexts. Stealth technology aims to render military targets challenging to detect, identify, track, and attack by manipulating their radar, infrared, and optical signatures, thereby enhancing the survivability, striking capability, and penetration ability of weapons and equipment amidst escalating military competition. As detection methods continue to evolve and become more accurate, stealth technology remains crucial. Simultaneously, the development of anti-stealth technology necessitates multi-spectrum studies to address the challenges posed by multi-detection scenarios. The adaptability and reversibility of electrochromic materials, allowing for controlled changes in color and emissivity, offer promising applications in intelligent stealth across the visible and infrared regions of the electromagnetic spectrum. This article provides an overview of electrochromism and its applications in intelligent stealth. Electrochromic materials have the remarkable ability to reversibly alter their properties, such as transmittance, reflectance, absorption, or emittance, under appropriate electrical potentials. Electrochromic devices (ECDs), typically fabricated as thin films, can be seamlessly integrated into devices to modulate optical properties.

A typical five-layer structure of an electrochromic device consists of an electrochromic layer, an ion storage layer, and an electrolyte layer sandwiched between two transparent conducting oxide layers. The electrochromic layer, crucial for device performance, undergoes a reversible reaction when exposed to an electrical potential, as exemplified by tungsten oxides (WO_3) reacting with lithium electrolytes. Upon application of the appropriate voltage, small cations such as Li^+ insert into WO_3, causing a change in valence state from W^{6+} to W^{5+}, resulting in a transition from transparent to dark blue, and vice versa, as described in Equation 11.3.

$$WO_3(\text{No color}) + xLi^+ + xe^- = Li_xWO_3(\text{blue}) \tag{11.3}$$

11.8.1.1 Applications of Electrochromic Materials

In recent years, there has been a concerted effort to develop visible light stealth technology, primarily focused on scattering light and camouflaging targets to minimize detection. By ensuring consistency in color and brightness between the object and its surroundings, the likelihood of detection by enemy optical detection equipment is significantly reduced. Traditional visible light stealth technology aims to eliminate contrast characteristics between the target and the background by applying camouflage coatings or nets matching the background color. While this method is effective when the object is stationary, it struggles to maintain stealth effectiveness during maneuvers, limiting its adaptability in dynamic combat scenarios. To address these limitations, researchers are actively exploring dynamic stealth technology to adapt to complex and changing combat environments. Intelligent stealth materials offer promising solutions, enabling dynamic camouflage on various backgrounds while saving maintenance time and reducing costs (Zhao et al., 2019).

Electrochromic materials play a crucial role in visible light camouflage and stealth due to their ability to change colors effectively. These materials, such as conductive polymers and vanadium oxides, offer a wide range of color options, making them ideal for visible light stealth applications. Studies have demonstrated their effectiveness in providing adaptive camouflage for weapons and equipment across different environmental backgrounds. For instance, flexible electrochromic devices using vanadium oxides and polyaniline can adjust colors based on environmental conditions, showcasing their suitability for smart equipment applications as shown in Figure 11.16.

Furthermore, electrochromic materials have found practical applications in military settings, such as in smart stealth films for aircraft and camouflage clothing for soldiers. These materials allow reversible color changes under different voltage controls, ensuring adaptability to diverse combat environments. Notably, the U.S. Army and Russia have utilized conductive polymer-based electrochromic materials to develop stealthy clothing and combat exoskeleton suits, respectively, demonstrating the feasibility and effectiveness of intelligent stealth technology in military applications.

Despite these advancements, challenges remain, including limited lifespan and insufficient brightness control of electrochromic materials. Nonetheless, ongoing

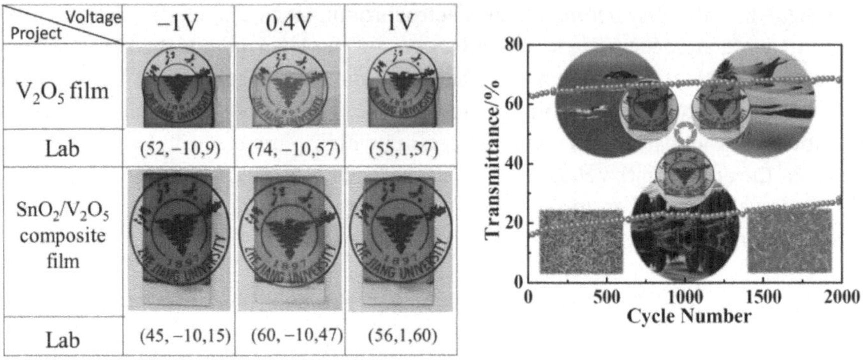

FIGURE 11.16 Electrochromic camouflage properties of SnO_2 /V_2O_5 electrochromic film. (Reprinted with permission from Zhao et al., 2019, Copyright © 2019, Royal Society of Chemistry.)

research efforts aim to overcome these limitations and further enhance the capabilities of visible light intelligent stealth materials, holding significant promise for future military camouflage applications.

Infrared stealth technology has also garnered considerable attention, driven by the need to evade modern infrared detection systems. These systems rely on temperature differences between targets and their backgrounds to detect and track them effectively. To reduce the probability of detection, adaptive infrared stealth technologies have emerged, including smart materials with adjustable emissivity, such as electrochromic and thermochromic materials.

Thermochromic materials alter the infrared emissivity of targets by changing their temperature, offering a relatively simple solution compared to electrochromic materials. However, challenges remain in terms of adjusting phase transition temperatures and achieving adequate temperature ranges for practical applications. In contrast, electrochromic stealth materials actively adjust infrared emissivity under the influence of electric fields or currents, making them a focal point in adaptive infrared camouflage research (Fu et al., 2023).

WO_3 and conductive polymers are among the most widely studied electrochromic materials for infrared stealth applications. These materials have shown promise in providing variable emissivity and have been incorporated into flexible electrochromic devices with excellent performance characteristics. For instance, improved porous polymer-doped PANI flexible electrochromic emissivity devices exhibit impressive emissivity ranges and cycle durability, making them suitable for infrared adaptive stealth and camouflage systems. Eclipse VEDTM, an all-solid-state electrochromic device based on WO_3, offers exceptional emissivity tuning capabilities, further expanding the potential applications of electrochromic materials in infrared adaptive stealth technologies (Tian et al., 2017).

In the realm of practical applications, various countries have made strides in developing electrochromic infrared adaptive systems to meet the infrared stealth requirements of tanks, vehicles, and other equipment across diverse environments.

FIGURE 11.17 "Black Fox" active adaptive infrared camouflage system. (Developed by Eltix company, Long et al., 2023)

These adaptive infrared stealth systems utilize electrochromic materials or devices capable of adjusting infrared emissivity, thereby enabling targets to achieve stealth in different backgrounds, particularly facilitating the transition between infrared camouflage during the day and night. One notable example is the ADAPTIV infrared stealth system developed by BAE Systems, which leverages electrochromic materials to modify the infrared radiation characteristics of the target. This system allows for the control of infrared signal characteristics to align with the background environment or mimic the infrared signatures of standard vehicles. Similarly, the "Black Fox" active adaptive infrared camouflage system, developed by Eltix company, has reached a high level of maturity, as depicted in Figure 11.17. This innovative camouflage system utilizes infrared cameras to capture the infrared radiation characteristics of the surrounding environment. Subsequently, the system adjusts the infrared radiation properties of the vehicle itself through a processor, effectively integrating it with the environment. The successful implementation of the "Black Fox" infrared adaptive stealth technology holds significant importance in addressing the challenge of nighttime infrared stealth for equipment such as tanks, marking a notable advancement in military stealth capabilities (Long et al., 2023).

11.8.2 Thermochromic Materials

Advancements in surveillance technology are progressing rapidly alongside the genuine evolution of imaging technology, particularly in hyperspectral and digital imaging. These advancements find applications for various purposes, both positive and negative, depending on the specific needs of various groups, primarily within defense sectors. Conversely, researchers and scientists are also directing their efforts toward developing anti-surveillance technology across different engineering fields, with chameleon camouflage textiles emerging as a significant area of interest. While existing literature reviews have acknowledged research and development endeavors in anti-surveillance technology, particularly in the realm of camouflage textiles, research on liquid crystal-based chameleon-type camouflage textiles is still in its early stages. Chameleon camouflage textiles possess the capability to adjust their color to blend with the surrounding environment, accommodating both positive and negative applications as required by different contexts. For example, combat environments such as snowy landscapes necessitate low-temperature adaptive camouflage, whereas desert settings call for high-temperature background camouflage (Degenstein et al., 2021).

11.8.2.1 Basics of Adaptive Camouflage Phenomenon

The fundamental principle of adaptive camouflage textiles revolves around the utilization of thermo chromic colorants, which fulfill the performance requirements of chromatic materials. Thermochromic liquid crystal (TLC) serves as a crucial mechanism for adaptive camouflaging, responding to temperature changes against various combat backgrounds (CB). Chromatic materials undergo color changes in response to external stimuli such as light (photochromic), heat (thermo chromic), electricity (electro chromic), liquid (solvatochromic), and electron beam (carsolchromic). Typically, the molecular structure of thermochromic colorants is altered, resulting in color changes above room temperature. However, thermochromic liquid crystals exhibit temperature changes within the range of 5–15°C, with cholesteric-type liquid crystals being particularly suitable for thermochromism. The pitch of the helical arrangement in liquid crystals is responsible for temperature-dependent reflection wavelength. Sudhakar and Gobi N proposed the concept of chameleon textiles, which can adapt to different CB environments through pH changes, electric or magnetic field effects, mechanochromism, bond breaking/making, and oxidation state changes. Materials used in camouflage textiles development include polymer binders/resins, pigments for controlling gloss, and heavy metal pigments such as lead, chromium, iron oxide, silica, and polyurethane. Despite advancements, thermochromic materials have limited applications in camouflage textiles due to their organic nature, short lifetimes, and restricted temperature ranges. Further research is needed to develop thermochromic materials with wider temperature ranges and longer lifetimes. Researchers emphasize the importance of chromophores with polymeric materials to enhance the stability of the color-changing process. Adaptive camouflage has been explored through color-changing leuco dyes, coupled with high stimuli-responsive studies. Heating mechanisms influence thermochromic systems for temperature-responsive camouflaging. Additionally, various studies have investigated adaptive camouflage devices and mechanisms based on electrical fields, mechanical systems, and temperature changes. The application of adaptive camouflage textiles extends beyond garments to include defense applications, offering versatile solutions for various scenarios and environments.

Selection of textile materials for adaptive camouflage formulation considers thermal conductivity, with wool and polyester exhibiting higher heat transfer rates compared to cotton. Thermal conductivity values for different materials, including cotton, wool, silk, PVC, nylon, polyester, and polypropylene, are crucial for temperature-responsive coloration in protective textiles. Thermochromic colorants find diverse applications in commercial products such as kitchen aprons, laboratory aprons, safety shoes, medical textiles, and flexible displays for clothing, transportation of sensitive goods, disease-detecting garments, firefighting garments, sports apparel, and building paints. These applications demonstrate the versatility and practicality of thermochromic colorants in various industries.

11.8.2.2 Thermochromic Coatings in Military Applications

Thermochromic materials, which change color in response to temperature variations, have several potential applications in military contexts, offering unique advantages in various scenarios. Some potential applications include:

Temperature-Sensitive Camouflage: Thermochromic materials can be integrated into camouflage fabrics or coatings to adjust their color and visibility based on temperature changes. In military operations, especially in diverse environments with fluctuating temperatures, adaptive camouflage can help soldiers blend into their surroundings more effectively, enhancing concealment and reducing the risk of detection by adversaries.

Temperature Monitoring and Warning Systems: Thermochromic materials can serve as indicators for monitoring equipment or systems subjected to temperature extremes. By incorporating these materials into critical components or surfaces, such as engine parts, weapon systems, or electronic devices, military personnel can visually detect overheating or temperature fluctuations, enabling timely intervention and preventive maintenance to avoid equipment failure or malfunctions.

Thermal Signature Reduction: Infrared (IR) signature reduction is crucial for military vehicles, aircraft, and personnel operating in hostile environments, as it minimizes the risk of detection by infrared sensors and targeting systems. Thermochromic coatings or materials with tailored thermal properties can help reduce the thermal signature of military assets by effectively dissipating or absorbing heat, thereby enhancing stealth capabilities and survivability on the battlefield.

Temperature-Responsive Signaling and Marking: Thermochromic materials can be utilized to create temperature-responsive signaling devices or markings for tactical communication, navigation, and identification purposes. By applying thermochromic paints, inks, or tapes to surfaces, military units can create temporary markings or indicators that change color in response to temperature variations, providing visual cues or signaling mechanisms in low-light conditions or during covert operations.

Thermal Camouflage for Vehicles and Equipment: Military vehicles and equipment, such as tanks, armored vehicles, and aircraft, can benefit from thermochromic coatings or materials that adapt their appearance to match ambient temperatures. By adjusting the surface coloration based on thermal gradients in the environment, thermochromic camouflage can help reduce the thermal contrast between military assets and their surroundings, making them less susceptible to detection by thermal imaging systems and infrared-guided weapons.

Thermal Management in Electronics and Optics: Thermochromic materials can play a role in thermal management systems for military electronics, optics, and sensor devices. By incorporating thermochromic coatings or films into heat sinks, enclosures, or optical components, military systems can regulate temperature fluctuations, dissipate excess heat, and maintain optimal operating conditions, thereby improving performance, reliability, and longevity.

Overall, the unique properties of thermochromic materials make them promising candidates for various military applications, ranging from camouflage and stealth technologies to temperature monitoring and thermal management systems. As research and development in this field continue to advance, the integration of thermochromic technologies into military platforms and equipment holds the potential to enhance operational effectiveness, survivability, and situational awareness for military forces.

11.9 DESALINATION AND WATER TREATMENT IN MILITARY APPLICATIONS

Water scarcity is a significant concern for military operations, particularly in arid, remote, or hostile environments where access to clean water is limited. Desalination and water treatment technologies play a critical role in providing safe drinking water to military personnel and supporting operational readiness. This section explores the challenges faced by military forces in water procurement, the importance of desalination and water treatment in military applications, key technologies utilized, and strategic implications for enhancing water resilience in military operations.

11.9.1 Challenges in Military Water Supply

11.9.1.1 Operational Requirements

The provision of an adequate and reliable water supply is fundamental to sustaining military operations in various environments. Operational requirements encompass the essential need for water to support troops, equipment, and various support operations. Troops require water for hydration, cooking, sanitation, and medical purposes. Additionally, equipment such as vehicles, machinery, and communication systems often rely on water for cooling and maintenance. Moreover, support operations, including field hospitals, logistics centers, and base camps, necessitate a steady water supply for their functioning.

However, meeting these operational water demands becomes particularly challenging in remote or contested areas where access to clean water sources is limited or compromised. In such environments, military personnel must contend with rugged terrain, extreme weather conditions, and the threat of hostile forces, all of which can impede efforts to source and transport water effectively. Remote areas may lack natural water sources, requiring logistical support to transport water over long distances. Furthermore, in contested areas, access to existing water sources may be restricted or unsafe due to the presence of enemy forces or contamination risks.

11.9.1.2 Contamination Risks

Contamination of water sources poses significant risks to military personnel, leading to adverse health effects and operational disruptions. Contaminants in water sources can originate from various sources, including industrial pollution, agricultural runoff, and biological agents. In addition, environmental hazards such as floods, hurricanes, and earthquakes can contribute to water contamination by disrupting infrastructure and spreading pollutants.

The importance of water purification and treatment cannot be overstated in mitigating these contamination risks. Purification processes such as filtration, disinfection, and chemical treatment are essential for removing or neutralizing contaminants, pathogens, and toxins from water sources. Military units often deploy portable water purification systems capable of treating water from diverse sources, including rivers, lakes, and groundwater wells.

Effective water purification not only safeguards the health and well-being of military personnel but also ensures the reliability and sustainability of water supplies during prolonged deployments. Moreover, access to clean and safe drinking water is essential for maintaining morale and operational readiness among troops. Hence, challenges in military water supply encompass meeting operational requirements for water in diverse environments while mitigating contamination risks to ensure the health and effectiveness of military personnel. Addressing these challenges requires a comprehensive approach that integrates logistical planning, technological innovation, and environmental stewardship to provide reliable and safe water supplies in all operational scenarios.

11.9.2 Importance of Desalination and Water Treatment in Military Applications

11.9.2.1 Operational Readiness

Clean water plays a critical role in maintaining the health, morale, and effectiveness of military personnel in various operational environments. Adequate access to clean water is essential for hydration, food preparation, personal hygiene, and medical treatment. Without a reliable supply of clean water, troops may suffer from dehydration, waterborne illnesses, and decreased physical and cognitive performance, ultimately compromising mission success.

The availability of water also directly impacts the success and sustainability of military missions. In arid or remote regions where natural water sources are scarce, desalination and water treatment technologies become indispensable for ensuring sufficient water supplies. These technologies enable military units to tap into alternative water sources such as seawater, brackish water, or contaminated groundwater, thereby expanding their operational capabilities and reducing reliance on vulnerable supply lines.

11.9.2.2 Logistics and Resilience

The integration of water treatment capabilities into military logistics planning is essential for enhancing operational resilience and sustainability. Military units must be equipped with the necessary infrastructure and resources to treat water efficiently and effectively, regardless of the operational environment. This requires the deployment of portable desalination plants, water purification systems, and advanced filtration technologies that can adapt to diverse water sources and quality conditions.

By incorporating water treatment capabilities into logistics planning, military forces can enhance their resilience against disruptions in water supply chains. This includes anticipating and mitigating risks such as infrastructure damage, contamination incidents, and supply line disruptions caused by natural disasters, enemy actions, or logistical constraints. By maintaining self-sufficiency in water treatment, military units can minimize their dependence on external resources and ensure continuity of operations in challenging environments.

Thus, the importance of desalination and water treatment in military applications lies in their critical role in maintaining operational readiness, enhancing mission success, and improving logistical resilience. By investing in these technologies and integrating them into operational planning, military forces can ensure access to clean and reliable water supplies, thereby safeguarding the health, morale, and effectiveness of personnel in all operational scenarios.

11.9.3 Desalination Technologies for Military Use

11.9.3.1 Portable Desalination Systems

Portable desalination systems are compact, lightweight solutions specifically designed for field deployment in military operations. These systems are engineered to be easily transportable and rapidly deployable, allowing military units to produce clean drinking water on-site, even in remote or austere environments. Key features of portable desalination systems include their small footprint, rugged construction, and compatibility with military operations.

These systems typically utilize advanced desalination technologies such as reverse osmosis or membrane distillation to remove salt and impurities from seawater or brackish water. They are equipped with integrated pumps, filters, and membranes that enable efficient water purification while minimizing energy consumption and maintenance requirements. Moreover, portable desalination systems may feature self-contained power sources, such as diesel generators or solar panels, to ensure operational independence.

11.9.3.2 Mobile Desalination Units

Mobile desalination units are modular systems designed for rapid setup and operation in expeditionary environments. These units consist of interconnected components housed within transportable containers or vehicles, allowing for flexible deployment and scalability to meet varying water demand requirements. Mobile desalination units are integral components of military logistics and support infrastructure, providing essential water treatment capabilities in forward operating bases, disaster relief missions, and other expeditionary scenarios.

The integration of mobile desalination units into military logistics enables rapid response and sustained operations in water-stressed environments. These units can be strategically positioned near water sources or deployed alongside military convoys to ensure continuous access to clean water throughout the duration of a mission. Additionally, their modular design allows for easy reconfiguration and relocation as operational needs evolve, enhancing adaptability and resource efficiency.

11.9.3.3 Solar-Powered Desalination

Solar-powered desalination systems harness renewable energy sources to power water treatment processes, offering distinct advantages for military operations. By utilizing solar energy to drive desalination processes, these systems reduce reliance on traditional fuel sources and minimize logistical burdens associated with fuel transportation and supply. This makes them particularly well-suited for off-grid operations and remote deployments where access to fuel may be limited or impractical.

Solar-powered desalination systems employ a variety of technologies, including solar stills, solar-assisted reverse osmosis, and solar thermal desalination, to convert sunlight into usable energy for water purification. These systems can operate autonomously, with minimal maintenance requirements, making them ideal for prolonged deployments in isolated or environmentally sensitive areas. Furthermore, their scalability and versatility allow for deployment in diverse geographic regions, providing military forces with a sustainable and environmentally friendly solution for producing potable water in challenging environments.

11.9.4 Water Treatment Innovations

11.9.4.1 Point-of-Use Purification

Point-of-use purification involves individual water purification kits designed for soldiers in the field, providing them with the means to produce potable water from various water sources. These kits typically incorporate a combination of filtration, chemical treatment, and ultraviolet (UV) disinfection technologies to remove contaminants and pathogens from water. Filtration systems remove physical impurities, while chemical treatments, such as chlorine or iodine tablets, neutralize microbial contaminants. UV disinfection utilizes ultraviolet light to inactivate bacteria, viruses, and other microorganisms, ensuring water safety.

11.9.4.2 Tactical Water Purification Systems

Tactical water purification systems are portable units deployed by military forces to treat water from natural sources or captured enemy supplies. These systems are designed for rapid deployment and operational flexibility, enabling troops to establish water purification capabilities in remote or hostile environments quickly. Tactical purification systems utilize advanced filtration, disinfection, and desalination technologies to produce clean, drinkable water from diverse water sources, including rivers, lakes, and groundwater. Their portability and versatility make them essential assets for ensuring water security in expeditionary missions.

11.9.5 Operational Significance and Strategic Considerations

11.9.5.1 Force Readiness and Sustainability

Water resilience plays a crucial role in maintaining operational effectiveness and troop morale in military operations. Access to clean and reliable water is essential for hydration, food preparation, personal hygiene, and medical care, directly impacting mission success and the well-being of personnel. Water availability influences troop morale, combat performance, and overall force readiness, highlighting the importance of prioritizing water security in military planning and operations.

11.9.5.2 Logistics and Supply Chain Resilience

The integration of water treatment capabilities into military logistics planning is essential for enhancing supply chain resilience and operational sustainability. By incorporating water treatment assets into logistical frameworks, military forces can mitigate the risks of water supply disruptions caused by environmental factors, infrastructure damage, or enemy actions. This includes pre-positioning water purification units, establishing alternative supply routes, and implementing contingency plans to ensure continuous access to clean water in challenging operational environments.

11.9.6 Future Directions and Challenges

11.9.6.1 Technological Advancements

Future research in water treatment innovation aims to improve the efficiency, reliability, and portability of desalination systems for military applications. Key priorities include developing advanced membrane technologies, enhancing energy efficiency, and optimizing system performance under diverse operating conditions. Additionally, there is potential for innovation in tailored water treatment technologies designed to meet the unique requirements of military operations, such as compact, deployable systems optimized for field use.

12.9.6.2 Training and Capacity Building

Training military personnel in water management and treatment is essential for ensuring operational readiness and sustainability. Building partnerships with industry and academia can facilitate knowledge exchange, technology transfer, and collaborative research efforts aimed at addressing emerging challenges and advancing water treatment capabilities. By investing in training and capacity building initiatives, military forces can enhance their resilience, effectiveness, and readiness in managing water resources in complex operational environments.

REFERENCES

Abdeen, D.H., El Hachach, M., Koc, M., Atieh, M.A., 2019. A review on the corrosion behaviour of nanocoatings on metallic substrates. Materials 12. https://doi.org/10.3390/ma12020210

Ahmad, I., Kan, C., 2016. A review on development and applications of bio-inspired superhydrophobic textiles. Materials 9. https://doi.org/10.3390/ma9110892

Atanasov, S.E., Oldham, C.J., Slusarski, K.A., Taggart-Scarff, J., Sherman, S.A., Senecal, K.J., Filocamo, S.F., McAllister, Q.P., Wetzel, E.D., Parsons, G.N., 2014. Improved cut-resistance of Kevlar® using controlled interface reactions during atomic layer deposition of ultrathin (<50 Å) inorganic coatings. J. Mater. Chem. A 2, 17371–17379. https://doi.org/10.1039/C4TA03662J

Bai, Y., Mora-Seró, I., De Angelis, F., Bisquert, J., Wang, P., 2014. Titanium dioxide nanomaterials for photovoltaic applications. Chem. Rev. 114, 10095–10130. https://doi.org/10.1021/cr400606n

Barthlott, W., Neinhuis, C., 1997. Purity of the sacred lotus, or escape from contamination in biological surfaces. Planta 202, 1–8. https://doi.org/10.1007/s004250050096

Bayda, S., Adeel, M., Tuccinardi, T., Cordani, M., Rizzolio, F., 2020. The history of nanoscience and nanotechnology: From chemical–physical applications to nanomedicine. Molecules 25. https://doi.org/10.3390/molecules25010112

Benzait, Z., Trabzon, L., 2018. A review of recent research on materials used in polymer–matrix composites for body armor application. J. Compos. Mater. 52, 3241–3263. https://doi.org/10.1177/0021998318764002

Bilisik, K., Karaduman, N.S., Bilisik, N.E., 2016. 3D fabrics for technical textile applications, in: Jeon, H.-Y. (Ed.), Non-Woven Fabrics. IntechOpen, Rijeka. https://doi.org/10.5772/61224

Birkett, M., Dover, L., Cherian Lukose, C., Wasy Zia, A., Tambuwala, M.M., Serrano-Aroca, Á., 2022. Recent advances in metal-based antimicrobial coatings for high-touch surfaces. Int. J. Mol. Sci. 23. https://doi.org/10.3390/ijms23031162

Caldona, E.B., De Leon, A.C.C., Thomas, P.G., Naylor, D.F.I., Pajarito, B.B., Advincula, R.C., 2017. Superhydrophobic rubber-modified polybenzoxazine/SiO_2 nanocomposite coating with anticorrosion, anti-ice, and superoleophilicity properties. Ind. Eng. Chem. Res. 56, 1485–1497. https://doi.org/10.1021/acs.iecr.6b04382

Chen, L., Bromberg, L., Lee, J.A., Zhang, H., Schreuder-Gibson, H., Gibson, P., Walker, J., Hammond, P.T., Hatton, T.A., Rutledge, G.C., 2010. Multifunctional electrospun fabrics via layer-by-layer electrostatic assembly for chemical and biological protection. Chem. Mater. 22, 1429–1436. https://doi.org/10.1021/cm902834a

Chen, L., Bromberg, L., Schreuder-Gibson, H., Walker, J., Alan Hatton, T., Rutledge, G.C., 2009. Chemical protection fabrics via surface oximation of electrospun polyacrylonitrile fiber mats. J. Mater. Chem. 19, 2432–2438. https://doi.org/10.1039/B818639A

Cheng, Y., Lu, S., Xu, W., Cao, K., Li, J., Zheng, Y., 2018. Controllable fabrication of superhydrophobic alloys surface on copper substrate for self-cleaning, anti-icing, anti-corrosion and anti-wear performance. Surf. Coat. Technol. 333, 61–70. https://doi.org/10.1016/j.surfcoat.2017.10.069

Chi, F., Liu, D., Wu, H., Lei, J., 2019. Mechanically robust and self-cleaning antireflection coatings from nanoscale binding of hydrophobic silica nanoparticles. Sol. Energy Mater. Sol. Cells 200, 109939. https://doi.org/10.1016/j.solmat.2019.109939

Chigondo, M., Chigondo, F., 2016. Recent natural corrosion inhibitors for mild steel: An overview. J. Chem. 2016, 6208937. https://doi.org/10.1155/2016/6208937

Degenstein, L.M., Sameoto, D., Hogan, J.D., Asad, A., Dolez, P.I., 2021. Smart textiles for visible and IR camouflage application: State-of-the-art and microfabrication path forward. Micromachines 12. https://doi.org/10.3390/mi12070773

Drexler, K.E., 1986. Engines of Creation: The Coming Era of Nanotechnology. Knopf Doubleday Publishing Group, Broadway, NY.

Dunne, C.P., C.A., Keinänen-Toivola, M.M., Kahru, A., Teunissen, B., Olmez, H., Gouveia, I., Melo, L., Murzyn, K., Modic, M., Ahonen, M., Askew, P., Papadopoulos, T., Crijns, F.R.L., 2017. Anti-microbial coating innovations to prevent infectious diseases

(AMiCI): Cost action ca15114. Bioengineered 8, 679–685. https://doi.org/10.1080/21655979.2017.1323593

Eduok, U., Ohaeri, E., Szpunar, J., 2020. 8 – Self-healing composite coatings with protective and anticorrosion potentials: Classification by healing mechanism, in: Khan, A., Jawaid, M., Raveendran, S.N., Asiri, A.M.A. (Eds.), Self-Healing Composite Materials, Woodhead Publishing Series in Composites Science and Engineering. Woodhead Publishing, Sawston, United Kingdom, pp. 123–162. https://doi.org/10.1016/B978-0-12-817354-1.00008-9

Ensikat, H.J., Ditsche-Kuru, P., Neinhuis, C., Barthlott, W., 2011. Superhydrophobicity in perfection: The outstanding properties of the lotus leaf. Beilstein J. Nanotechnol. 2, 152–161. https://doi.org/10.3762/bjnano.2.19

Farooq, S.A., Raina, A., Mohan, S., Arvind Singh, R., Jayalakshmi, S., Irfan Ul Haq, M., 2022. Nanostructured coatings: Review on processing techniques, corrosion behaviour and tribological performance. Nanomaterials 12. https://doi.org/10.3390/nano12081323

Faruk, O., Yang, Y., Zhang, J., Yu, J., Lv, J., Lv, W., Du, Y., Wu, J., Qi, D., 2023. A comprehensive review of ultrahigh molecular weight polyethylene fibers for applications based on their different preparation techniques. Adv. Polym. Technol. 2023, 6656692. https://doi.org/10.1155/2023/6656692

Faust, O., Acharya, U.R., Ng, E.Y.K., Hong, T.J., Yu, W., 2014. Application of infrared thermography in computer aided diagnosis. Infrared Phys. Technol. 66, 160–175. https://doi.org/10.1016/j.infrared.2014.06.001

Feng, X., Feng, L., Jin, M., Zhai, J., Jiang, L., Zhu, D., 2004. Reversible super-hydrophobicity to Super-hydrophilicity transition of aligned ZnO nanorod films. J. Am. Chem. Soc. 126, 62–63. https://doi.org/10.1021/ja038636o

Fihri, A., Bovero, E., Al-Shahrani, A., Al-Ghamdi, A., Alabedi, G., 2017. Recent progress in superhydrophobic coatings used for steel protection: A review. Colloids Surf. Physicochem. Eng. Asp. 520, 378–390. https://doi.org/10.1016/j.colsurfa.2016.12.057

Frank, S.N., Bard, A.J., 1977. Heterogeneous photocatalytic oxidation of cyanide and sulfite in aqueous solutions at semiconductor powders. J. Phys. Chem. 81, 1484–1488. https://doi.org/10.1021/j100530a011

Fu, H., Zhang, L., Dong, Y., Zhang, C., Li, W., 2023. Recent advances in electrochromic materials and devices for camouflage applications. Mater Chem Front 7, 2337–2358. https://doi.org/10.1039/D3QM00121K

Fujihira, M., Satoh, Y., Osa, T., 1981. Heterogeneous photocatalytic oxidation of aromatic compounds on TiO_2. Nature 293, 206–208. https://doi.org/10.1038/293206a0

Fujishima, A., Honda, K., 1972. Electrochemical photolysis of water at a semiconductor electrode. Nature 238, 37–38. https://doi.org/10.1038/238037a0

Gadow, R., Von Niessen, K., 2006. Lightweight ballistic with additional stab protection made of thermally sprayed ceramic and cermet coatings on aramide fabrics. Int. J. Appl. Ceram. Technol. 3, 284–292. https://doi.org/10.1111/j.1744-7402.2006.02088.x

Gao, Y., Gereige, I., El Labban, A., Cha, D., Isimjan, T.T., Beaujuge, P.M., 2014. Highly transparent and UV-resistant superhydrophobic SiO_2-coated ZnO nanorod arrays. ACS Appl. Mater. Interfaces 6, 2219–2223. https://doi.org/10.1021/am405513k

Ghosh Chaudhuri, R., Paria, S., 2012. Core/shell nanoparticles: Classes, properties, synthesis mechanisms, characterization, and applications. Chem. Rev. 112, 2373–2433. https://doi.org/10.1021/cr100449n

González-García, Y., Mol, J.M.C., Muselle, T., Graeve, I.D., Assche, G.V., Scheltjens, G., Mele, B.V., Terryn, H., 2011. A combined mechanical, microscopic and local electrochemical evaluation of self-healing properties of shape-memory polyurethane coatings. Electrochim. Acta 56, 9619–9626. https://doi.org/10.1016/j.electacta.2011.03.081

Grandcolas, M., Louvet, A., Keller, N., Keller, V., 2009. Layer-by-layer deposited titanate-based nanotubes for solar photocatalytic removal of chemical warfare agents from textiles. Angew. Chem. Int. Ed. 48, 161–164. https://doi.org/10.1002/anie.200802932

Hare, E.F., Shafrin, E.G., Zisman, W.A., 1954. Properties of films of adsorbed fluorinated acids. J. Phys. Chem. 58, 236–239. https://doi.org/10.1021/j150513a011

Hassan, T.A., Rangari, V.K., Jeelani, S., 2010. Synthesis, processing and characterization of shear thickening fluid (STF) impregnated fabric composites. Mater. Sci. Eng. A 527, 2892–2899. https://doi.org/10.1016/j.msea.2010.01.018

Heyse, P., Vilder, I.D., Vanneste, M., 2016. 4 – Smart durable and self-healing textile coatings, in: Hu, J. (Ed.), Active Coatings for Smart Textiles, Woodhead Publishing Series in Textiles. Woodhead Publishing, Sawston, United Kingdo, pp. 55–80. https://doi.org/10.1016/B978-0-08-100263-6.00004-6

Hossain, M.T., Shahid, M.A., Limon, M.G.M., Hossain, I., Mahmud, N., 2024. Techniques, applications, and challenges in textiles for a sustainable future. J. Open Innov. Technol. Mark. Complex. 10, 100230. https://doi.org/10.1016/j.joitmc.2024.100230

Hu, J.L., Lu, J., 2016. 2 – Memory polymer coatings for smart textiles, in: Hu, J. (Ed.), Active Coatings for Smart Textiles, Woodhead Publishing Series in Textiles. Woodhead Publishing, Sawston, United Kingdom, pp. 11–34. https://doi.org/10.1016/B978-0-08-100263-6.00002-2

Huang, Q., Yang, Y., Hu, R., Lin, C., Sun, L., Vogler, E.A., 2015. Reduced platelet adhesion and improved corrosion resistance of superhydrophobic TiO_2-nanotube-coated 316L stainless steel. Colloids Surf. B Biointerfaces 125, 134–141. https://doi.org/10.1016/j.colsurfb.2014.11.028

Jahid, M.A., Hu, J., Wong, K., Wu, Y., Zhu, Y., Sheng Luo, H.H., Zhongmin, D., 2018. Fabric coated with shape memory polyurethane and its properties. Polymers 10. https://doi.org/10.3390/polym10060681

Kausar, A., Ahmad, I., Maaza, M., Bocchetta, P., 2023. Self-healing nanocomposites—Advancements and aerospace applications. J. Compos. Sci. 7. https://doi.org/10.3390/jcs7040148

Khan, Y., Sadia, H., Ali Shah, S.Z., Khan, M.N., Shah, A.A., Ullah, N., Ullah, M.F., Bibi, H., Bafakeeh, O.T., Khedher, N.B., Eldin, S.M., Fadhl, B.M., Khan, M.I., 2022. Classification, synthetic, and characterization approaches to nanoparticles, and their applications in various fields of nanotechnology: A review. Catalysts 12. https://doi.org/10.3390/catal12111386

Kim, H.-A., 2021. Moisture vapor permeability and thermal wear comfort of ecofriendly fiber-embedded woven fabrics for high-performance clothing. Materials 14. https://doi.org/10.3390/ma14206205

Kosashvili, Y., Hiss, J., Davidovic, N., Lin, G., Kalmovic, B., Melamed, E., Levy, Y., Blumenfeld, A., 2005. Influence of personal armor on distribution of entry wounds: Lessons learned from urban-setting warfare fatalities. J. Trauma Acute Care Surg. 58, 1236–1240.

Krogman, K.C., Lowery, J.L., Zacharia, N.S., Rutledge, G.C., Hammond, P.T., 2009. Spraying asymmetry into functional membranes layer-by-layer. Nat. Mater. 8, 512–518. https://doi.org/10.1038/nmat2430

Kumar, A., Singh, S., 2018. Development of coatings for radar absorbing materials at X-band. IOP Conf. Ser. Mater. Sci. Eng. 330, 012006. https://doi.org/10.1088/1757-899X/330/1/012006

Kumar, N., Dixit, A., 2019. Camouflage and stealth technology based on nanomaterials, in: Nanotechnology for Defence Applications. Springer International Publishing, Cham, pp. 155–203. https://doi.org/10.1007/978-3-030-29880-7_5

Lee, J.A., Krogman, K.C., Ma, M., Hill, R.M., Hammond, P.T., Rutledge, G.C., 2009. Highly reactive multilayer-assembled TiO_2 coating on electrospun polymer nanofibers. Adv. Mater. 21, 1252–1256. https://doi.org/10.1002/adma.200802458

Lee, J.-S. [Ji-Sun], Kim, H.-W., Lee, J.-S. [Jun-Seo], An, H.-S., Chung, C.-M., 2021. Microcapsule-type self-healing protective coating that can maintain its healed state upon crack expansion. Materials 14. https://doi.org/10.3390/ma14206198

Lee, Y.S., Wetzel, E.D., Wagner, N.J., 2003. The ballistic impact characteristics of Kevlar woven fabrics impregnated with a colloidal shear thickening fluid. J. Mater. Sci. 38, 2825–2833. https://doi.org/10.1023/A:1024424200221

Li, J., Wan, H., Ye, Y., Zhou, H., Chen, J., 2012. One-step process to fabrication of transparent superhydrophobic SiO_2 paper. Appl. Surf. Sci. 261, 470–472. https://doi.org/10.1016/j.apsusc.2012.08.034

Liu, S., Latthe, S.S., Yang, H., Liu, B., Xing, R., 2015. Raspberry-like superhydrophobic silica coatings with self-cleaning properties. Ceram. Int. 41, 11719–11725. https://doi.org/10.1016/j.ceramint.2015.05.137

Long, X., Lei, X., Du, H., Mehmood, A., Li, X., Liu, X., 2023. Recent research progress in electrochromic materials for intelligent stealth (visible and infrared): A review. J. Phys. Conf. Ser. 2440, 012005. https://doi.org/10.1088/1742-6596/2440/1/012005

Mahadevan, L., 2001. Non-stick water. Nature 411, 895–896. https://doi.org/10.1038/35082164

Mahadik, S.A., Mahadik, S.S., 2021. Surface morphological and topographical analysis of multifunctional superhydrophobic sol-gel coatings. Ceram. Int. 47, 29475–29482. https://doi.org/10.1016/j.ceramint.2021.07.115

Malik, S., Muhammad, K., Waheed, Y., 2023. Nanotechnology: A revolution in modern industry. Molecules 28. https://doi.org/10.3390/molecules28020661

Md. Abdul Hannan, S.M.F.K., Papia Haque, Rahman, M.M., 2020. Scope of sustainable pre-treatment of cotton knit fabric avoiding major chemicals. J. Nat. Fibers 17, 623–634. https://doi.org/10.1080/15440478.2018.1518742

Mo, C., Zheng, Y., Wang, F., Mo, Q., 2015. A simple process for fabricating organic/TiO_2 super-hydrophobic and anti-corrosion coating. Int. J. Electrochem. Sci. 10, 7380–7391. https://doi.org/10.1016/S1452-3981(23)17357-3

Mobin, M., Zehra, S., 2023. Chapter 13 – Corrosion control by cathodic protection, in: Aslam, J., Verma, C., Hussain, C.M. (Eds.), Electrochemical and Analytical Techniques for Sustainable Corrosion Monitoring. Elsevier, New York, pp. 265–279. https://doi.org/10.1016/B978-0-443-15783-7.00006-2

Narendra Kumar, Sunita Kumbhat, 2018. Concise Concepts of Nanoscience and Nanomaterials, 1st ed. Scientific Publishers, India.

National Research Council, 1989. Materials Science and Engineering for the 1990s: Maintaining Competitiveness in the Age of Materials. The National Academies Press, Washington, DC. https://doi.org/10.17226/758

Neinhuis, C., Barthlott, W., 1997. Characterization and distribution of water-repellent, self-cleaning plant surfaces. Ann. Bot. 79, 667–677. https://doi.org/10.1006/anbo.1997.0400

Nik Md Noordin Kahar, N.N.F., Osman, A.F., Alosime, E., Arsat, N., Mohammad Azman, N.A., Syamsir, A., Itam, Z., Abdul Hamid, Z.A., 2021. The versatility of polymeric materials as self-healing agents for various types of applications: A review. Polymers 13. https://doi.org/10.3390/polym13081194

Nishino, T., Meguro, M., Nakamae, K., Matsushita, M., Ueda, Y., 1999. The lowest surface free energy based on –CF3 alignment. Langmuir 15, 4321–4323. https://doi.org/10.1021/la981727s

Obayi, C.S., Nnamchi, P.S., 2020. Exploits, advances and challenges in characterizing self-healing materials, in: Tasaltin, N., Nnamchi, P.S., Saud, S. (Eds.), Advanced Functional Materials. IntechOpen, Rijeka. https://doi.org/10.5772/intechopen.93031

Ohko, Y., Saitoh, S., Tatsuma, T., Fujishima, A., 2001. Photoelectrochemical anticorrosion and self-cleaning effects of a TiO_2 coating for type 304 stainless steel. J. Electrochem. Soc. 148, B24. https://doi.org/10.1149/1.1339030

Pagliero, M., Bottino, A., Comite, A., Costa, C., 2020. Novel hydrophobic PVDF membranes prepared by nonsolvent induced phase separation for membrane distillation. J. Membr. Sci. 596, 117575. https://doi.org/10.1016/j.memsci.2019.117575

Pan, S., Wang, N., Xiong, D., Deng, Y., Shi, Y., 2016. Fabrication of superhydrophobic coating via spraying method and its applications in anti-icing and anti-corrosion. Appl. Surf. Sci. 389, 547–553. https://doi.org/10.1016/j.apsusc.2016.07.138

Parvate, S., Dixit, P., Chattopadhyay, S., 2020. Superhydrophobic surfaces: Insights from theory and experiment. J. Phys. Chem. B 124, 1323–1360. https://doi.org/10.1021/acs.jpcb.9b08567

Pesika, N.S., Hu, Z., Stebe, K.J., Searson, P.C., 2002. Quenching of growth of ZnO nanoparticles by adsorption of octanethiol. J. Phys. Chem. B 106, 6985–6990. https://doi.org/10.1021/jp0144606

Phoenix, S.L., Porwal, P.K., 2003. A new membrane model for the ballistic impact response and V50 performance of multi-ply fibrous systems. Int. J. Solids Struct. 40, 6723–6765. https://doi.org/10.1016/S0020-7683(03)00329-9

Pilate, F., Toncheva, A., Dubois, P., Raquez, J.-M., 2016. Shape-memory polymers for multiple applications in the materials world. Eur. Polym. J. 80, 268–294. https://doi.org/10.1016/j.eurpolymj.2016.05.004

Pogrebnjak, A., Sukhodub, L. [Ludmila], Sukhodub, L. [Leonid], Bondar, O., Kumeda, M., Shaimardanova, B., Shaimardanov, Z., Turlybekuly, A., 2019. Composite material with nanoscale architecture based on bioapatite, sodium alginate and ZnO microparticles. Ceram. Int. 45, 7504–7514. https://doi.org/10.1016/j.ceramint.2019.01.043

Polizos, G., Jang, G.G., Smith, D.B., List, F.A., Lassiter, M.G., Park, J., Datskos, P.G., 2018. Transparent superhydrophobic surfaces using a spray coating process. Sol. Energy Mater. Sol. Cells 176, 405–410. https://doi.org/10.1016/j.solmat.2017.10.029

Qing, Y., Yang, C., Hu, C., Zheng, Y., Liu, C., 2015. A facile method to prepare superhydrophobic fluorinated polysiloxane/ZnO nanocomposite coatings with corrosion resistance. Appl. Surf. Sci. 326, 48–54. https://doi.org/10.1016/j.apsusc.2014.11.100

Radwan, A.B., Mohamed, A.M.A., Abdullah, A.M., Al-Maadeed, M.A., 2016. Corrosion protection of electrospun PVDF–ZnO superhydrophobic coating. Surf. Coat. Technol. 289, 136–143. https://doi.org/10.1016/j.surfcoat.2015.12.087

Rahman, Md.M., Rahaman, Md.S., Islam, Md.R., Rahman, F., Mithi, F.M., Alqahtani, T., Almikhlafi, M.A., Alghamdi, S.Q., Alruwaili, A.S., Hossain, Md.S., Ahmed, M., Das, R., Emran, T.B., Uddin, Md.S., 2022. Role of phenolic compounds in human disease: Current knowledge and future prospects. Molecules 27. https://doi.org/10.3390/molecules27010233

Rezayi, T., Entezari, M.H., 2016. Toward a durable superhydrophobic aluminum surface by etching and ZnO nanoparticle deposition. J. Colloid Interface Sci. 463, 37–45. https://doi.org/10.1016/j.jcis.2015.10.029

Rogalski, A., 2019. Graphene-based materials in the infrared and terahertz detector families: A tutorial. Adv. Opt. Photon. 11, 314–379. https://doi.org/10.1364/AOP.11.000314

Rosen, B.A., Laufer, C.H.N., Kalman, D.P., Wetzel, E.D., 2007. Multi-threat performance of Kaolin-based shear thickening fluid (STF)-treated fabrics. In: Proceedings of SAMPE 2007. Vol. 52. Baltimore, MD; p. 018. 3-7 June 2007.

Ruckdashel, R.R., Venkataraman, D., Park, J.H., 2021. Smart textiles: A toolkit to fashion the future. J. Appl. Phys. 129, 130903. https://doi.org/10.1063/5.0024006

Sajanlal, P.R., Samal, A.K., Sreeprasad, T.S., Pradeep, T., 2011. Anisotropic nanomaterials: Structure, growth, assembly, and functions. Nano Rev. 2, 5883. https://doi.org/10.3402/nano.v2i0.5883

Schreuder-Gibson, H.L., Truong, Q., Walker, J.E., Owens, J.R., Wander, J.D., Jones, W.E., 2003. Chemical and biological protection and detection in fabrics for protective clothing. MRS Bull. 28, 574–578. https://doi.org/10.1557/mrs2003.168

Sewell, B.T., Bouloukos, C., von Holt, C., 1984. Formaldehyde and glutaraldehyde in the fixation of chromatin for electron microscopy. J. Microsc. 136, 103–112. https://doi.org/10.1111/j.1365-2818.1984.tb02550.x

Sharma, R., Agarwala, R.C., Agarwala, V., 2008. Development of radar absorbing nano crystals by microwave irradiation. Mater. Lett. 62, 2233–2236. https://doi.org/10.1016/j.matlet.2007.11.076

Sharma, R., Agarwala, R.C., Agarwala, V., 2009. Development of electroless (Ni-P)/ $BaNi_{0.4}Ti_{0.4}Fe_{11.2}O_{19}$ nanocomposite powder for enhanced microwave absorption. J. Alloys Compd. 467, 357–365. https://doi.org/10.1016/j.jallcom.2007.11.141

Sharma, V., Sharma, V., Goyat, M.S., Hooda, A., Pandey, J.K., Kumar, A., Gupta, R., Upadhyay, A.K., Prakash, R., Kirabira, J.B., Mandal, P., Bhargav, P.K., 2020. Recent progress in nano-oxides and CNTs based corrosion resistant superhydrophobic coatings: A critical review. Prog. Org. Coat. 140, 105512. https://doi.org/10.1016/j.porgcoat.2019.105512

Shri Prakash, B., Balaraju, J.N., 2024. Chromate (Cr^{6+})-free surface treatments for active corrosion protection of aluminum alloys: A review. J. Coat. Technol. Res. 21, 105–135. https://doi.org/10.1007/s11998-023-00831-1

Sikdar, P., Dip, T.M., Dhar, A.K., Bhattacharjee, M., Hoque, Md.S., Ali, S.B., 2022. Polyurethane (PU) based multifunctional materials: Emerging paradigm for functional textiles, smart, and biomedical applications. J. Appl. Polym. Sci. 139, e52832. https://doi.org/10.1002/app.52832

Silva, V.A., Folgueras, L. de C., Cândido, G.M., Paula, A.L. de, Rezende, M.C., Costa, M.L., 2013. Nanostructured composites based on carbon nanotubes and epoxy resin for use as radar absorbing materials. Mater. Res. 16, 1299–1308. https://doi.org/10.1590/S1516-14392013005000146

Singh, P., Ali, S.W., Kale, R.D., 2023. Antimicrobial nanomaterials as advanced coatings for self-Sanitizing of textile clothing and personal protective equipment. ACS Omega 8, 8159–8171. https://doi.org/10.1021/acsomega.2c06343

Singh, V.K., Shukla, A., Patra, M.K., Saini, L., Jani, R.K., Vadera, S.R., Kumar, N., 2012. Microwave absorbing properties of a thermally reduced graphene oxide/nitrile butadiene rubber composite. Carbon 50, 2202–2208. https://doi.org/10.1016/j.carbon.2012.01.033

Soethe, V.L., Nohara, E.L., Fontana, L.C., Rezende, M.C., 2011. Radar absorbing materials based on titanium thin film obtained by sputtering technique. J. Aerosp. Technol. Manag. 3, 279–286. https://doi.org/10.5028/jatm.2011.03030511

Srivastava, A., Majumdar, A., Butola, B.S., 2011. Improving the impact resistance performance of Kevlar fabrics using silica based shear thickening fluid. Mater. Sci. Eng. A 529, 224–229. https://doi.org/10.1016/j.msea.2011.09.021

Srivastava, A.K., Deepa, M., Bahadur, N., Goyat, M.S., 2009. Influence of Fe doping on nanostructures and photoluminescence of sol–gel derived ZnO. Mater. Chem. Phys. 114, 194–198. https://doi.org/10.1016/j.matchemphys.2008.09.005

Starr, C.L., Cao, H., Peksoz, S., Branson, D.H., 2015. Thermal effects of design and materials on QuadGard™ body armor systems. Cloth. Text. Res. J. 33, 51–63. https://doi.org/10.1177/0887302X14556151

Sun, L.-L., Xiong, D.-S., Xu, C.-Y., 2013. Application of shear thickening fluid in ultra high molecular weight polyethylene fabric. J. Appl. Polym. Sci. 129, 1922–1928. https://doi.org/10.1002/app.38844

Sun, X.-G., Gao, M., Li, C., Wu, Y., 2011. Microwave absorption characteristics of carbon nanotubes, in: Yellampalli, S. (Ed.), Carbon Nanotubes. IntechOpen, Rijeka. https://doi.org/10.5772/16514

Sundarrajan, S., Chandrasekaran, A.R., Ramakrishna, S., 2010. An update on nanomaterials-based textiles for protection and decontamination. J. Am. Ceram. Soc. 93, 3955–3975. https://doi.org/10.1111/j.1551-2916.2010.04117.x

Tian, Y., Zhang, X., Dou, S., Zhang, L., Zhang, H., Lv, H., Wang, L., Zhao, J., Li, Y., 2017. A comprehensive study of electrochromic device with variable infrared emissivity based on polyaniline conducting polymer. Sol. Energy Mater. Sol. Cells 170, 120–126. https://doi.org/10.1016/j.solmat.2017.05.053

Vaidya, M.Y., McBain, A.J., Butler, J.A., Banks, C.E., Whitehead, K.A., 2017. Antimicrobial efficacy and synergy of metal ions against *Enterococcus faecium*, *Klebsiella pneumoniae* and *Acinetobacter baumannii* in planktonic and biofilm phenotypes. Sci. Rep. 7, 5911. https://doi.org/10.1038/s41598-017-05976-9

Varghese, T.V., Kumar, H.A., Anitha, S., Ratheesh, S., Rajeev, R.S., Rao, V.L., 2013. Reinforcement of acrylonitrile butadiene rubber using pristine few layer graphene and its hybrid fillers. Carbon 61, 476–486. https://doi.org/10.1016/j.carbon.2013.04.104

Wang, H., Xue, Y., Ding, J., Feng, L., Wang, X., Lin, T., 2011. Durable, self-healing superhydrophobic and superoleophobic surfaces from fluorinated-decyl polyhedral oligomeric silsesquioxane and hydrolyzed fluorinated alkyl silane. Angew. Chem. Int. Ed. 50, 11433–11436. https://doi.org/10.1002/anie.201105069

Wang, L., Jia, X., Li, Y., Yang, F., Zhang, L., Liu, L., Ren, X., Yang, H., 2014. Synthesis and microwave absorption property of flexible magnetic film based on graphene oxide/carbon nanotubes and Fe_3O_4 nanoparticles. J. Mater. Chem. A 2, 14940–14946. https://doi.org/10.1039/C4TA02815E

Wei, C., Wang, G., Cridland, M., Olson, D.L., Liu, S., 2018. Chapter 25 – Corrosion protection of ships, in: Kutz, M. (Ed.), Handbook of Environmental Degradation of Materials (Third Edition). William Andrew Publishing, Norwich, NY, pp. 533–557. https://doi.org/10.1016/B978-0-323-52472-8.00026-5

Wilhelm, M., Bir, C., 2008. Injuries to law enforcement officers: The backface signature injury. Forensic Sci. Int. 174, 6–11. https://doi.org/10.1016/j.forsciint.2007.02.028

Wu, B., Lyu, J., Peng, C., Jiang, D., Yang, Jun, Yang, Jinshui, Xing, S., Sheng, L., 2020. Inverse infusion processed hierarchical structure towards superhydrophobic coatings with ultrahigh mechanical robustness. Chem. Eng. J. 387, 124066. https://doi.org/10.1016/j.cej.2020.124066

Wu, S., Sikdar, P., Bhat, G.S., 2023. Recent progress in developing ballistic and anti-impact materials: Nanotechnology and main approaches. Def. Technol. 21, 33–61. https://doi.org/10.1016/j.dt.2022.06.007

Xiao, H.-M., Liu, X.-M., Fu, S.-Y., 2006. Synthesis, magnetic and microwave absorbing properties of core-shell structured $MnFe_2O_4/TiO_2$ nanocomposites. Compos. Sci. Technol. 66, 2003–2008. https://doi.org/10.1016/j.compscitech.2006.01.001

Xiao, L., Ma, H., Liu, J., Zhao, W., Jia, Y., Zhao, Q., Liu, K., Wu, Y., Wei, Y., Fan, S., Jiang, K., 2015. Fast adaptive thermal camouflage based on flexible VO2/graphene/CNT thin films. Nano Lett. 15, 8365–8370. https://doi.org/10.1021/acs.nanolett.5b04090

Yan, D., Li, Y., Liu, Y., Li, N., Zhang, X., Yan, C., 2021. Antimicrobial properties of chitosan and chitosan derivatives in the treatment of enteric infections. Molecules 26. https://doi.org/10.3390/molecules26237136

Yoshikawa, N., 2010. Fundamentals and applications of microwave heating of metals. J. Microw. Power Electromagn. Energy 44, 4–13. https://doi.org/10.1080/08327823.2010.11689772

Zhang, Z., Ge, B., Men, X., Li, Y., 2016. Mechanically durable, superhydrophobic coatings prepared by dual-layer method for anti-corrosion and self-cleaning. Colloids Surf. Physicochem. Eng. Asp. 490, 182–188. https://doi.org/10.1016/j.colsurfa.2015.11.049

Zhao, G., Wang, W., Wang, X., Xia, X., Gu, C., Tu, J., 2019. A multicolor electrochromic film based on a SnO_2/V_2O_5 core/shell structure for adaptive camouflage. J. Mater. Chem. C 7, 5702–5709. https://doi.org/10.1039/C9TC01073D

Zhao, X., Soper, S.A., Murphy, M.C., 2021. A high-adhesion binding strategy for silica nanoparticle-based superhydrophobic coatings. Colloids Surf. Physicochem. Eng. Asp. 625, 126810. https://doi.org/10.1016/j.colsurfa.2021.126810

Zhou, K., Deng, J., Yin, L., Ma, S., Gao, S., 2007. Microwave absorbing properties of $La_{0.8}Ba_{0.2}MnO_3$ nano-particles. Trans. Nonferrous Met. Soc. China 17, 947–950. https://doi.org/10.1016/S1003-6326(07)60205-2

Zhu, H., Li, Q., Zheng, C., Hong, Y., Xu, Z., Wang, H., Shen, W., Kaur, S., Ghosh, P., Qiu, M., 2020. High-temperature infrared camouflage with efficient thermal management. Light Sci. Appl. 9, 60. https://doi.org/10.1038/s41377-020-0300-5

12 Biomedical Applications of Nanotechnology

There are many applications of nanotechnology and nanomaterials in biomedical applications. The application of nanotechnology in biotechnology leads to a field of interest called nanobiotechnology. Nanotechnology finds vast applications in biomedical engineering/biotechnology; for instance, diagnostics, treatment, care, drug delivery systems, wound-healing prostheses, and implants are a few significant applications. Nanomaterials of organic and inorganic origin are extensively used for biomedical applications. Metal, ceramic, carbon, polymer nanomaterials, liposomes, etc., are used in biomedical applications. Many organic molecules are attached to the surface of the nanoparticles to induce specific biological properties for antibody-antigen interactions, targeted delivery of drugs, DNA hybridization, etc. Tuning of nanomaterials with the required surface chemistry, physics, and thermodynamics can enhance the functionality and efficacy of the respective application.

12.1 TISSUE ENGINEERING

The primary focus of tissue engineering is to enable the natural healing of damaged tissues. The fundamental feature of a synthetic scaffold for tissue engineering applications is its ability to mimic the extracellular matrix (ECM) architecture in terms of physical, chemical, and biological characteristics. In another way, the synthetic scaffolds must have good porosity with even pore size distribution, high surface area, non-toxicity, biocompatibility, biodegradability, structural integrity, and wettability among cells. Eventually, the synthetic scaffolds should mimic the biological characteristics of the specific tissue.

The extracellular matrix (ECM) has a complex role in regulating the behavior of the cells in contact with it, influencing their survival, development, migration, proliferation, shape, and functions. The extracellular matrix has a complex molecular composition. ECM has a system of non-living tissue made of cellulose in plants, proteins, minerals, and certain carbohydrates in animals. The macromolecules that constitute the extracellular matrix are mainly produced locally by cells in the matrix, such as fibroblast cells (Figure 12.1). The synthetic scaffolds are made of biodegradable and biocompatible materials. The advantage of using biodegradable material

DOI: 10.1201/9781003317272-12

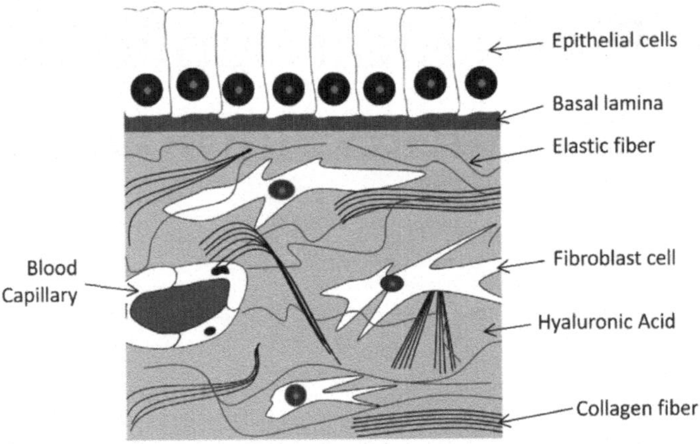

FIGURE 12.1 Connective tissue underlying an epithelium contains an extracellular matrix.

is that a second surgery is not involved in removing the scaffold. Simultaneously, biocompatibility omits the chances of the immune system rejecting it from the body.

Inorganic nanomaterials such as clays, carbon-based nanomaterials, metal oxides, and metallic nanoparticles are studied for tissue engineering applications. These nanomaterials are combined with organic/biomaterials to form synthetic scaffolds with biological properties suitable for tissue engineering. Gold and titanium dioxide (TiO_2) nanoparticles are extensively studied for tissue engineering applications. These nanoparticles enhance cell proliferation rates during various tissue regeneration. Gold nanoparticles are biocompatible, and their surface can be easily modified. Gold nanoparticles are identified as excellent candidates for bone tissue regeneration (Boisselier and Astruc, 2009). TiO_2 nanoparticles support the enhanced proliferation of cardiomyocyte cells in cardiac muscles. The TiO_2 nanoparticles in embedded PLGA are a potential candidate for bone regeneration (Jawad et al., 2008). Carbon nanomaterials are an essential candidate for developing synthetic scaffolds in tissue engineering. Incorporating carbon nanomaterials, such as carbon nanotubes, graphene, graphene oxide, etc., in polymers and inorganic materials improves the mechanical properties of the composites (Ku et al., 2013).

Electrospun nanofibrous ECM is widely studied as a temporary scaffold in tissue engineering applications. The electrospun nanofibers mimic the physical structure and morphological similarities of proteins in ECM. The electrospun nanofibers can provide a porous 3D structure suitable for synthetic scaffolds. Nanofibers of biocompatible and biodegradable materials can be fabricated using the electrospinning process. If the material is a biodegradable polymer but not biocompatible, then the biocompatibility of the nanofibers can be improved by blending it with biological materials like albumin or protein. The nanofibers of biological materials such as protein can be spun as nanofibers at a reasonably low cost and can be directly used for ECM without further treatment. Additionally, incorporating bioactive materials like nano-hydroxyapatite (HA), tricalcium phosphate (TCP), etc., to electrospun nanofibrous scaffolds can support cell adhesion, proliferation, and differentiation. Moreover, one

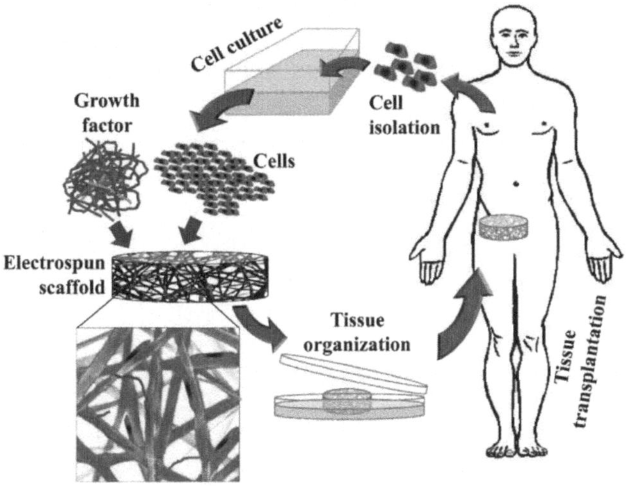

FIGURE 12.2 Schematic representation of steps involved in tissue engineering.

can tune the degradation time of the synthetic scaffolds to match the new tissue formation. The various stages of tissue engineering are shown in Figure 12.2.

Ceramic and polymeric nanofibers are generally used in tissue engineering applications because they have good biocompatibility, are lightweight, are noncorrosive, and have bonding properties. Naturally occurring polymers such as silk are also used to electrospin tissue scaffolds. Numerous studies have been performed on the potential of electrospun fibers as tissue scaffolds. Electrospun fibers are explored as synthetic scaffolds of bones, muscle, nerve, cartilage, blood vessels, skin, etc.; a few electrospun polymers used in tissue engineering applications are listed in Table 12.1.

12.2 DRUG DELIVERY

Controlled release and targeted delivery of drugs are critical in medical treatment. In conventional treatments, the concentration of drugs in the blood is always kept above the minimum practical level to achieve the desired cure. Each drug has its own degradation time in the biological system, which leads to the weakening of its concentration below the minimum practical level in blood after a few hours of intake. To surmount the problem due to degradation, drugs are consumed at higher doses at regular intervals. The blood transports these drugs, which are at a high level, to all parts of the body and can impart toxicity to the healthy regions of the body, typically known as side effects. Controlled release of drugs can retain the desired drug level in the blood for a longer duration of time without reaching a toxic level or dropping below the minimum practical level. If the drugs are delivered at the specific site of the tissue or the infections, the efficacy of the drugs will improve remarkably with no side effects. The site-specific delivery of the drugs is called targeted drug delivery.

TABLE 12.1
Different Polymers for Tissue Engineering Applications

Nanofiber	Applications
Chitosan, PCL, PLGA, PLLA, PU	Vascular grafts
PVP, PEG, Human tropoelastin, PLGA	Skin patch
PLGA, PLLA, PLA, PEG	Tendon
PVA/PEG, PLLA	Cartilage tissue
PLLA, PCL, PEG-PLA diblock copolymer, PANI/PCL, CA, PEO	Muscle
PMGI, PEG–PDLLA diblock polymer, Chitosan, PCL Poly(1,4-dioxan-2-one) (PDO), PU	Cardiac tissue
Surface-modified PLLA, Tussah silk fibroin (TSF), PLLA blended with PANI, Poly(L-lactic)-*co*-PCL	Nerve Regeneration
PAN-MA, PVDF– TrFE, PES	Neural tissue
PVDF- TrFE	Spinal cord
PCL, PVA, PLGA, Silk fibroin, PBS, Polyamide-6, chitosan	Bone

Many nanostructured materials are proposed for the controlled release and targeted delivery of drugs. The large surface area of the nanomaterials enables them to modify with drug molecules and site-specific markers for targeted delivery and controlled release of drugs. As a result, nanocarriers with the ability to deliver drugs with high specificity and release the drugs in a controlled manner can improve the effectiveness of the drugs, thus reducing the dosage, side effects, and cooperation of the patients towards the treatment as the frequency of administration is reduced. Different nanomaterials are proposed as nanocarriers with tailored compositions, morphology, and surface characteristics for the application. Nanomaterials can be tuned for drug loading and controlled release at specific sites without any side effects while maintaining the concentration of the drug locally above the minimum dosage levels. The drug is loaded into nanostructured materials in different forms; they are schematically represented below.

Nanostructured drug carriers must be designed with the kinetics desired for the selected applications; therefore, one has to understand the mechanisms behind drug retention and drug release. The mechanism of drug release from nanostructured drug carriers is a function of composition (drug, polymer, and additives), their fraction and the interactions among them, and the synthesis technique. The mechanism of drug release can be classified into diffusion-controlled, solvent-controlled, degradation-controlled, chemical reaction, and stimuli controlled. In a diffusion-controlled release, the drug is diffused through the drug carrier due to the difference in concentration. In a solvent-controlled release, the solvent (water in body fluids) diffuses into the drug carrier to swell and release the drugs. In the degradation-controlled release, the drug is released on the degradation of the nanocarriers. In a simulation-controlled process, internal or external stimuli such as temperature, pH, ionic strength, sound, and electric or magnetic fields activate the release of drugs (Lee and Yeo, 2015).

Biodegradable polymers, liposomes, micelles, dendrimers, nanogels, porous silica, etc., are nanostructured drug carriers. Biodegradable polymers with measured degradability can be used to control drug release. Biodegradable polymers and non-biodegradable polymers are used in controlled drug delivery. In the case of biodegradable polymers, drugs are released using diffusion and degradation; conversely, in non-biodegradable polymers, it is through diffusion alone. Liposomes are essential in drug delivery, encapsulating hydrophilic and hydrophobic drugs within their lipid bilayer structures. Modifications of liposomes with polyethylene glycol (PEG) led to the creation of pH-sensitive and redox-sensitive systems intended to release the drug to specific environments like a tumor site in a controlled way. Liposomes are critical in targeted therapy, combining longevity, specificity, and responsive release mechanisms.

Due to their core-shell structure, polymeric micelles created from amphiphilic block copolymers can support drug delivery and encapsulate hydrophobic and hydrophilic drugs. These micelles are designed to release drugs in response to specific triggers such as pH changes or external stimuli like heat and ultrasound, enhancing targeted therapy. Nanogels are nanoscale hydrogel particles that can release drugs through swelling, degradation, and diffusion when an external stimulus such as pH, temperature, and redox changes are applied. These particles can be engineered to react to specific environments to trigger the release of drugs. For instance, the temperature or pH varies in the infected regions, which triggers the release of drugs from these particles. Chitin-based nanogels are an example of stimuli-responsive drug carriers.

Dendrimers are efficient in carrying and releasing drugs through covalent conjugation or physical entrapment. The capacity for drug loading and the drug release rate can be finely tuned by adjusting the dendrimer's generation and surface characteristics, enabling applications ranging from targeted cancer therapy to the delivery of nucleic acids. Porous silica nanoparticles (NPs) are highlighted for their ability to provide gradual and sustained drug release, thanks to their stable, tunable structures that can encapsulate drugs within their pores, with surface modifications enhancing drug loading and controlling release rates. Using "gatekeepers" such as organic molecules or inorganic precipitates can further refine release profiles, allowing for stimuli-responsive drug delivery mechanisms that respond to changes in pH, light, or redox potential. For instance, calcium phosphate (CaP) capped on mesoporous silica NPs can selectively release drugs at acidic pH levels, illustrating the potential for targeted therapy applications. Moreover, azobenzene-functionalized silica NPs demonstrate light-triggered drug release, showcasing the versatility and adaptability of porous silica NPs for controlled drug delivery.

Hollow metal nanocarriers, including those based on gold (Au), platinum (Pt), or palladium (Pd), offer potential for drug delivery applications. For instance, hollow Au nanospheres are utilized as drug carriers due to their photothermal activity, allowing for triggered drug release upon near-infrared (NIR) laser irradiation. Titanium oxide (TiO_2) nanotubes with amphiphilic interiors have also been explored, offering controlled drug release mechanisms facilitated by surface modifications and UV illumination. Additional coating with polyelectrolyte complexes further refines drug release control from these hollow metal nanoparticles. Carbon-based nanocarriers, such as carbon nanotubes and fullerenes, are also gaining attention for their controllable size, large surface area, and unique geometry. Surface modifications of carbon-based nanocarriers enhance drug loading and prevent aggregation, contributing to improved drug delivery efficiency. Functionalized fullerenes, like C60 modified with PEI and folic acid, have shown promise as drug carriers for compounds like DTX, exhibiting enhanced cellular uptake and controlled release kinetics. Similarly, amphiphilic C60 modified with carboxylate-terminated dendritic moieties has demonstrated comparable cytotoxicity to conventional formulations of PTX. These diverse nanocarriers offer versatile platforms for targeted and controlled drug delivery, showing potential for various biomedical applications (Prato et al., 2008).

Electrospun nanofibers are the easiest way to produce nanostructured drug carriers. The fibrous nature will enhance the solubility of drugs while protecting them from gastric acids and enzymes. It is possible to load the drugs into the electrospun nanofibers during the electrospinning process or as a template for generating drug-loaded nanotubes, or it can be in the form of core-sheath fibers. Drugs loaded into the fibers during processing are released from the fibers as the polymer fiber degrades. Because of the controlled degradation of the polymer fibers, the drugs are released proportionally to the polymer degradation. The sensitivity of the polymers to the nature of the gastrointestinal fluids (acidic or basic) aids the targeted delivery of drugs.

Tuning nanoparticles to carry drugs directly to the site of injury or disease can make treatments more effective and reduce side effects, known as targeted drug delivery. Targeted delivery of drugs is achieved through active or passive strategies. In active targeting, the drug carriers are conjugated to tissue- or cell-specific ligands. However, in passive targeting, therapeutic agents are incorporated into macromolecules or nanoparticles to reach the target organ passively. Nanoparticles encapsulating drugs or coupled to macromolecules can exploit the enhanced permeability and retention (EPR) effect to target tumors passively. Alternatively, catheters can administer nanoparticles directly to the target organ or tissues. For example, localizing drug-loaded nanoparticles at sites of vascular restenosis may enable sustained drug release at specific arterial wall sites.

Liposomes can be used as drug carriers to deliver precisely to the various parts of the body. They have site markers that allow them to attach to specific cells and provide the drug to the cell's outer layer. The targeted drug delivery ensures that the drug is released in a controlled way, enhancing its efficacy by prolonging its presence at the intended site of action. As a result, liposome-based targeted delivery minimizes side effects and maximizes therapeutic outcomes. Nanoparticles smaller than liposomes are promising for targeted delivery of drugs for tumors, especially in brains, where traditional drug delivery methods fail. By engineering nanoparticles, new methodologies are developed for more effective and precise drug delivery to the brain. Modifying the nanoparticles with sugars and other specialized materials to enhance the transport of drugs to the brain is one example, and such approaches can improve future treatment outcomes and patient care. Various nano-drug carriers are shown in Figure 12.3.

12.3 WOUND DRESSING

There is always a continual emergence in demand for skin regeneration products. Wounds on the periphery of the dermis are capable of self-regeneration. Still, sadly, the body cannot heal deep injuries extended to the connective tissues in such cases

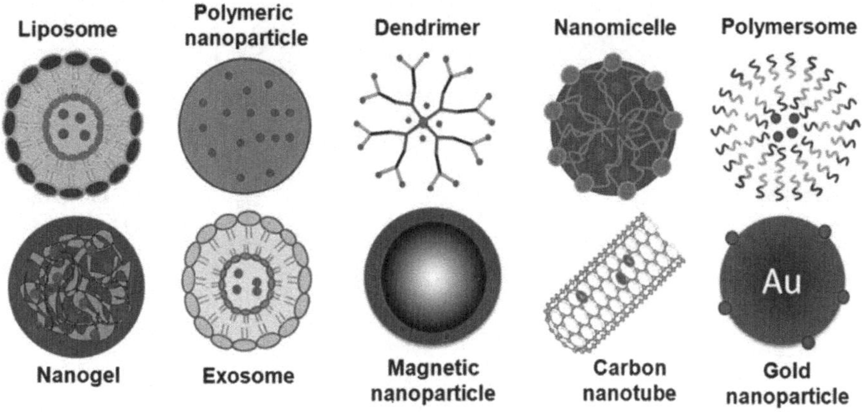

FIGURE 12.3 Nanosized drug carriers for drug delivery. (Meng et al., 2017, reproduced with permission of Springer-Nature.)

where there are no remaining sources of cells for regeneration except at the wound periphery. Complete rehabilitation, therefore, requires extended time and ample care. Also, the possibility of scare marks cannot be omitted. In the case of severe wounds or burning, survival of the patient or complete recovery is possible only by the quick closure of the wounds. The introduction of bioengineered skin changed the scenario to a large extent. However, the production of bioengineered skin involves tedious processes, and it also depends on the availability of the donor site. The output of bioengineered skin requires experienced personnel and involves large-scale investment.

Wound healing is a combined action of different steps: step1: hemostasis-sealing of wounds; step 2: inflammation—remove debris, bacteria, and damaged cells; step 3: proliferation—replacement of damaged tissues with new tissues; and step 4: remodeling—the new tissue undergo remodeling to strengthen and remove the scar. These steps involve different types of cells, such as keratinocytes, fibroblasts, immune cells, endothelial cells, and progenitor cells, and molecules, such as growth factors, chemokines, and cytokines. The different conventional and nanotechnology-based methods for wound dressing are shown in Figure 12.4.

Chronic wounds, such as wounds associated with diabetics and ulcers, are long-lasting and necessitate novel therapies. Nanomaterials of different origins, sizes, shapes, and compositions are explored for their application in wound healing. Nanomaterials promote fast wound healing by helping cell migration, proliferation, and differentiation while regulating the inflammatory response and helping the formation of new tissues. Modern dressings have been designed to create a conducive microenvironment for healing by managing wound moisture, absorbing excess exudate, and targeting bacterial load and protease levels. Nanomaterials are smaller than the pores on the cell walls, and they can penetrate cells to control biological processes at the cellular and molecular levels. Moreover, the tunable surface properties of nanomaterials enhance these characteristics of the nanoparticles.

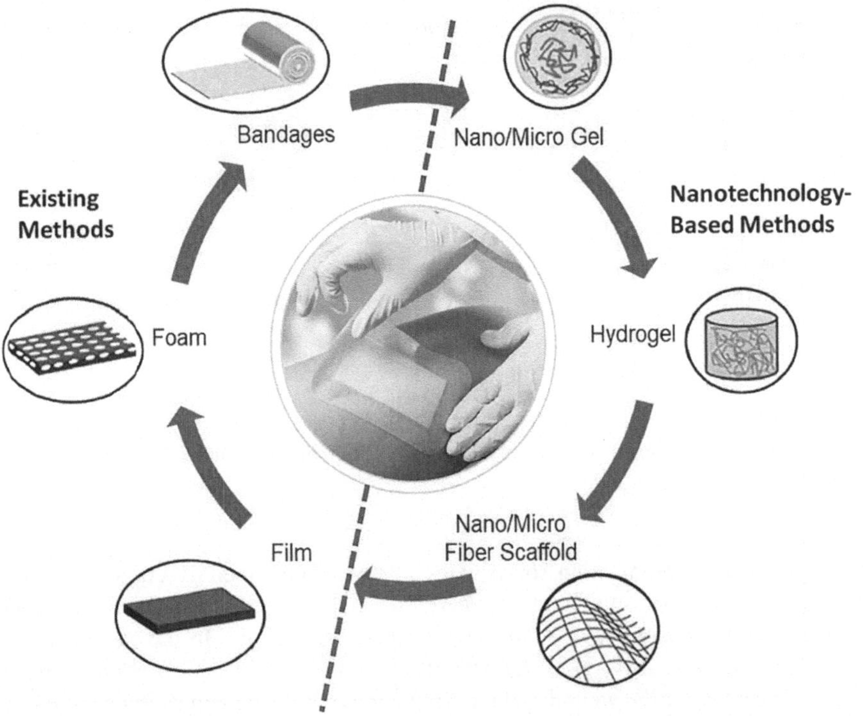

FIGURE 12.4 Methods for wound dressings (left) existing techniques, and (right) nanotechnology applied methods. (Teixeira et al., 2020)

Therefore, nanomaterials open up many choices to provide flexible wound-healing treatments, including wound care, tissue regeneration, infection control, targeted medication administration, and scar reduction (Nandhini et al., 2024). The multifunctionality of nanomaterials, such as antibacterial characteristics, controlled drug release, targeted delivery of drugs, etc., can maximize the speed of the wound-healing process. Additionally, nanostructured materials promote cell growth by supporting cell adhesion, proliferation, and migration, and they direct the growth of cells. The wound-healing process thus combines aspects of tissue engineering, anti-bacterial activity, and targeted controlled drug delivery.

Many organic, inorganic, and metallic nanoparticles are studied for wound-healing applications. The nanomaterial with antibacterial properties is primarily investigated in wound healing because of its ability to prevent the multiplication of pathogens in the wounds. Silver, gold, copper, $TiO_{2,}$ and ZnO nanoparticles exhibit superior anti-bacterial properties, preventing the growth of infectious pathogens in the wound, thus accelerating the healing process. Nanostructured biocompatible and biodegradable polymeric nanofibers and hydrogels are identified as synthetic scaffolds for the regen-eration of tissues in deep wounds. As discussed earlier, the drug's targeted and con-trolled delivery improves its efficacy for a prolonged period. Some nanomaterials, such as HA, TCP, silica, graphene, etc., increase cell proliferation, growth of blood vessels, regeneration of tissue, and, ultimately, rapid recovery from a wound (Pei et al., 2023).

Electrospinning can play a crucial role in overcoming the complexities of the bioengineered skin. The porous nature of the non-woven electrospun fibrous mats can allow cells, drugs, and genes to make them ideal for drug delivery. Electrospun fibers have architecture and morphologies similar to the skin's ECM. Electrospinning can fabricate nanofibers from materials that are biodegradable, biocompatible, and commercially available at low cost. All these properties enhance the proliferation of cells through it. The incorporation of nanoparticles like silver can keep the microorganisms away. At the same time, the loading of nanosized drugs accelerates the healing rate. Artificial dermal analogs often fail because they lack flexibility, permeability, high thickness, etc.

12.4 NANOTECHNOLOGY IN ORTHOPEDICS

Nanotechnology can solve many challenging orthopedics problems, including diagnostics, treatment, and care. The unique properties of the materials at the nanoscale helped the researchers and clinicians to develop effective treatments for injury, wear, and diseases affecting the musculoskeletal system. The various applications of nanotechnology in orthopedics are shown in Figure 12.5.

The nanomaterials can help improve bone and cartilage repair, regeneration, and replacement. The nanofibrous scaffolds and nanocomposites with morphology

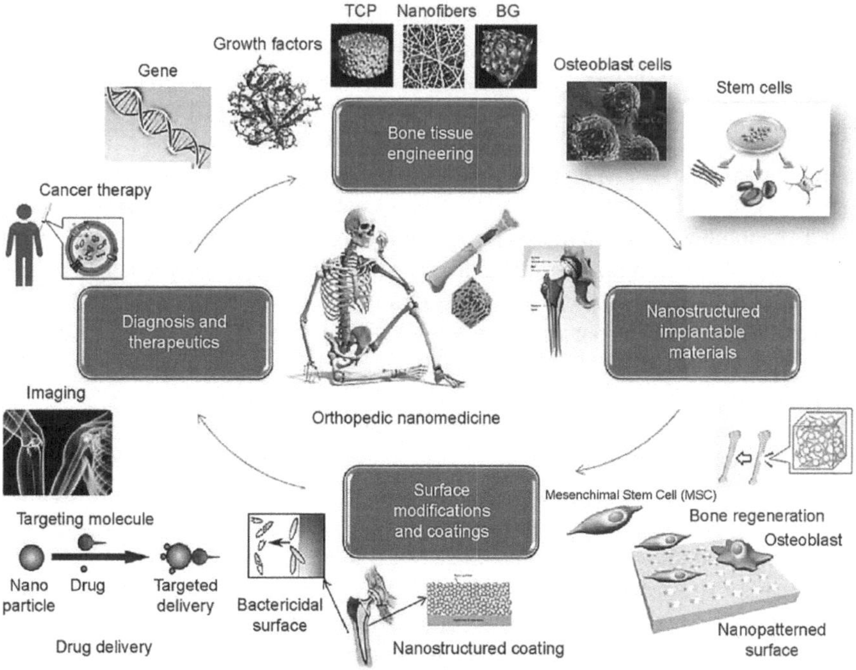

FIGURE 12.5 Applications of nanotechnology in orthopedics. (Mazaheri et al., 2015, International Journal of Nanomedicine 2015-10 6039-6054 "Originally published by and used with permission from Dove Medical Press Ltd.")

analogous to the extracellular matrix can provide support and optimal conditions for the rapid growth of new bone tissue. The biocompatibility of the materials is essential when considering a material for orthopedic implants. The materials must support the proliferation of bone marrow and fibroblast cells on its surface. The above qualities are necessary to avoid any inflammation or toxicity by the implants at the bone-implant interface. The electrospun nanofibrous scaffolds made of biodegradable and biocompatible polymers and reinforced with nanoparticles like hydroxyapatite (HA) and tricalcium phosphate (TCP), minerals in bone, are effective for rapid repair of damaged bone tissues. HA and TCP alone exhibit good biocompatibility but have inferior structural properties. Therefore, one has to integrate them with biocompatible polymers such as poly-glycolic acid (PGA) and poly-lactic acid (PLLA). The composite structure can support and help in the proliferation of bone cells and guide the formation of new bone. Furthermore, treating these biomaterials with growth factors, such as bone sialoproteins (BSP) and bone morphogenic proteins (BMP), can enhance osteointegration further.

Another important application of nanotechnology in orthopedics is the surface modification of orthopedic implants, such as hip and knee replacements. The failure of osteointegration is the major problem associated with prosthetic joints. The treatment for surface irregularity can enhance the growth of the cells, and the cellular interactions become smooth. However, fibrous instead of bony ingrowth occurs, and one has to expect a premature failure. The nanostructured surface of the implants can resolve the issue by promoting osteoblastic cell activity. The increased surface area of the nanotextured surface of the implants creates a thorough connection between the surface of the implant and the host bone, resulting in good osteointegration and lifespan of implants. For instance, coatings made from calcium phosphate nanoparticles can improve osseointegration, securing it more firmly and reducing the risk of implant failure. Additionally, incorporating nanoparticles with antibacterial properties into these coatings can help prevent post-surgical infections, a common complication that can lead to severe outcomes. The nanotexturing of the implants includes using Ti nanotubes, nano-silver, and polypeptide thin film loaded with drugs. Such an approach allows a regulated release of antibiotics, and thus, bacterial adhesion and colonization are weakened. Nanotechnology can improve the strength of the existing implant materials; for instance, the mechanical properties of ultra-high molecular weight polyethylene (UHMWPE) are significantly enhanced by reinforcing it with carbon nanotubes (Puértolas and Kurtz, 2014).

Nanomaterials are helpful for the targeted delivery of drugs in orthopedics. The targeted delivery of antibiotics to an infection site or an anti-inflammatory drug to inflamed joints can provide quick relief to patients without any side effects. Studies show that nanoparticles of gold are suitable carriers of drugs to cure tendinopathy—a tendon illness and injury (Dohnert et al., 2012). Nanosized gold enhances the efficacy of drugs, and nanostructured PLLA can deliver a growth promoter, BMP, to accelerate the growth of osteoblasts and cure bone damage (Schofer et al., 2011).

Nanotechnology has the potential to diagnose bone and joint disorders. Nanoparticles can be used as contrast agents in imaging techniques, providing more precise and detailed images than currently possible. This capability can significantly aid in the early detection and treatment of conditions such as osteoporosis and osteoarthritis, potentially preventing severe damage and disability.

12.5 NANOTECHNOLOGY IN DENTISTRY

The potential of nanomaterials in dentistry spans a wide range: nanocomposites for fillings, nanocoatings for dental prosthetics, and dental care products like toothpaste, mouthwashes, bleaching gels, etc. The incorporation of nanotechnology into dentistry not only improves the efficacy of dental treatments but also improves patient outcomes and comfort. The various applications of nanotechnology in dentistry are shown in Figure 12.6.

The human teeth have a nanostructure composed of different tissues. The outermost layer of teeth is enamel, which is made of hydroxyapatite. Enamel is the hardest and strongest tissue in the human body, and it is intended to act as an insulating barrier and protect the rest of the tooth from injuries. Dentin is the next layer, which acts as a support to prevent any cracking of enamel. Dental caries or cavities are the most common diseases associated with the teeth. It occurs due to the demineralization of enamel by acid, as the microorganisms convert glucose to acid. The cavities are filled with restorative materials. The restorative materials must be nontoxic and antimicrobial, restore the damaged tooth, have comparable mechanical properties to enamel, and have an excellent aesthetic appearance. Nanocomposites of acrylic resin, poly(methyl methacrylate) (PMMA), bisphenol-A diglycidylether methacrylate (Bis-GMA), and glass ionomer (resin and glass powder mixture). These polymer nanocomposites are designed in such a way that they meet the requirements of dental fillings listed above. Nanomaterials can act as multifunctional fillers in these composites; for instance, ZnO, Ag, etc., can improve the strength and impart antimicrobial properties of the fillings.

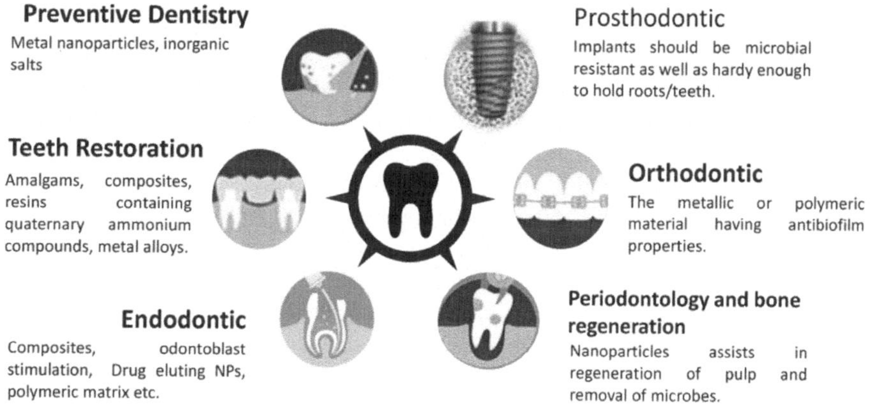

FIGURE 12.6 Schematic representation of various applications of nanotechnology in dentistry. (Sreenivasalu et al., 2022)

The dental adhesive system determines the effective sealing of tooth structures and the durability of sealing in restorative dentistry. The dental adhesives are composed of etchants, primers, and adhesives. Etchants demineralized collagen network, allowing the primer to penetrate the tooth structure effectively. Primers bond the hydrophilic tooth structures with the hydrophobic adhesive, with a methacrylate group for bonding with composites, a spacer group for flexibility, and a reactive group to bond with dentine. The antibacterial properties of the adhesives can be enhanced by Ag nanoparticles and 12-methacryloyloxy-dodecylpyridinium bromide (MDPB). Integrating nanostructured calcium phosphate (NACP) and Ag nanoparticles can enhance re-mineralization, reduce bacterial colonization, and improve bond strength. Adding ZrO_2 and SiO_2-BaO nanofillers in the additive resin improves mechanical properties such as tensile strength, elastic modulus, and bonding strength.

Nanoparticles are used in endodontics, where pulp disorders are treated. The nanoparticles can be used as drug carriers for targeted and controlled release of drugs to the infected regions, as discussed in the previous section on drug delivery. Nanoparticles such as ZnO and Ag can improve the antibacterial properties of endodontic sealers. Nanoparticle-reinforced composites are used for prosthodontics (Prosthetics of dentistry). Periodontics is another field of dentistry where nanotechnology is employed. The remedy includes the treatment using antibiotics, antiseptics, and probiotic agents to prevent the colonization of microorganisms from spreading the disease. Dental implants and orthodontics require biocompatible special coatings that can promote cell proliferation. Such coatings are discussed under the topic of coatings.

12.6 HYPERTHERMIA THERAPY

Hyperthermia therapy is a recent technique for treating diseases, notably cancer, thermal therapy, or thermotherapy. Hyperthermia therapy is based on the increase in local temperature to 40–45 °C near the tumor region to destroy cancer cells and shrink tumors. The rise in temperature during the therapy can damage cells, proteins, structures, and blood vessels inside tumors. As a result, the tumor size will be significantly reduced. Additionally, hyperthermia therapy is based on nanotechnology, and such techniques open opportunities to develop new therapeutic strategies for various diseases. Hyperthermia therapy can treat tumors without ablative surgery and increase the effectiveness of radio- and chemotherapies. The therapy can improve the immune response to fight cancer by activating immune cells (killer cells, dendritic cells, and cytotoxic T cells), reducing immune suppression and drug resistance, altering cancer cell-surface molecules, etc. Hyperthermia therapy not only improves the cancer treatment but reduces the side effects of current therapy used for cancer treatment. Hyperthermia heating can be achieved by radio frequency waves, microwaves, alternating magnetic fields or near-infrared lasers, and magnetic or photothermal nanoparticles. Hyperthermia therapy based on a magnetic field is shown in Figure 12.7. If a light source generates heat, then the therapy is called photothermal therapy.

The nanosized magnetic materials can heat cancer-affected areas with the help of alternating magnetic fields without any surgical intervention (Kulkarni et al. 2015).

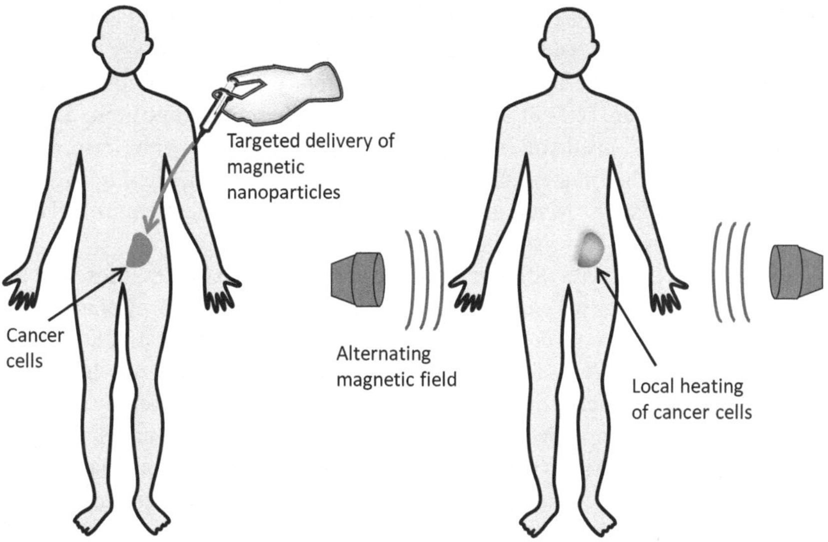

FIGURE 12.7 Hyperthermia therapy using magnetic nanoparticles.

The nanoparticles for hyperthermia therapy are chosen based on their magnetic properties, and the most popular nanoparticle is iron oxide. The nanomaterials that can absorb electromagnetic waves and convert them to heat energy are suitable for hyperthermia applications. Curie temperature (T_c) determines the nature of magnetization in materials. Above T_C, the material becomes purely paramagnetic. In a hypothermia application, the heating of the magnetic particles is possible up to Curie temperature, and by selecting the magnetic particles with appropriate Curie temperature, one can achieve self-controlled heating and avoid damage to unaffected regions. Additionally, engineering nanoparticles to have a specific size and shape, surface modification, etc., can impact the targeted delivery of the nanoparticles to the tumor site. Targeted delivery of the nanoparticles to the tumor site is critical in hyperthermia, as good cells can be damaged. Targeting nanoparticles in the tumor area can be done by functionalizing nanomaterials with molecules like antibodies, peptides, or ligands that specifically bind to receptors or antigens on the cancer cells. This specificity enhances the concentration of nanoparticles in the tumor tissue, improving the selectivity and efficiency of hyperthermia treatment. Additionally, the nanoparticles can accumulate in cancer-affected regions than normal tissues due to the enhanced permeability and retention (EPR) effect. Tumor blood vessels are more permeable than normal blood vessels, allowing nanoparticles to pass through and accumulate in the tumor.

Many nanostructured materials are identified as perfect candidates for hyperthermia treatment. Iron oxide nanoparticles (Magnetite Fe_3O_4 and Maghemite γ–Fe_2O_3) are the most widely used nanoparticles for magnetic hyperthermia. They can be controlled and heated by applying an alternating magnetic field, causing them to generate heat through Neel and Brownian relaxation mechanisms. $CoFe_2O_4$ is also a potential candidate for hyperthermia application as they have high coercivity and magnetic anisotropy, but toxicity studies are needed.

Perovskite-structured magnetic nanomaterials can exhibit hyperthermia properties. Perovskite structures with manganese in the B-site are extensively studied as a potential candidate for the hyperthermia application, e.g., $LaMnO_3$, owing to their low Curie temperature (40–50 °C), ideal for hypothermia application. $LaMnO_3$, an antiferromagnetic insulator, is transformed into metallic with ferromagnetic ordering by doping the trivalent A-site with divalent Sr, Ag, Na, etc. $La_{1-x}Ag_xMnO_3$, $La_{1-y}Na_yMnO_3$, and $La_{1-x}Sr_xMnO_3$ are attractive candidates for self-controlled hyperthermia applications.

Gold nanostructures with different architectures, such as particles, rods, and core-shells, exhibit photothermal characteristics. Due to their tunable plasmonic properties, they effectively convert near-infrared light (NIR) into heat. Carbon-based nanostructures such as CNTs, carbon quantum dots, graphene, and graphene oxides can also convert NIR to heat, making them useful for photothermal therapy. Bimetallic Ni-Cu and Ni–Mo nanoparticles exhibit superior heating effects under fluctuating magnetic fields. Gold-iron oxide (Au/Fe_3O_4) nanoparticles exhibit magnetic and plasmonic properties; these nanoparticles can be heated by alternating magnetic fields and NIR light, offering dual-mode hyperthermia potential. Semiconducting quantum dots and nanoparticles, such as CuS, Ag2Te, etc., can absorb NIR light and convert it into heat.

12.7 COATINGS

We come across a wide range of coatings in our daily lives. Coatings for industrial applications are also equally important as those in household applications. The nanostructured composite coatings with selective barrier properties can be used in chemical industries, war fields, mines, etc., which have remarkable breathability while preserving maximum protection against chemicals. Many coatings are used in biomedical applications as well. The antimicrobial coatings are the primary applications of nanostructured coatings. The antibacterial coatings are effective measures against the growth of pathogens on various surfaces, thus preventing the spread of infectious diseases. Antimicrobial nanoparticles such as copper, silver, ZnO, CuO, and TiO_2-reinforced polymer coatings can control microbial growth by releasing ions, damaging the cell walls of the pathogens, and generating reactive oxygen species. Such coatings are used for wound healing, implants, and stents. Chitosan-based nanocoatings exhibit substantial antibacterial properties and are promising for biomedical applications.

Biocompatibility of the materials is very important in biomedical implants, and stainless steel, cobalt alloys, titanium alloys, hydroxyapatite, tricalcium phosphates (CAPs), and many polymers are used for making implants. The coatings on implants can be classified as passive or active. Passive coatings, such as HA, promote adhesion, while active coatings release interactive molecules/ions like antibacterial agents. HA and TCP constitute bone tissue, and they are biocompatible and support osteoconductive properties, making them desirable for dental and orthopedic implants. Porous HA coatings on metallic surfaces can reduce corrosion rates and enhance bioactivity. ZnO nanolayers support osteoblast proliferation besides the

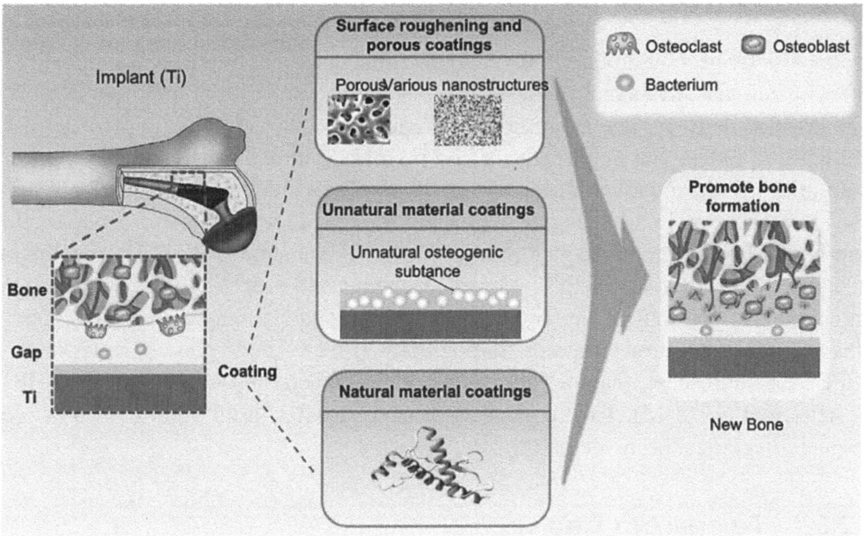

FIGURE 12.8 A representative multifunctional nanocoating on implants. (Lu et al., 2021)

antibacterial properties. The use of nanocoatings on biomedical implants improves patient outcomes. A schematic representation of multifunctional nanocoatings on implants is shown in Figure 12.8.

Nanocoating on stents is crucial in reducing thrombogenicity (tendency to clot blood) by minimizing blood protein adsorption and platelet accumulation on the stents. Nanoporous coatings on stents have a nanosized surface texture, which can support muscle cell and endothelial cell adherence and proliferation. The nano surface texture also affects the platelet interaction with the stents. Polymeric coatings loaded with drugs offer sustained drug release and biocompatibility.

Surgical instruments are coated with nanostructured diamond coatings, and they exhibit low physical adhesion to materials or tissues, chemical/biological inertness, and a low friction coefficient, reducing the penetration force required. Novel manufacturing methods can produce surgical blades with cutting edges as small as 5 nm to 1 μm, using diamond scalpels at the cutting edges with few atoms wide. These instruments are used for high-precision surgeries such as eye, neurosurgery, etc., where minimal invasion is allowed.

Multifunctional nanocoatings with self-cleaning and antifouling properties are essential for microfluidic devices, biosensors, filtration, implants, therapeutic delivery systems, etc. In biomedical applications, nanocoatings are expected to exhibit biocompatible and biodegradable properties. TiO_2-OA nanocomposite coatings exhibit tunable wetting characteristics with high water contact angles and superhydrophobicity. These advancements reveal the versatility and potential of nanocoatings in addressing various challenges across various biomedical applications.

12.8 IMAGING

Nanoparticles become an essential tool for imaging in various biomedical applications, offering unique advantages such as tunable optical properties, multifunctionality, and targeted delivery. The imaging of biological processes and tissues provides data similar to a biopsy but noninvasively and performed in real time. Therefore, a suitable imaging technique with high selectivity and specificity can be effectively used for monitoring and early diagnosis of many critical illnesses. These nanoparticles can be used in different imaging modes, providing enhanced sensitivity, resolution, and specificity for visualizing biological structures and processes. The conventional modalities for biological imaging are X-ray computed tomography (CT), optical imaging (OI), positron emission tomography (PET), single-photon emission CT (SPECT), fluorescent imaging, ultrasound, and magnetic resonance imaging (MRI) (Estelrich et al., 2015). The application of nanoparticles in different modes of biological imaging is discussed below.

12.8.1 Fluorescence Imaging

Fluorescence imaging using nanostructured materials exhibits high special resolution for diagnosing various diseases at a microscopic level. There are different approaches to fluorescent imaging using nanoparticles.

Near-infrared fluorescence (NIRF) imaging uses nanoparticles that exhibit fluorescence in near-infrared (NIR) radiation. Unlike visible light, NIR radiation penetrates deeply into tissues and avoids background noise signals such as auto-fluorescence from the tissues. However, the penetration depth of NIRF is limited, and some auto-fluorescence and scattering by various tissues remain a challenge, in addition to the poor sensitivity to abnormalities. The use of fluorescent nanomaterials can overcome such challenges. For instance, the fluorescent dye or fluorophores used for NIRF is injected into the bloodstream, and the image is taken as the blood reaches the tissue; a representative image is shown in Figure 12.9. The amount of dye reaching the tissue is usually minimal. Suppose the fluorescent dye molecules are loaded to nanoparticle carriers and targeted actively or passively to the respective tissue. In that case, the dye reaching the tissue can be increased significantly, and imaging quality can be improved by obtaining more signals. Moreover, the nanoparticles can be tuned to prevent quenching of NIR fluorescence and upconversion (converting low-energy photons to high energy photons, i.e., infrared light with longer wavelength to UV and visible light with shorter wavelength), which in turn reduces blink and photobleaching of dyes.

The nanoparticle-based fluorescence imaging uses many strategies. The plasmonic/magnetic nanoparticles are conjugated fluorescent dye molecules; the dye emits light for fluorescent imaging, while nanoparticles ensure the targeted delivery of dyes for high concentrations. The core-shell/hollow nanoparticles are also used for the same purpose structure: the fluorescent dye is the core, and the shell is labeled with antibodies, ligands, and peptides. Liposomes, micelles, dendrimers, and polymer nanoparticles are commonly employed for the abovementioned purpose. This design prevents the loss of fluorophores by rapid degradation and excretion. In some designs, a fluorescent quencher is used to quench the emission from the dye; it can be a

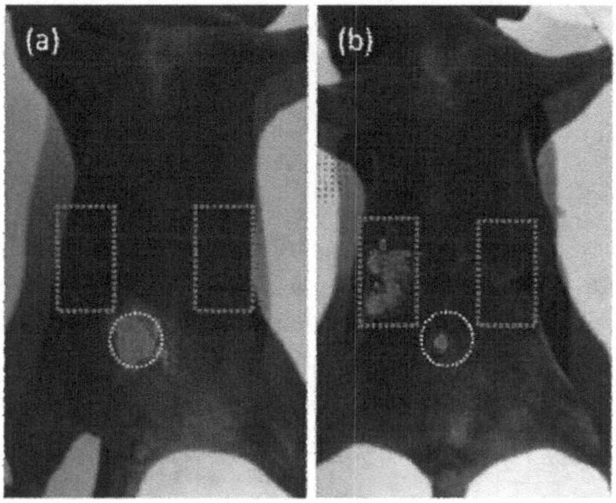

FIGURE 12.9 Fluorescence images of a mouse injected with upconversion nanoparticles labeled with mouse mesenchymal stem cells (mMSCs) taken (a) right after injection and (b) 6h after injection for the detection of stem cells. (Chen et al., 2016)

nanoparticle itself or a second molecule for quenching. The quenchers are degraded at the imaging site, enabling the dye to emit the light freely; such systems have an on/off mechanism. CdSe/ZnS quantum dots are used for such purposes.

Forster resonance energy transfer (FRET) between the donor and acceptor molecules is effectively utilized for fluorescent imaging with the help of nanoparticles. In FRET, the energy from the donor is transferred to the acceptor molecule, which is released from the acceptor molecules radiatively, resulting in fluorescence. Since the low-energy infrared light is used for excitation in upconversion nanoparticles, tissue damage can be avoided while increasing tissue penetration depth. In the spectral range of 650 to 1300 nm, the absorption and scattering of the light by the tissues is minimal.

Nanoparticle-based fluorescence imaging is efficient in detecting genes, protein analysis, enzyme activity, early-stage disease diagnosis, identification of tumors, and study of the effect of drug administration. Nanoparticles of different morphology, size, and composition are used for fluorescent imaging. Quantum dots (QDs) and organic and inorganic nanoparticles are used for this purpose. The band gap is the energy gap between the valence and conduction bands of the nanoparticles. The difference in sizes makes a significant difference in the band gaps; the difference in band gaps leads to different excitation energies to excite the nanoparticles. QDs of carbon, graphene, and semiconducting materials exhibit a broad absorption spectrum and tunable fluorescence emission. Most fluorescent QDs of semiconductors are made of heavy metals that lead to cell poisoning, so sufficient care must be taken while designing imaging systems based on them.

Nanoparticles with quenching effects, such as gold nanoparticles (AuNP), graphene oxide, graphene, carbon nanotubes, etc., are helpful for an on/off mechanism in fluorescent imaging. They are coupled with fluorophores for imaging.

Upconversion nanoparticles (UCNPs) are synthesized by doping the host lattice with lanthanide ions. UCNP nanoparticles exhibit a FRET-type emission in which the absorber elements are La, Nb, Gd, Y, and Yb ions, while emitter elements are Tm, Er, Yb, or Ho. They are prepared in core-shell structure to improve photostability, biocompatibility, and bright fluorescence.

12.8.2 Magnetic Resonance Imaging

Magnetic Resonance Imaging (MRI) is a widely used imaging technique in biomedical applications due to its high resolution. However, its application is limited due to its low specificity and sensitivity, especially in cancer detection. MRI works based on aligning the spin of hydrogen protons in a single direction by an external magnetic field. The sensitivity and specificity of MRI can be enhanced using paramagnetic and superparamagnetic nanoparticles as cell markers, which can be called contrast agents. The disease-affected region absorbs these particles and will contrast the region under study with the unaffected area. Superparamagnetic nanoparticles enable slower or faster water-proton relaxation, thus disturbing the magnetic fields and MRI imaging (Kačenka et al., 2010).

In MRI, the nanomaterials can be used as contrast agents to enhance the resolution of the internal structure of the tissues during the scan by significantly distinguishing the relaxation times of hydrogen protons in water molecules. These agents are classified as T_1-weighted and T_2-weighted categories, as they affect protons' relaxation times (T_1 and T_2). T_1 agents shorten the relation time, thus providing positive contrast and leading to brighter images, e.g., Gd^{3+}-based paramagnetic nanostructures. T_2 agents create negative contrast by shortening T_2 relaxation time, resulting in darker images, e.g., superparamagnetic iron oxide (Fe_3O_4 or Fe_2O_3) nanoparticles (resulting in darker images). However, dual-weighted agents mean materials that can impact T_1 and T_2 relaxation times, e.g., Gd_2O_3 nanoparticles. The abundance of hydrogen in the body, such as water, is a significant drawback, leading to background noise; however, imaging based on fluorine is helpful in tumor and lung imaging. The contrast change in MR images by injecting iron oxide nanoparticles is shown in Figure 12.10.

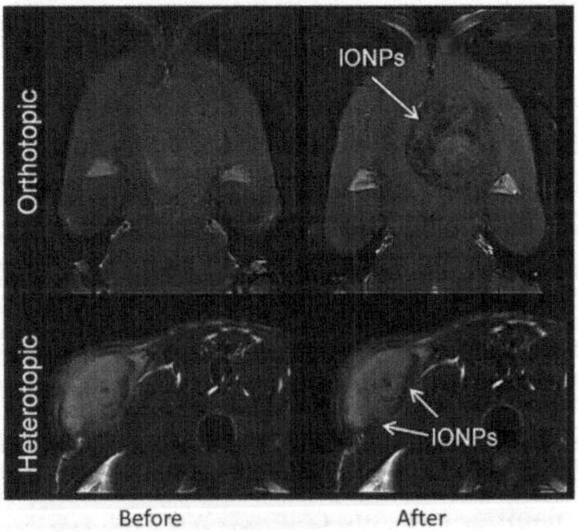

FIGURE 12.10 Representative MR images with contrast enhancement by the injection of iron oxide nanoparticles (IONPs) for brain tumor detection in mouse. (Avasthi et al., 2020)

Directly using paramagnetic or superparamagnetic nanoparticles, surface modified with target markers, is helpful for MRI imaging. The core-shell nanoparticles with a magnetic core and an organic or inorganic shell are also studied as contrast agents; the organic shell includes liposomes, micelles, and dendrimers. The core-shell structure allows the controlled delivery of the magnetic metallic ions at the target. The core can be a perovskite structure with a superparamagnetic material, $La_{1-x}Sr_xMnO_3$, and the shell can be an organic biocompatible molecule such as bovine serum albumin (BSA) or dextran sulfate or an inorganic inactive material such as SiO_2. Fluorescent imaging of the targeted region is also possible at the same time if the nanoparticles are embedded with fluorescent molecules/ions.

12.8.3 Computed Tomography (CT)

Computed tomography (CT) work is based on the difference in the absorption of X-rays, called X-ray attenuation, by various tissues. This variation can also be due to the thickness of the tissues, which results in the generation of three-dimensional images. CT generates high-resolution images quickly at a low cost. Like MRI, contrast agents can play a vital role in CT imaging for better differentiation of tissues with similar attenuation coefficients. Iodine is a common CT contrast agent, but iodine is cleared quickly from the body, and there are chances of renal toxicity. Nanostructured materials are introduced as contrast agents to overcome the limitations of iodine.

The nanostructured contrast agents can be classified as iodine-based and metal-based. In iodine-based contrast agents, nanoparticles are used as shells to load iodine to form a core-shell structure so that it can be delivered to the target in a controlled way. The shell can be liposomes, dendrimers, micelles, etc. Metal-based contrast agents, metals, and their oxides with high X-ray attenuation coefficients, such as gold, tantalum oxide, zirconium dioxide, etc., are used. Gold nanoparticles are commonly used for CT scans. Metal particles are used as cores or shells in core-shell designs, such as gold nanospheres coated with indocyanine-loaded mesoporous silica shells (Zeng et al., 2016). and nanoparticles incorporating zirconium dioxide in the shell with PPy and doxorubicin in the core (Tan et al., 2016). Contrast enhancement by gold nanoparticles in CT scans is shown in Figure 12.11.

12.8.4 Ultrasound Imaging

Ultrasound is the most common medical imaging process due to its portability, versatility, noninvasiveness, high spatial resolution, low cost, and real-time imaging. The contrast agents for ultrasound imaging differentiate the target tissues from the surrounding tissues. The currently used commercial technique applies microbubbles, which can change the contrast of blood pool to cellular uptake. Additionally, they have a larger size and low lifetime and stability. To reflect acoustic waves, the nanoparticles must be larger than the ones used for CT or MRI contrast imaging. Nanoparticles of size >200 nm are used in ultrasound contrast imaging. The nanoparticles are labeled with markers/ligands to match the target tissues. Nanoparticles such as liposomes, micelles, silica, and various polymer nanoparticles are used for ultrasound contrast imaging. Nanoparticles with high acoustic wave scattering are apt for this application.

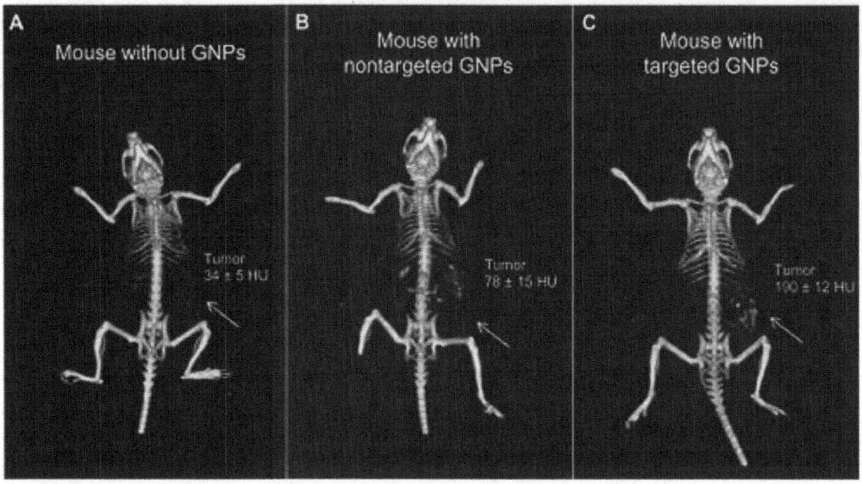

FIGURE 12.11 X-ray computed tomography (CT) images of a mouse (A) before injection of gold nanoparticles (GNPs), (B) 6 hours after injecting nontargeted GNPs, and (C) 6 hours after injecting targeted GNPs modified with epidermal growth factor receptor revealing the head and neck tumor. (Reuveni et al., 2011, International Journal of Nanomedicine 2011 6 2859-2864. "Originally published by and used with permission from DoveMedical Press Ltd.")

12.8.5 PET/SPECT Imaging

Positron emission tomography (PET) is a nuclear imaging technique that can penetrate tissues deeply, has prominent sensitivity, and helps quantify the analysis in real time, even at the molecular level. Single-photon emission computed tomography (SPECT) is another nuclear medicine imaging similar to PET imaging. Still, it is used to detect changes in biochemical functions in anatomy in case of a disease. These techniques are costly, and radioactive materials are used for the analysis. Radioactive atoms with long half-life are used for PET/SPECT imaging. These nuclides are injected into the veins, and their traces are recorded in PET/SPET. Currently, Fluorine-18 is used for PET imaging, which has a half-life of 109.8 minutes, which is very short as we consider the time for preparation of nanoparticles using it and cell uptake. Other choices are copper-64 (half-life of 12.7 h), Indium-111 (half-life of 2.8 days), and Iodine-124 (half-life of 4.2). The nanoparticle adopts a structure similar to other contrast agents discussed earlier, and the most common one is the core-shell type. Radionuclides in SPECT require half-lives more than PET, and currently, Technetium-99m (half-life of 6 h) is used. The radionuclides, including Indium-111, Iodine-123, 125 and 131, and Gold-198, 199, are also used in nanoparticles. Copper-64 infused iron nanoparticles for PET imaging enhance the contrast of tumor region in mice, as shown in Figure 12.12.

PET/SPECT imaging is commonly used for tumor detection. Tumor detection is possible by attaching the specific binding agents to target the tumor cells/tissue.

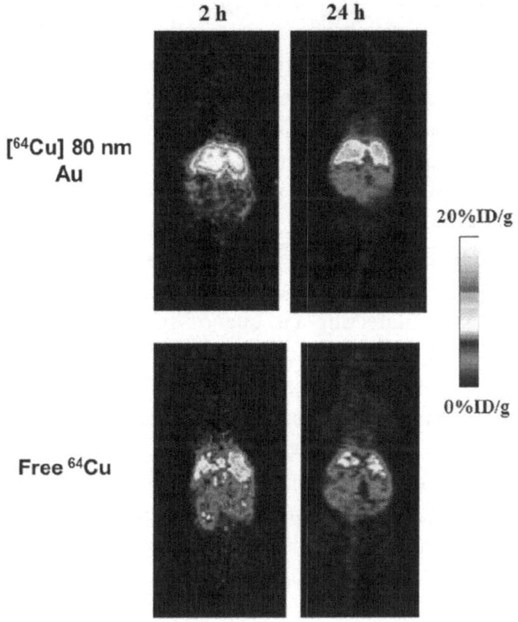

FIGURE 12.12 Representative whole-body coronal PET images of mice at 2 and 24 h after intravenous injection of [^{64}Cu]80 nm Au (upper) as well as free ^{64}Cu (lower). (Sun et al., 2014)

Polymeric nanoparticles such as poly(4-vinyl phenol) (PVPh) labeled using radio nuclei and surface-coated antibodies targeting the tumor tissue is a model used in PET imaging.

12.8.6 Multimodality Imaging

The single biomedical imaging technique is insufficient to diagnose a disease, as every method has many advantages and limitations. For instance, NIRF imaging has high spatial resolution but poor tissue penetration. On the other hand, PET is sensitive with adequate tissue penetration and provides data at a molecular level, but lacks spatial resolution. CT has good spatial resolution, but the ionizing radiation is harmful. Therefore, multimodality imaging is introduced. Multimodal imaging is possible by multifunctional nanoparticles. Nanoparticles carrying contrast agents, fluorescent molecules, and targeting agents can help image the tissues using NIRF imaging and CT, MRI, or PET imaging. For example, fluorescein isothiocyanate (FI) conjugated with PEG-modified Au nanoparticles can be used for fluorescence and CT imaging. Some nanoparticles, such as Gd2O3, show fluorescent, CT, and MRI imaging properties with appropriate modifications. The multimodality combinations are MRI-fluorescence, MRI-PET, CT-fluorescence, CT-PET, MRI-US, MRI-CT, MRI-CT-fluorescence, MRI-PET-fluorescence, US-CT-MRI-fluorescence and PET-fluorescence, etc.

12.9 BIOSENSORS

Nanostructured materials are extensively used to detect biological molecules and physical quantities like pressure, temperature, etc. Nanomaterials are capable of detecting minute concentrations of the analyte. The nanosensors require a bioreceptor and a transducer. As the analyte diffuses to the sensor's surface, the physicochemical properties of the transducer can change the various properties of the surface of the transducer. This change is converted into a measurable signal, mainly an electrical signal, recorded. With nanotechnology's help, biosensing has advanced significantly in healthcare, diagnostics, environmental monitoring, and food safety. Biosensors. These sensors are crucial in detecting and quantifying biological analytes with high specificity and sensitivity. Nanotechnology has significantly enhanced the performance of biosensors by providing novel materials, miniaturization capabilities, and precise manipulation at the nanoscale. These sensors can rapidly and accurately detect biomarkers associated with cancer, infectious diseases, diabetes, and cardiovascular disorders. Additionally, nano biosensors facilitate real-time patient monitoring of therapeutic drug levels, optimizing treatment efficacy and minimizing side effects.

The nanostructure offers a large surface area-to-volume ratio, enhancing interaction with the analyte molecules. Nanoparticles such as gold, silver, quantum dots, carbon nanomaterials, transition metal oxides, and magnetic nanoparticles exhibit excellent optical, electrical, and magnetic properties suitable for biosensing in different modalities. Functionalizing nanomaterials with biomolecules like antibodies, enzymes, DNA, and aptamers enables selective detection of target analytes. Beyond healthcare, nanotechnology-based biosensors are crucial in environmental monitoring by detecting air, water, and soil pollutants, toxins, and pathogens. They contribute to food safety by identifying contaminants, allergens, and pollutants in food products, ensuring quality control and consumer safety. Furthermore, these biosensors are utilized in biodefense to rapidly detect biothreat agents, enhancing preparedness against bioterrorism and infectious disease outbreaks.

The biosensors can be categorized based on how they detect the analytes, such as the varying parameters for the analyte concentration. They are;

Acoustic or Piezoelectric Sensors: Response is measured as a change in sound waves or vibrations.

Resistive Sensors: Response is the changes in electrical resistance due to external factors.

Fluorescent Sensors: response is the change in fluorescence of the materials.

Optical Sensors: response changes in light color or optical properties.

Electrochemical or Amperometric Sensors: Response is the change in the electrochemical properties, measured by current flow

12.9.1 ACOUSTIC/PIEZOELECTRIC SENSORS

Acoustic sensors use piezoelectric materials for the detection of analytes. In acoustic sensors, the electromechanical properties of the piezoelectric materials change as the analyte interacts with the receptors. Another approach is measuring the changes in acoustic waves generated by piezoelectric materials, and the response can be a change in velocity, frequency, or amplitude. For the fabrication of acoustic /piezo-electric sensors, quartz, lithium tantalate ($LiTaO_3$), and lithium niobate ($LiNbO_3$) are employed as substrates, and piezoelectric thin films are deposited on the substrate using PVD. The piezoelectric materials are lead zirconate titanate (PZT), barium zirconate titanate (BZT), aluminum nitride (AlN), and zinc oxide (ZnO).

The detection of analytes is possible using microgravimetric analysis in which the variation in mass is monitored as the sensor response due to the interaction of biomolecules to active sensing materials. Micro-electro-mechanical systems (MEMS) can measure minute changes in mass. The MEMS oscillators are often used to determine the change in mass due to the absorption of biomolecules. QCM is a MEMS module that measures changes in mass more accurately and is more sensitive than other sensors at low temperatures. In general, the bioreceptor is coated on the surface of the QCM oscillator to adsorb and interact with the analyte. The bioreceptor layer can be an antigen, antibody, aptamers, nucleic acids, etc., which makes the QCM sensors selective and specific to adsorb the target analyte. In QCM, an AC voltage is supplied to the electrodes in the oscillator, a piezoelectric crystal, and the oscillator oscillates at a particular frequency. The construction of a QCM is shown in Figure 12.13. When the gas is adsorbed/interacts with the QCM oscillator, the frequency of oscillation changes, and the shift in frequency measures the concentration of VOCs/gases.

Piezoelectric strain sensors have also found extensive applications in biomedical engineering. The sensing mechanism converts mechanical energy (mechanical stress) into electrical energy. As the stress is applied, the ions in the piezoelectric materials are displaced from their crystal lattice site, and as a result, an electric potential is developed across the material. The magnitude of electric potential is directly proportional to the applied stress. Piezoelectric sensors are advantageous due to their sensitivity, rapid response, and self-power. They can be used to monitor the blood pressure in the bold vessels. Piezoelectric sensors are incorporated into wearable devices to monitor the patient's activity levels, vital signs, or specific body movements.

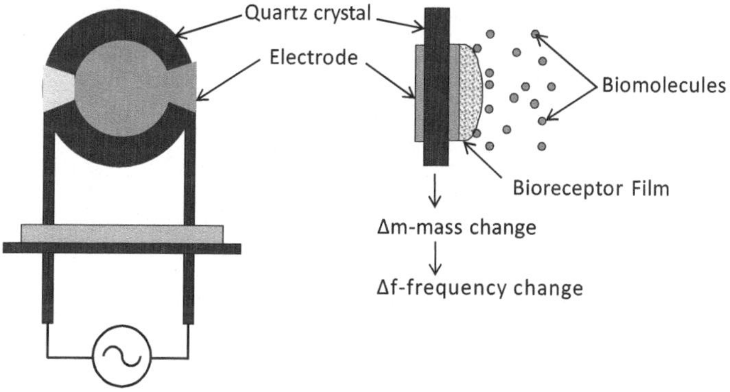

FIGURE 12.13 Construction of a quartz crystal microbalance sensor.

Piezoelectric sensors measure the flow rate of respiratory gases in ventilators and anesthesia machines.

The nanostructured materials are highly sought after for piezoelectric strain sensors. Nanostructured ZnO, especially nanorods and nanowires, are beneficial for piezoelectric sensors. These nanostructures can be quickly grown on any substrate, and the 1-D structure exhibits excellent piezoelectric properties. Barium Titanate ($BaTiO_3$) Nanoparticles are known for their significant piezoelectric properties. Its nanoparticles can be used to create composites or thin films that are highly sensitive due to the piezoelectric effect. Polyvinylidene Fluoride (PVDF) and its copolymers exhibit piezoelectric properties when polarized. Nanofibers made from PVDF can be especially effective in piezoelectric applications due to their high flexibility, making them ideal for wearable technology.

12.9.2 Resistive Sensors

The resistive sensors rely on the change in the conductivity of the bioreceptor, as the analyte is adsorbed to the surface of it. The construction of a resistive sensor is shown in Figure 12.14. Resistive sensors are popular due to their low cost, simple operation, linear response, fast response/recovery time, and better stability. This type of sensor is used for analyzing different gases/vapors such as acetone, ammonia, CO, etc., and the sensing elements are metal-oxides-based semiconductors (e.g., SnO_2, ZnO, TiO_2, WO_3, NiO, Co_3O_4, etc.). These oxides are n-type or p-type, depending on the oxygen vacancy. If the receptor is an n-type material, the adsorbed gases capture (reducing gases) or release (oxidizing gases) electrons from/to the conduction band of the oxide surface, resulting in a decrease/increase in the conductivity.

The resistive sensors can be used for the detection strain, usually made of conductive nanocomposite films, which exhibit a change in electrical resistance when stretched, twisted, or bent. Therefore, the matrix used must be elastic, and the reinforcement must be conductive. The nanocomposites for resistive sensors must have interconnected

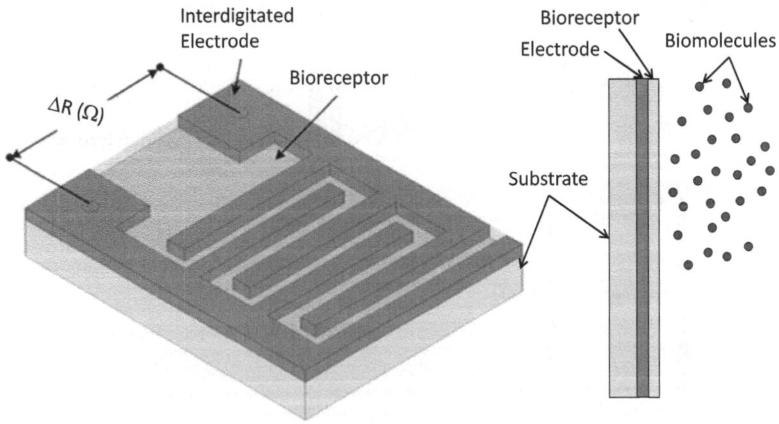

FIGURE 12.14 Construction of a resistive biosensor. (Khan and Kang, 2015)

nanofillers to make the composite conductive under no load conditions. The concentration of the nanofillers at which the nanomaterials are touching each other is called the percolation threshold; at this point, the composites transform from an insulator to a conductor. Polymers such as polydimethylsiloxane (PDMS), polyurethane (PU), polyethylene (PE), and polyvinylidene fluoride (PVDF) are commonly used as resistive strain sensors. Silicon rubber and hydrogels are also used as matrices. Nanofillers such as carbon nanotubes (CNTs), graphene, and nanoparticles of metals like silver (Ag), gold (Au), and copper (Cu), carbon nanofibers (CNFs). The metal oxide nanoparticles zinc oxide (ZnO), titanium dioxide (TiO_2), and tin dioxide (SnO_2) are incorporated to enhance the sensitivity of the sensor.

12.9.3 Fluorescent Sensors

When exposed to the target biomolecules, a fluorescent sensor measures the sensing element's fluorescence. The fluorescent sensors, otherwise called fluorescent probes, are susceptible and can detect deficient concentrations, have high selectivity, fast response, and are non-destructive. Nanomaterials exhibiting fluorescence and affinity to the target biomolecules are fluorescent probes. The fluorescent nanomaterials are surface functionalized with specific ligands or receptors to selectively detect the target analytes. Depending on the application requirements, this functionalization can be tailored to recognize biomolecules, ions, gases, or pathogens. Various fluorescent quantum dots, carbon dots, gold nanoparticles, graphene dots, nanostructured fluorescent polymers, inorganic nanoparticles, etc., are sensing platforms to detect biomolecules. The quantum dots exhibit tunable fluorescent emission, biocompatibility, and surface functionalization capabilities.

The detection mechanism of the fluorescent probes can be fluorescence intensity, lifetime of the emission, or the emission wavelength upon interaction with the target analyte. The change in fluorescence intensity is by quenching, during which non-radiative energy transfer occurs between the fluorescent probe and the analyte molecules. This transfer of energy depends on the concentration of the analyte molecules. The fluorescent probes detect pH, temperature, metallic ions, glucose, proteins, nucleic acids, and neurotransmitters.

12.9.4 Optical Sensors

The basic principles of optical sensors are the spectrophotometric changes due to the interaction of biomolecules and active sensing centers. The optical sensors are formed by reinforcing baroreceptor nanoparticles in a suitable polymer matrix, such as graphene reinforced in the polydiacetylene matrix. The change in color, refractive index, etc., of the optical sensors in the presence of biomolecules is measured as the response. The latter is called a surface plasmon resonance (SPR) sensor. The polydiacetylene/graphene composite sensor exhibits color change as the sensor responds when exposed to analytes. Optical sensors are widely preferred due to their passive nature, compactness, and electromagnetic stability; however, their sensitivity and reproducibility are still challenging. In SPR optical sensors, a monochromatic light with a constant intensity is allowed through the above fiber-optic system, as

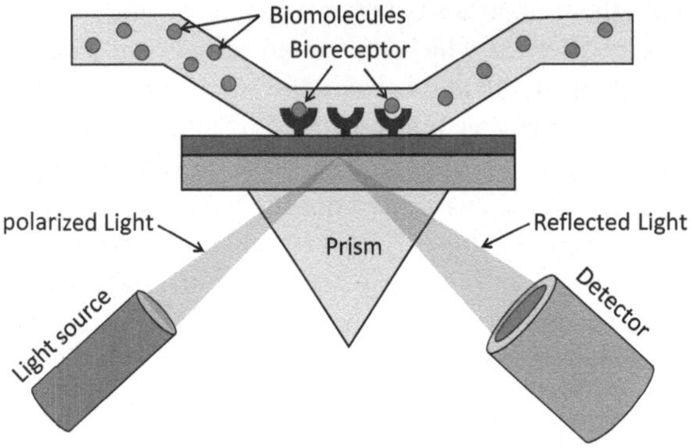

FIGURE 12.15 Construction of a surface plasmon resonance (SPR) sensor.

shown in Figure 12.15. As the sensitive layer absorbs the analyte, the refractive index of the layer changes and results in a dip in the transmitted light intensity, which is the response towards the analyte.

12.9.5 Electrochemical or Amperometric Sensors

Electrochemical/amperometric sensors are extensively used to analyze biomolecules such as glucose, viruses, DNA, proteins, lipids, etc. These sensors are helpful for clinical monitoring and early diagnosis of diseases. The nanotechnology sensors exhibit high sensitivity, selectivity, and rapid response towards analytes. Nanoparticles of different origins, such as metal nanoparticles, carbon nanomaterials, and metal oxide nanoparticles, exhibit electrocatalytic properties desirable for electrochemical sensors. The sensing techniques used for the detection of biomolecules require expensive reagents. The electrochemical sensors can overcome many drawbacks of conventional methods with fast response, cost-effectiveness, and high sensitivity and selectivity toward target biomolecules.

Nanomaterials with tailored morphology, composition, crystal structure sizes, surface charges, etc., are used for electrochemical sensing of biomolecules. The surface modification of these nanomaterials functionalized with polymers, enzymes, or bioactive molecules enhances the selectivity and sensitivity of the sensors. Biocompatibility and targeting capabilities make them highly effective for biosensing applications. CNTs, graphene, nanoparticles of Au, Pt, Co_3O_4, NiO, WO_3, Fe_3O_4, CuO, etc., are strong candidates for electrochemical sensors for biomolecules non-enzymatically. The above materials are reported for the detection of hydrogen peroxide and glucose. Non-enzymatic detection of H_2O_2 and glucose is more accurate and reliable than enzymatic detection, owing to susceptibility to poisonous chemicals, temperature, pH, humidity, etc.

Sensing neurotransmitters like dopamine is essential to screen the mental wellness of humans, so it can help us detect neural diseases like brain cancers, mental illness, and Parkinson's disorder. However, due to their interference, sensing dopamine is very difficult in the presence of uric acid (UA) and ascorbic acid (AA). Hence, finding a sensor with high selectivity to sense dopamine is essential, especially in living cells. Nanostructured materials can be combined/modified/tuned to detect biomolecules such as dopamine and several other complex biomolecules.

12.10 BIOCATALYSTS OR ENZYME CARRIERS

The nanoparticles are used as carriers for enzymes to act as biocatalysts. Nanomaterials offer high enzyme loading, increase in mass transfer, stabilization, high catalytic property, reusability, etc. The enzymes can be conjugated to different nanomaterial surfaces, which is called immobilization of enzymes. The enzyme immobilization can be done through physical adsorption (Attaching an enzyme to a nonreactive material), covalent bonding (by forming covalent bonds), and entrapment (Binding enzyme hydrogen bonds, ionic interactions, and Van der Waals forces, etc.), as represented in Figure 12.16.

Many nanomaterials can effectively mimic the catalytic activities of enzymes. Some oxides with transition elements exhibit the characteristics of the enzyme peroxidase, which catalyzes the oxidation of a substrate in the presence of peroxide. The peroxidase-like activity of perovskite materials originates from the *d*-electron population of the e_g (σ^*) antibonding orbitals of the transition metal sites and the highest activity is observed when e_g is ~1.2, and the lowest when e_g Is 0 or 2.

12.11 COSMETICS

As shown in Figure 12.17, nanomaterials are extensively used to address various skincare and beauty concerns. The addition of nanomaterials to cosmetics enabled the development of advanced formulations that enhance the performance and efficacy of the cosmetic materials. The nanomaterials in cosmetics create products with improved texture, stability, and targeted delivery of active ingredients, resulting in superior skincare benefits.

The most used application of nanotechnology in cosmetics is the incorporation of nano-TiO_2 and ZnO particles in sunscreen lotions. These nanoparticles can absorb a broad UV spectrum, thus protecting the users from UV exposure while being lightweight and having a transparent finish. The earlier version of sunscreen lotion had a white appearance due to the presence of white-colored minerals in it. As particle size is in the nanoscale regime, sunscreen lotions become more cosmetically pleasing and comfortable, encouraging regular use for adequate sun protection.

Additionally, nanostructured materials are used as a template for encapsulating the active ingredients to release them in a controlled way at the targeted layers of the skin. Liposomes, nanoemulsions, polymer nanocapsules, dendrimers, etc., are potential candidates for the abovementioned applications. Such formulation allows penetration of the active ingredients deep into the skin, thus preventing hydration, aging, and hyperpigmentation more effectively. Overall, the presence of nanocrystals in cosmetics improves the stability and bioavailability of active ingredients, thereby increasing their efficacy in improving skin texture and appearance.

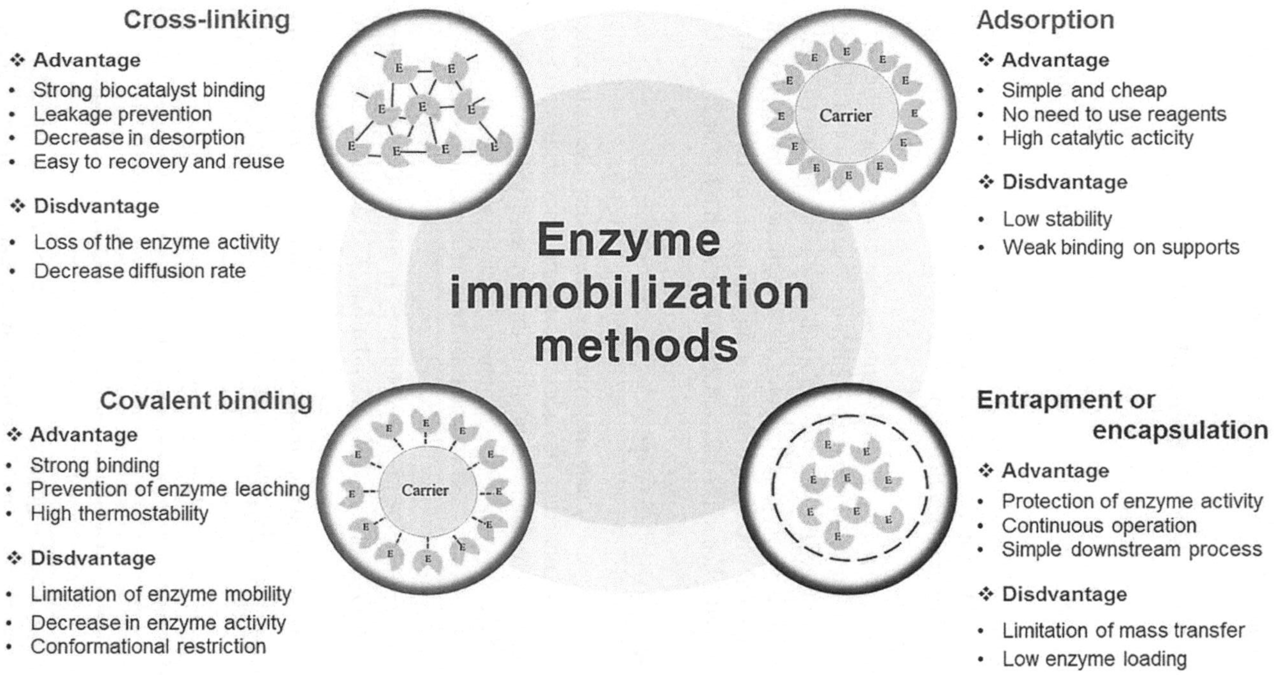

FIGURE 12.16 Different mechanisms of enzyme immobilization. (Kim et al., 2018)

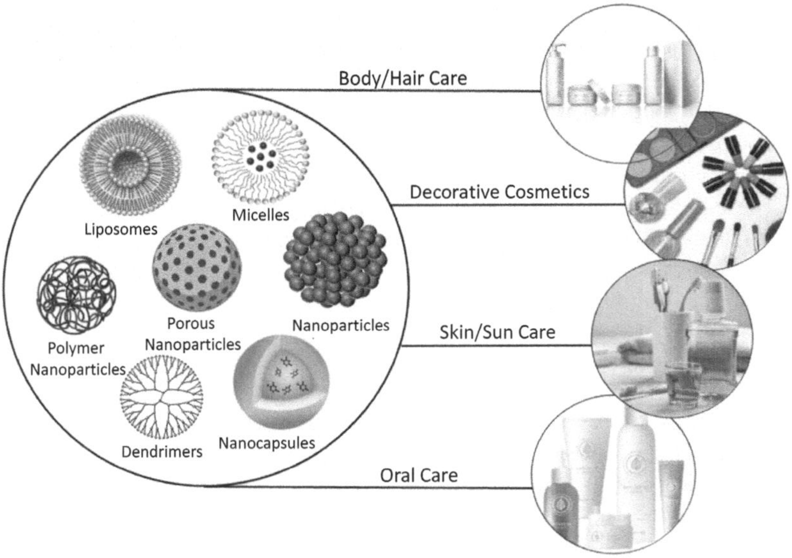

FIGURE 12.17 Nanostructures for various cosmetic applications.

Though nanomaterials are game changers in the cosmetic industry, providing cutting-edge solutions to skincare and beauty, they are also associated with a safety concern. Studies raise concerns about the long-term effects of nanoparticles on human health and the environment, revealing the need for risk assessment and regulation. Nowadays, regulatory agencies strictly monitor the use of nanomaterials in cosmetics.

12.12 NANOSTRUCTURED MATERIALS FOR POWER GENERATION

Many nanotechnology approaches enable the development of energy harvesting methods from different renewable sources to power biomedical devices. They can harness energy from mechanical motion, thermal gradient, and light. These devices are highly demanding for implant and wearable health monitoring systems because they generate power from the body's motion, temperature, etc. A detailed discussion of these topics is available in Chapter 5. When selecting these devices, the material used for the fabrication must be biocompatible and inert. Piezoelectric materials can convert body movements into electrical energy and are used in wearable devices and implants. Zinc oxide (ZnO) nanowires and polyvinylidene fluoride (PVDF) and its copolymer nanofibers are functional piezoelectric materials that can be made flexible and have high energy harvesting ability. Triboelectric nanogenerators (TENGs) can generate electricity from the mechanical interaction of different materials, such as skin contact or vibrations. Nanotextured polymers and their composites effectively harvest triboelectric energy from human motions or bodily fluid flows. Thermoelectric materials generate electricity through the Seebeck effect, producing a voltage across a temperature gradient. The nanostructuring of materials can increase the thermoelectric properties significantly. Bismuth telluride (Bi_2Te_3) nanoparticles and silicon germanium (SiGe) nanostructures are popular thermoelectric generators that generate electricity from body temperature. Photovoltaic materials such as quantum dots and nanostructured silicon can absorb the light falling on the skin or penetrate it and convert it to electricity. A recent invention for powering implants is enzymatic biofuel cells. These fuel cells are designed to convert chemical and electrical energy from the body. The nanostructured electrodes are made of carbon nanotubes or graphene, on which enzymes for splitting the glucose are immobilized; as the enzymes react to the blood glucose, electrical energy is generated.

REFERENCES

Avasthi, A., Caro, C., Pozo-Torres, E., Leal, M.P., García-Martín, M.L., 2020. Magnetic nanoparticles as MRI contrast agents. Top. Curr. Chem. (Z) 378, 40. https://doi.org/10.1007/s41061-020-00302-w

Boisselier, E., Astruc, D., 2009. Gold nanoparticles in nanomedicine: preparations, imaging, diagnostics, therapies and toxicity. Chem. Soc. Rev. 38, 1759–1782. https://doi.org/10.1039/B806051G

Chen, H., Wang, Y., Wang, T., Shi, D., Sun, Z., Xia, C., Wang, B., 2016. Application prospective of nanoprobes with MRI and FI dual-modality imaging on breast cancer stem cells in tumor. J. Nanobiotechnol. 14, 52. https://doi.org/10.1186/s12951-016-0195-8

Dohnert, M.B., Venâncio, M., Possato, J.C., Zeferino, R.C., Dohnert, L.H., Zugno, A.I., De Souza, C.T., Paula, M.M., Luciano, T.F., 2012. Gold nanoparticles and diclofenac diethylammonium administered by iontophoresis reduce inflammatory cytokines expression in Achilles tendinitis. Int. J. Nanomedicine 7, 1651–1657. https://doi.org/10.2147/IJN.S25164

Estelrich, J., Sánchez-Martín, M.J., Busquets, M.A., 2015. Nanoparticles in magnetic resonance imaging: from simple to dual contrast agents. Int. J. Nanomedicine 10, 1727–1741. https://doi.org/10.2147/IJN.S76501

Jawad, H., Lyon, A.R., Harding, S.E., Ali, N.N., Boccaccini, A.R., 2008. Myocardial tissue engineering. Br. Med. Bull. 87, 31–47. https://doi.org/10.1093/bmb/ldn026

Kačenka, M., Kaman, O., Kotek, J., Falteisek, L., Černý, J., Jirák, D., Herynek, V., Zacharovová, K., Berková, Z., Jendelová, P., Kupčík, J., Pollert, E., Veverka, P., Lukeš, I., 2010. Dual imaging probes for magnetic resonance imaging and fluorescence microscopy based on perovskite manganite nanoparticles. J. Mater. Chem. 21, 157–164. https://doi.org/10.1039/C0JM01258K

Khan, M.R.R., Kang, S.-W., 2015. Highly sensitive multi-channel IDC sensor array for low concentration taste detection. Sensors 15, 13201–13221. https://doi.org/10.3390/s150613201

Kim, K.H., Lee, O.K., Lee, E.Y., 2018. Nano-immobilized biocatalysts for biodiesel production from renewable and sustainable resources. Catalysts 8, 68. https://doi.org/10.3390/catal8020068

Ku, S.H., Lee, M., Park, C.B., 2013. Carbon-based nanomaterials for tissue engineering. Adv. Healthcare Mater. 2, 244–260. https://doi.org/10.1002/adhm.201200307

Kulkarni, V.M., Bodas, D., Paknikar, K.M., 2015. Lanthanum strontium manganese oxide (LSMO) nanoparticles: a versatile platform for anticancer therapy. RSC Adv. 5(74), 60254–60263. https://doi.org/10.1039/C5RA02731D

Lee, J.H., Yeo, Y., 2015. Controlled drug release from pharmaceutical nanocarriers. Chem. Eng. Sci. 125, 75–84. https://doi.org/10.1016/j.ces.2014.08.046

Lu, X., Wu, Z., Xu, K., Wang, X., Wang, S., Qiu, H., Li, X., Chen, J., 2021. Multifunctional coatings of titanium implants toward promoting osseointegration and preventing infection: Recent developments. Front. Bioeng. Biotechnol. 9. https://doi.org/10.3389/fbioe.2021.783816

Mazaheri, M., Eslahi, N., Ordikhani, F., Tamjid, E., Simchi, A., 2015. Nanomedicine applications in orthopedic medicine: State of the art. Int. J. Nanomedicine 10, 6039–6054. https://doi.org/10.2147/IJN.S73737

Meng, J., Agrahari, V., Youm, I., 2017. Advances in targeted drug delivery approaches for the central nervous system tumors: The inspiration of nanobiotechnology. J. Neuroimmune Pharmacol. 12, 84–98. https://doi.org/10.1007/s11481-016-9698-1

Nandhini, J., Karthikeyan, E., Rajeshkumar, S., 2024. Nanomaterials for wound healing: Current status and futuristic frontier. Biomed. Technol. 6, 26–45. https://doi.org/10.1016/j.bmt.2023.10.001

Pei, Z., Lei, H., Cheng, L., 2023. Bioactive inorganic nanomaterials for cancer theranostics. Chem. Soc. Rev. 52, 2031–2081. https://doi.org/10.1039/D2CS00352J

Prato, M., Kostarelos, K., Bianco, A., 2008. Functionalized carbon nanotubes in drug design and discovery. Acc. Chem. Res. 41, 60–68. https://doi.org/10.1021/ar700089b

Puértolas, J.A., Kurtz, S.M., 2014. Evaluation of carbon nanotubes and graphene as reinforcements for UHMWPE-based composites in arthroplastic applications: A review. J. Mech. Behav. Biomed. Mater. 39, 129–145. https://doi.org/10.1016/j.jmbbm.2014.06.013

Reuveni, T., Motiei, M., Romman, Z., Popovtzer, A., Popovtzer, R., 2011. Targeted gold nanoparticles enable molecular CT imaging of cancer: An in vivo study. Int. J. Nanomedicine 6, 2859–2864. https://doi.org/10.2147/IJN.S25446

Schofer, M.D., Roessler, P.P., Schaefer, J., Theisen, C., Schlimme, S., Heverhagen, J.T., Voelker, M., Dersch, R., Agarwal, S., Fuchs-Winkelmann, S., Paletta, J.R.J., 2011. Electrospun PLLA nanofiber scaffolds and their use in combination with BMP-2 for reconstruction of bone defects. PLoS ONE 6, e25462. https://doi.org/10.1371/journal.pone.0025462

Sreenivasalu, P.K.P., Dora, C.P., Swami, R., Jasthi, V.C., Shiroorkar, P.N., Nagaraja, S., Asdaq, S.M.B., Anwer, M.K., 2022. Nanomaterials in dentistry: Current applications and future scope. Nanomaterials 12, 1676. https://doi.org/10.3390/nano12101676

Sun, X., Huang, X., Yan, X., Wang, Y., Guo, J., Jacobson, O., Liu, D., Szajek, L.P., Zhu, W., Niu, G., Kiesewetter, D.O., Sun, S., Chen, X., 2014. Chelator-free ^{64}Cu-integrated gold nanomaterials for positron emission tomography imaging guided photothermal cancer therapy. ACS Nano 8, 8438–8446. https://doi.org/10.1021/nn502950t

Tan, L., Liu, T., Fu, C., Wang, S., Fu, S., Ren, J., Meng, X., 2016. Hollow ZrO_2/PPy nanoplatform for improved drug delivery and real-time CT monitoring in synergistic photothermal-chemo cancer therapy. J. Mater. Chem. B 4, 859–866. https://doi.org/10.1039/C5TB02205C

Teixeira, M.A., Paiva, M.C., Amorim, M.T.P., Felgueiras, H.P., 2020. Electrospun nanocomposites containing cellulose and its derivatives modified with specialized biomolecules for an enhanced wound healing. Nanomaterials 10, 557. https://doi.org/10.3390/nano10030557

Zeng, C., Shang, W., Liang, X. [Xiaoyuan], Liang, X. [Xiao], Chen, Q., Chi, C., Du, Y., Fang, C., Tian, J., 2016. Cancer diagnosis and imaging-guided photothermal therapy using a dual-modality nanoparticle. ACS Appl. Mater. Interfaces 8, 29232–29241. https://doi.org/10.1021/acsami.6b06883

Index